THE CHINESE TYPEWRITER

Studies of the Weatherhead East Asian Institute, Columbia University

The Studies of the Weatherhead East Asian Institute of Columbia University were inaugurated in 1962 to bring to a wider public the results of significant new research on modern and contemporary East Asia.

THE CHINESE TYPEWRITER

A HISTORY

THOMAS S. MULLANEY

THE MIT PRESS   CAMBRIDGE, MASSACHUSETTS   LONDON, ENGLAND

This book was set in ITC Stone Sans Std and ITC Stone Serif Std by Toppan Best-set Premedia Limited. Printed and bound in the United States of America.

Library of Congress Cataloging-in-Publication Data

Names: Mullaney, Thomas S. (Thomas Shawn), author.
Title: The Chinese typewriter : a history / Thomas S. Mullaney.
Description: Cambridge, MA : The MIT Press, [2017] | Series: Studies of the
    Weatherhead East Asian Institute, Columbia University | Includes bibliographical
    references and index.
Identifiers: LCCN 2016050197 | ISBN 9780262036368 (hardcover : alk. paper)
Subjects: LCSH: Typewriters, Chinese--History. | Typewriters--History. | Chinese
    language--Writing--History. | Written communication--Technological
    innovations--China--History. | Information technology--China--History. |
    Communication and technology--China--History.
Classification: LCC Z49.4.C4 M85 2017 | DDC 681.61--dc23 LC record available at
    https://lccn.loc.gov/2016050197

10 9 8 7 6 5 4 3 2 1

Per Chiara

漢字無罪
Chinese characters are innocent.
Zhou Houkun
1915

CONTENTS

ACKNOWLEDGMENTS   ix

INTRODUCTION: THERE IS NO ALPHABET HERE   1

1   INCOMPATIBLE WITH MODERNITY   35

2   PUZZLING CHINESE   75

3   RADICAL MACHINES   123

4   WHAT DO YOU CALL A TYPEWRITER WITH NO KEYS?   161

5   CONTROLLING THE KANJISPHERE   195

6   QWERTY IS DEAD! LONG LIVE QWERTY!   237

7   THE TYPING REBELLION   283

CONCLUSION: TOWARD A HISTORY OF CHINESE
COMPUTING AND THE AGE OF INPUT   315

TABLE OF ARCHIVES   323
BIOGRAPHIES OF KEY HISTORICAL PERSONS   325
CHARACTER GLOSSARY   329
NOTES   337
BIBLIOGRAPHY OF SOURCES   401
INDEX   457

ACKNOWLEDGMENTS

What is your problem? Out of all possible questions you could be asking in this life, which one defines the flight path of your mental holding pattern? What is the question that you are always asking, even when to others (and perhaps to yourself) you seem to be changing the subject constantly? And why this problem? To answer these questions is exhausting and wonderful. It takes everything you have.

But to the outside world, it is socially and financially essential to make this process seem orderly and composed. To seem like we have things under control. At the outset of a new project, we say what needs saying in cover letters and funding proposals, at conferences and cocktail parties. But the truth knows better. The first thousand miles are propelled by faith, not in the answer you will find, but in the joy of the labor itself, and a fragile but persistent sense that something of value awaits if only you can hold on.

Without a well-timed and balanced mixture of criticism and faith, close friends and colleagues, it is all too easy to lose touch with this inner sense. Alex Cook continues to be one of my most cherished friends. Without his insight, kindness, and sobering humor, I have no idea where I'd be. I also wish to express my gratitude to Matt Gleeson, who is not only a dear friend, brilliant writer, and longtime music partner, but also the most talented developmental editor I've ever worked with. As before,

I thank Mattie Zelin for her continued support and mentorship—once your advisee, always your advisee. And I thank my families, both by kinship and by kindred spirit: Tom, Merri, Sonia, the late and beloved Giancarlo (IK3IES), Speranza, Scott, Mojgan, Cameron, Laura, Samantha, Mario, Fabiana, Alessio, Andy, Salley, Olivia, Kari, Annelise, Katie, Ruben, Sarah, Dennis, and Kelley.

My colleagues at Stanford may not know this, but a substantial part of my decision to set out on this book project was inspired by informal lunchtime conversations that followed my mid-term reappointment. Senior colleagues I deeply admire took me aside and subjected my entire intellectual being to dissection and scrutiny. It was unsettling, but I like unsettling. Although the idea of the Chinese typewriter had already taken hold of me by this point, it was in large part thanks to these conversations that I decided to be true to myself—unapologetically, radically, and uncompromisingly true to that inner sense of which I just spoke.

In particular, I wish to thank the generosity and criticism of Kären Wigen, Tamar Herzog, Paula Findlen, and Matt Sommer—you have been harbormasters to me, helping me navigate the unseen mountain ranges that lie beneath the surface. Thank you also to Gordon Chang, Londa Schiebinger, Robert Proctor, Zephyr Frank, Jessica Riskin, Steve Zipperstein, Estelle Freedman, Richard Roberts, and Fred Turner for innumerable hallway conversations you likely don't remember (but which I will not forget). Thank you as well to Jim Campbell for inspiring me to try my hand at curating an exhibition, and to Becky Fischbach for showing me how it's done. I am also deeply grateful to intellectual fellow travelers Miyako Inoue, Yumi Moon, Jun Uchida, and Matt for reading the full early draft of this book, as well as to Haiyan Lee and Monica Wheeler. I am deeply thankful to my students as well, for this project particularly Gina Tam, Andrew Elmore, Ben Allen, and Jennifer Hsieh.

Archivists and librarians are the very best of humanity and of the academy. There should be a bank holiday just in their honor. I cannot hope to repay my debt, both in general and in particular, and so allow me instead to offer thanks to Zhaohui Xue, Regan Murphy Kao, Grace Yang, Charles Fosselman, Jidong Yang, Lisa Nguyen, Hsiao-ting Lin, and Carol Leadanham at Stanford University; Liwei Yang and the late Bill Frank at the Huntington Library, Jacques Perrier at the Musée de la Machine à Écrire,

David Baugh at the Philadelphia Records Office, Ann Marie Linnabery at the Niagara County Historical Society, Cyrille Foasso at the Musée des Arts et Métiers, Trina Yeckley and Kevin Bailey at the National Archives and Records Administration, Ulf Kyneb and Anette Jensen at the Danish National Archives, Taos-Hélène Hani at the Bibliothèque Nationale de France, David Kessler at the Bancroft Library, Rene Stein at the National Cryptologic Museum, Enrico Bandiera, Arturo Rolfo, and Marcello Turchetti at the Archivio Storico Olivetti, Maria Mayr at the Mitterhofer Schreibmaschinenmuseum, Nancy Miller at the University of Pennsylvania Archives, Wang Hsien-chun at the Academia Sinica, Craig Orr, Cathy Keen, and David Haberstich at the Smithsonian Institution, John Moffett at the Needham Research Institute, Stacy Fortner at the IBM Corporate Archives, Diane Kaplan at the Yale University Library Manuscripts and Archives division, Ben Primer at the Princeton University Library Rare Books and Special Collections, Geoff Alexander at the Academic Film Archive of North America, John Strom at the Carnegie Institution, Patrick Dowdey at Wesleyan University, Thomas Russo at the Museum of Business History and Technology, Jan Shearsmith at the Museum of Science and Industry, Alan Renton and Charlotte Dando at the Porthcurno Telegraph Museum, John Hutchins at the Machine Translation Archive, G.M. Goddard at the University of Sheffield, Charles Aylmer at the Cambridge University Library, Deb Boyer at PhillyHistory, Lucas Clawson and Carol Lockman at the Hagley Museum and Library, Frank Romano at the Museum of Printing, Ming-sun Poon at the Library of Congress, Donna Rhodes at Pearl S. Buck International, Henning Hansen in Copenhagen, Rolf Heinen in Drolshagen, Victoria West at the V&A Archive, Rory Cook at the Science Museum (London), Remi Dubuisson at the United Nations Archive (New York), Sherman Seki at the University of Hawai'i at Manoa, Rebecca Johnson Melvin at the University of Delaware, Jim Bowman at FEBC International, Myles Crowley at MIT, Katherine Fox at the Harvard Business School, Raymond Lum at Harvard University, Paul Robert at the Virtual Typewriter Museum, Dag Spicer, Hansen Hsu, David Brock, Marguerite Gong Hancock, and Poppy Haralson at the Computer History Museum, and colleagues at the Beijing Municipal Archives, the Shanghai Municipal Archives, the Shanghai library, the Tianjin Municipal Archives, Tsinghua University, and Fudan University. Given the state of affairs in

Chinese archives at present, and the sensitivities surrounding access and personnel, I opt not to list your names.

Historians of the nineteenth and twentieth centuries enjoy a rare pleasure that many of our colleagues do not: conversing with family members and descendants of the historical personages in our work, not to mention the individuals themselves. In the research and writing of this book, I have been graced in this respect more times than anyone could reasonably expect, and for this I am particularly thankful. I am grateful to Ruth Johnson and Kellogg S. Stelle, grandniece and great-grandson of Devello Sheffield, inventor of the first Chinese typewriter; Shu Chonghui, grandson of Shu Zhendong, co-inventor of China's first mass-manufactured Chinese typewriter; Yu Shuolin, son of Yu Binqi, inventor and manufacturer of the Yu-Style Chinese Typewriter; John Marshall, Stan Umeda, and Christine Umeda for sharing so much with me about the late Hisakazu Watanabe and his Japanese typewriters; Andrew Sloss, son of Robert Sloss; James Yee and Pastor Joy Yee, who reached out to me and gave me what became the first Chinese typewriter in my collection (a story I retell briefly in the book); and Pastor Tony Kuriyama and Hideko Kuriyama for donating their beautiful Japanese Panwriter. I have also had the bracing experience, on more than one occasion now, of returning to emails in my inbox sent by individuals who passed away during the course of researching and writing this book. Special thanks and farewell go to Chan Yeh, inventor of the IPX system, who kindly spoke to me on multiple occasions.

Many scholars made critical contributions to early drafts of the book, and to the research process as a whole. While of course taking full responsibility for all errors or shortcomings that remain, I am most grateful to Annelise Heinz, Miyako Inoue, Rob Culp, Michael Gibbs Hill, Kariann Yokota, and Mara Mills for reading and offering invaluable feedback on the entire book manuscript; and to Christopher Reed, Lisa Gitelman, Endymion Wilkinson, Sigrid Schmalzer, Eugenia Lean, Roy Chan, Rebecca Slayton, Andrew Gordon, and Raja Adal for reading substantial portions thereof. I am also deeply grateful to Geof Bowker, Mark Elliott, Jeff Wasserstrom, Erik Baark, Wen-hsin Yeh, Nick Tackett, Victor Mair, Chris Leighton, Fa-ti Fan, Glenn Tiffert, John Kelly, Shumin Zhai, Ingrid Richardson, Cyril Galland, Paul Feigelfeld, Kurt Jacobsen, Kim Brandt,

Jim Hevia, Judith Farquhar, Denise Ho, Joan Judge, Josh Fogel, Li Chen, Michael Schoenhals, Kaushik Sunder Rajan, Geremie Barmé, Emma Teng, Melissa Brown, Michael Fischer, Toby Lincoln, Nicole Barnes, Claire-Akiko Brisset, Jacob Eyferth, Brian Rotman, Stefan Tanaka, Cynthia Brokaw, Lydia Liu, Bill Kirby, Lisa Onaga, Ramekon O'Arwisters, John Williams, Tae-Ho Kim, Zev Handel, Steve Harrell, Pat Ebrey, Marlon Zhu, Ken Lunde, Joe Katz, Kees Kuiken, Elize Wong, Yung-O Biq, Cao Nanping, Ann Blair, Jana Remy, Stijn Vanorbeek, Wolfgang Behr, Jean-Louis Ruijters, and three anonymous outside reviewers. For their assistance on the data visualizations of tray beds, I would like to thank Alberto Pepe, Rui Lu, Li Meng, and Lanna Wu. My thanks as well to a community of talented research assistants at Stanford, including Samantha Toh, Youjia Li, Mona Huang, Chuan Xu, Anna Polishchuk, Truman Chen, and Yuqing Luo. Special thanks go to Suzanne Moon and her associates at *Technology and Culture*, both for their early support of this project and for permission to reprint those parts of chapter 7 that appeared in the October 2012 issue of the journal under the title "The Moveable Typewriter: How Chinese Typists Developed Predictive Text during the Height of Maoism." Thanks also go to Ben Elman and Jing Tsu for permission to reprint those parts of chapter 2 that appeared in the edited volume *Science and Republican China* (Leiden: Brill, 2014) under the title "Semiotic Sovereignty: The 1871 Chinese Telegraph Code in Global Historical Perspective." Finally, my thanks go to Jeff Wasserstrom and Jennifer Munger for permission to include those parts of chapter 5 that appeared in the August 2016 issue of the *Journal of Asian Studies* under the title "Controlling the Kanjisphere."

Research for this book was possible thanks to the generous support of many institutions. The author wishes to express thanks to the Hellman Faculty Fund, the Stanford University Freeman Spogli Institute China Fund, the National Science Foundation, the Stanford University Center for East Asian Studies, and sabbatical support from Stanford University and the Department of History. In particular, I wish to offer my heartfelt thanks to Fred Kronz at the NSF, whose patience and encouragement saw me through what otherwise would have been a discouraging process of revision and resubmission (not to mention one bizarre political witch hunt). I also wish to thank the MIT Press, and especially Katie

Helke. I am thankful to Amy Brand, Katie Hope, Michael Sims, Matthew Abbate, Colleen Lanick, Justin Kehoe, Yasuyo Iguchi, and David Ryman. At the Weatherhead Institute, special thanks go to Carol Gluck and Ross Yelsey.

And now let me begin. Why we authors wait until the end of our acknowledgments to thank the ones who mean the most is a convention I have never understood—and yet here I go again. When I try to remember writing this book—to really remember—all I can summon to mind are thousands of fragmentary moments strewn between the true episodes of my life. I wrote this book as a conversation with you, Chiara, somewhere in between samphire and St. Ives, scala quaranta and Sudtirol, Battlestar and Black Bear Inn, Mission Pie and Mendocino, pie shops and Project Snow, road trips and rose milk tea, birthday cakes and Beijing, Drolshagen and dandelion, Lincoln and Lego models, City Hall and *Cuochi e fiamme*, Torcello and *Tempesta d'amore*, Purple Bamboo and Puccini, belonging to you and being your husband. This book is for you. You teach me bravery and balance. You shelter friends and slay vampires. I love you beyond all comprehension—including my own. Despite what strangers sometimes think, I honestly have no idea what I'm doing in life, and I'm anxious all the time. But when I'm by your side, I simply don't care. You run my hurt away.

INTRODUCTION: THERE IS NO ALPHABET HERE

We Chinese wish to say that the privilege of a mere typewriter is not tempting enough to make us throw into waste our 4000 years of superb classics, literature and history. The typewriter was invented to suit the English language, not the English language the typewriter.
—"Judging Eastern Things from Western Point of View," 1913

As we mark the spectacular rise of the People's Republic of China, the opening ceremonies of the Olympic Games in 2008 have become a new node on our timeline. Observers of China were already familiar with the country's economic achievements over the preceding two decades, and perhaps with its advances in science, medicine, and technology. Never before had the world witnessed the full scale of China's twenty-first-century strength and self-confidence all in one sitting, however. August 8 was a theater of superlatives. The ceremony culminated the longest torch relay in Olympic history to that point (eighty-five thousand miles over 129 days), enrolled some fifteen thousand performers, and boasted a production budget of 300 million US dollars, all for the opening day pageantry alone.[1] If we include the games as a whole, and the massive infrastructural build-up in Beijing and other cities, the total budgetary outlay was somewhere in the neighborhood of 44 billion dollars.[2]

When we consider the towering cost of the spectacle—the cast, electricity bills, catering, costume design, construction crews, director Zhang

Yimou's paycheck, and more—it might seem curious to suggest that its one and perhaps only truly revolutionary moment was its least expensive and most easily overlooked. This was the Parade of Nations, the procession of national teams around the grounds of the Bird's Nest.

The first team to enter the Bird's Nest was Greece, as per Olympic tradition. Greece is the perpetual *ur*-host of the games, an event historically rooted in veneration of ancient Greek society and its esteemed place as the fount of Western democracy, science, reason, and humanism. The parade pays homage to Greece in a second, subtler way as well: the alphabetic order by which national teams enter the field of play. In his *Origins of Western Literacy*, Eric Havelock nominated Greek alphabetic script as a revolutionary invention that surpassed all prior writing systems, including the Phoenician from which it and all alphabets originate.[3] For historian, philosopher, and former president of the Modern Language Association Walter Ong, the Greek adoption and adaptation of the Phoenician alphabet was a force for democratization, as "little children could acquire the Greek alphabet when they were very young and their vocabulary limited."[4] Still others have ventured into dubious neurological claims, arguing that the invention of the Greek alphabet activated a hitherto dormant left hemisphere of the human brain and thereby inaugurated a new age of human self-actualization.[5] Greece gave us "Our Glorious Alphabet," and so every two years, we honor it in the opening ceremonies of both the summer and winter games.

The rules governing the Parade of Nations were first set down in writing in 1921 by the International Olympic Committee.[6] "Each contingent participating in the games," the regulations read, "must be preceded by a sign bearing the name of its country, to be accompanied by the national flag." Following this was a parenthetical note: "(countries proceed in alphabetical order)."[7] Such phrasing carried through to 1949, when the charter was adjusted slightly to take on the distinctly cosmopolitan form it maintains to this day. The revised regulations stipulated that it was the prerogative of the host country to organize the opening parade according to *alphabetic order as it functioned in the host country's language*.[8] With this adjustment, the IOC had taken clear steps to relativize, and thus universalize, the rules governing this international ceremony.

In the Tokyo Olympics of 1964, global television audiences might have been exposed to a non-Western and nonalphabetic script for the first time, were it not for Japan's decision to use English alphabetic order rather than kanji—the subset of the Japanese written language based upon Chinese characters—or kana—the syllabic part of Japanese writing, encompassing hiragana and katakana. Instead, it was not until the Seoul Olympics of 1988 that the world first witnessed a non-Western alphabet applied to this venerable Olympic tradition. Here in Korea, where *ga* (가) is the first syllable in Korean hangul, Greece was followed by Ghana (가나 *Gana*) and then Gabon (가봉 *Gabong*).[9]

In 2008, with the Greek national team entering the Bird's Nest—the architectural wonder designed in part by Chinese artist Ai Weiwei—the parade in Beijing was following a conventional script. Television commentators Bob Costas, Matt Lauer, Tom Brokaw, and others droned on in an unbroken stream of synthesis, as their roles demanded. They riffed on subjects as diverse as Confucianism, Tang dynasty cosmopolitanism, *taiji*, the Ming dynasty eunuch seafarer and explorer Zheng He, calligraphy, Buddhist iconography of the Dunhuang cave complex in northwest China, and the colorful diversity of China's non-Han ethnic minority peoples, among a mash-up of others.[10] The synthesis stumbled from time to time, tripping up in awkward turns of phrase (references to China's "long, *long* march" and "great *step* forward" come to mind). Endearing lapses notwithstanding, these play-by-play commentators were in rare form.

This constant hum of commentary contrasted sharply, however, with a forty-five-second span of complete exegetical breakdown that ensued when the *second* national team took the field: Guinea. Suddenly, Costas and his colleagues lost their groove.

COSTAS: Guinea follows them in. There is no alphabet here, so, y'know, if you're expecting one nation to follow the other the way they generally do at an opening ceremony, think again.

LAUER: Yeah, you're out of luck. It goes based on the number of strokes in the Chinese character [a shimmer of quiet laughter] that represents the country's name, so you could easily see a country that starts in 'A' followed by a country that starts in 'R' or vice versa. So we're gonna have the graphics at the bottom of the screen which will give you an idea … which countries are approaching the tunnel.

Greece, Guinea, Guinea-Bissau, Turkey, Turkmenistan, Yemen, Maldives, Malta.

G, T, Y, M?

There is no alphabet here.

If Costas was at a loss for words, one can hardly blame him. 2008 was the first time in history that the Olympic Games had been hosted in a country that did *not* organize the Parade of Nations according to alphabetic order of one sort or another—because it was the first time that the games had been held in a country whose language possessed no alphabet at all.

For over a century, IOC regulations had only *appeared* to the world as capacious, embracing of cultural difference, or in a word, *universal*. In 2008, the bylaws of the International Olympic Committee were unmasked as false pretenders to the throne of universalism. Predicated on the idea of choice and cultural relativism, the regulation's foundation in the idea of "alphabetical order as it functioned in the host country's language" brought the Olympic Games and their Chinese hosts to an embarrassing impasse. IOC regulations afforded China "permission" to undertake something that was, by definition, a logical impossibility: to organize the parade according to a "Chinese alphabet," which does not in fact exist.

The 2008 parade was not sequenced at random, however. There was a Chinese *dao* to match the Greek *logos*, one that functioned according to a two-part organizational system well known in China. In the first of these, Chinese characters are ordered according to the number of pen- or brushstrokes needed to compose them, an organizational scheme that had been a mainstay in China for centuries. The three-character Chinese name for *Guinea* (几内亚 *Jineiya*)—the first country to follow Greece—begins with one of the simplest characters in the written language, orthographically speaking: 几 *ji*, composed of only two strokes. By comparison, the three-character Chinese name for *Turkey* (土耳其 *Tu'erqi*), begins with the character 土 *tu*, whose composition requires three strokes in all. Consequently, Guinea preceded Turkey in the parade.

By itself, stroke count is not enough to arrive at an unambiguous order for the simple reason that many characters are composed of the same number of strokes. For example, the Chinese name for *Yemen* (也门 *Yemen*) begins with another three-stroke Chinese character, 也 *ye* (figure

I.1). Who would enter the Bird's Nest first, then: the Turkish national team or the Yemeni?

The second level of organization is based upon a centuries-old principle of Chinese calligraphy dating back at least to the Jin dynasty calligrapher Wang Xizhi (303–361). According to this principle, all characters in Chinese are composed of eight fundamental types of brushstrokes, ranked in a simple hierarchy: the *dian* (dot), *heng* (horizontal), *shu* (vertical), *pie* (left-falling diagonal), *na* (right-falling diagonal), *tiao* (rising), *zhe* (bending downward/rightward), and *gou* (hook) (figure I.2). When we return to the question of Turkey and Yemen, then, we find that "tu" (土)

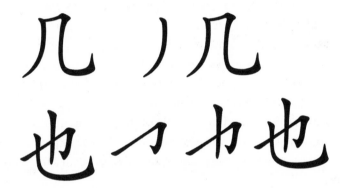

I.1 Stroke order of *ji* and *ye*

I.2 The eight fundamental strokes of the character *yong* (eternity)

of *Tu'erqi* (Turkey) is composed of the strokes *horizontal/vertical/horizontal*, or 2-3-2 in terms of the ranking of each stroke; while the "ye" (也) of *Yemen* is composed of the sequence *bending downward/vertical/bending downward*, or 7-3-7. The sequence 2-3-2 outranks 7-3-7, and so Turkey entered the Bird's Nest before Yemen.

Being unfamiliar with Chinese orthographic tradition, some in the Western viewing audience resorted instead to conspiracy theory. "Did NBC Alter the Olympics' Opening Ceremony?," user *techmuse* posted on *Slashdot* on the evening of August 9, 2008, triggering a cascade of just under 500 responses over the course of 48 hours.[11] A working thesis quickly formed that the sequence of national delegations—which so clearly violated anything remotely resembling an orderly procession—must have been garbled and resequenced as part of a profit-driven decision by television executives. Anticipating that American viewers would tune out following the appearance of the US team, the conspiracy theory held, NBC had cut up the original sequence and reorganized it so as to place the US delegation toward the end of the procession, and in doing so ensured a more enduring viewing audience. "American media alters the truth to boost ratings! Movie at 11," quipped *kcbanner* just moments after the opening salvo by *techmuse*.

Notwithstanding a scattered few who attempted to highlight the obvious—that the Chinese language has no alphabet, and thus that there might be an alternate explanation—online commentary traipsed further into the abyss of speculation, as if taking Costas's admonishment to "think again" with the utmost seriousness. Some believed the theory yet pardoned it, drawing upon a kind of gritty, world-weary cynicism. "Olympic Events have always been rearranged when on a Tape delay," interjected *wooferhound*. "I expect it, and why not? It is not even displayed in correct order when it's hosted in the USA." More extreme and jocular speculations surfaced in a comment by *Minwee*, likening NBC's supposed act to "the practice started with the 1936 Berlin Olympics when the German newsreels showed only negatives of all of the track and field events, so that a white Jesse Owens could be seen beating the pants off of all the black athletes."

It was not until the second day of commentary that the fraudulence of the original theory began to receive treatment. NBC had not doctored

Table I.1

SEQUENCE OF THE 2008 OLYMPIC GAMES PARADE OF NATIONS
(FIRST TEN COUNTRIES)

Order in parade	Country	Name in Chinese	Pinyin	Stroke count of first and second characters in Chinese name
1	Greece	希腊	Xīlà	7, 12
2	Guinea	几内亚	Jǐnèiyǎ	2, 4
3	Guinea-Bissau	几内亚比绍	Jǐnèiyǎ Bǐshào	2, 4
4	Turkey	土耳其	Tǔ'ěrqí	3, 6
5	Turkmenistan	土库曼斯坦	Tǔkùmànsītǎn	3, 8
6	Yemen	也门	Yěmén	3, 3
7	Maldives	马尔代夫	Mǎěrdàifū	3, 5
8	Malta	马耳他	Mǎěrtā	3, 6
9	Madagascar	马达加斯加	Mǎdájiāsījiā	3, 6
10	Malaysia	马来西亚	Mǎláixīyà	3, 7

the 2008 Parade of Nations; the sequence had simply followed a different organizational logic. What started with fury and excited speculation, then, ended limping, with one final rhetorical exclamation from *smitth1276*: "it doesn't bother any of you that this is an entirely inaccurate claim? The order wasn't changed at all, and whoever alleged that it was is smoking crack." And so the storm ended exactly two days after it began, on the evening of August 11.

Compared to the pageantry of August 8, 2008, 8:08 pm—a decadent onslaught of wireworks, fireworks, synchronized shouting, levitating LCD screens, Han Chinese toddlers donning non-Han minority outfits, an intense cardiovascular workout in the form of human-powered Chinese movable type, and child model Lin Miaoke lip synching "Ode to the Motherland" over the angelic, prerecorded voice of her more talented but apparently less aesthetically acceptable counterpart, Yang Peiyi—the nonalphabetic sequence of the Parade of Nations was more like an astute, Banksy-esque prank carefully crafted to perplex and subvert:

Greece, Guinea, Guinea-Bissau, Turkey, Turkmenistan, Yemen, Maldives, Malta.

G, T, Y, M.

There is no alphabet here.

Beijing's prank is even more delicious when we consider that China could have easily played along with IOC mythology by organizing the proceedings according to the Latin alphabet. For four decades or more, scarcely any Chinese dictionary, reference work, or indexing system on the mainland has employed the stroke-count organization used in the parade. To the contrary, in the 1950s mainland China developed and promulgated a Latin alphabet–based phoneticization system known as *Hanyu pinyin*, or pinyin for short. Designed by Chinese linguists shortly after the Communist revolution of 1949, pinyin is now ubiquitous in China, functioning as a paratextual technology that runs alongside and supports character-based Chinese writing, but does not replace it. Pinyin is not a "Chinese alphabet," thus, but China's *use* of the Latin alphabet toward a variety of ends. When Chinese toddlers first learn to read and write Chinese characters, for example, they learn pinyin at the outset in order to assist them with the memorization of standard, nondialect pronunciation. When computer users in mainland China sit down at their laptops, moreover, the keyboard they use is of the standard QWERTY variety, but is used to produce screen output entirely in Chinese characters (more on this subject later).[12]

Beijing could have spared Costas and Lauer their embarrassment, and avoided bewildering the global viewing audience, and yet chose not to. Clearly, Chinese organizers did not *want* to spare us, and herein resides Beijing's subtle act of defiance—the one truly revolutionary moment in the 2008 ceremonies, and perhaps the only one that did not contribute to their towering budget.

CHINESE IN THE AGE OF THE ALPHABET

This is the first of two books charting out a global history of modern Chinese information technology. Divided into seven chapters, this book moves across roughly one century, from the advent of telegraphy in the 1840s to the advent of computing in the 1950s. In the forthcoming book, we will carry this history into the present age of Chinese computing and new media. Over the course of this history, we will see that the encounter between Chinese script and the International Olympic Committee in 2008 was but one of its many encounters with false alphabetic universalisms of one form or another. Whether Morse code, braille, stenography, typewriting, Linotype, Monotype, punched-card memory, text encoding, dot matrix printing, word processing, ASCII, personal computing, optical character recognition, digital typography, or a host of other examples from the past two centuries, each of these systems was developed *first* with the Latin alphabet in mind, and only later "extended" to encompass non-Latin alphabets—and perhaps nonalphabetic Chinese.

As these information technologies spread around the world, a process of globalization greatly facilitated by European colonialism and later American global dominance, they came to be viewed by many as language-agnostic, neutral, and "universal" systems—systems that worked for everyone and every tongue. In truth, however, such myths of "universality" held up only to the degree that Chinese script was effaced or erased from the story. Remington and Olivetti, as we will see, proudly declared universality on behalf of their typewriters, as did Mergenthaler Linotype and Monotype on behalf of their composing machines, all despite the fact that not one of these companies ever succeeded in breaking into the Chinese-language market—a rather substantial omission in an otherwise triumphal story. Every time Chinese script did show up, as in the 2008 opening ceremonies, it could only lead to awkward situations. Just as engineers in China and elsewhere reconciled Chinese script with one or another of these technologies, moreover, the invention and circulation of a new alphacentric technology rebooted the struggle, again placing Chinese script at risk of being denied entry to and participation in the "next big thing" as it further transformed the worlds of economics, politics, warfare, statecraft, science, and more. Threaded together, then, we are confronted with a 150-year history of "Chinese information crisis, redux."

Throughout our exploration, we will pay close attention to engineers, linguists, entrepreneurs, language reformers, and everyday practitioners who struggled to usher character-based Chinese writing into a modern age of global information, and who subscribed to a common belief that, as one of our historical actors will phrase it, "Chinese characters are innocent" (*hanzi wu zui*).[13] For these individuals, the responsibility for China's technolinguistic challenges during the modern period rested squarely, not with Chinese characters per se, but with *people*—engineers who had yet to discover the key to what, in their estimation, was an eminently solvable puzzle, and perhaps everyday users of the language who, if Chinese writing was to survive in the modern age, would need to be willing to engage with this language in unprecedented and perhaps radically new ways. This puzzle would have to be solved quickly, though, for it constituted nothing short of a civilizational trial by which to judge once and for all whether Chinese script was compatible with Modernity with a capital M.

What name shall we give to this long history of false universalisms? *Linguistic imperialism* leaps to mind, and at first seems to fit the bill. The history of these encounters is inseparable, after all, from the broader history of China's engagement with Euro-American imperialism. Beginning in the nineteenth century, as we will see, Chinese script was enmeshed within a novel global information order whose infrastructure depended increasingly upon something China did not possess, and could not simply "adopt": namely, an alphabet. Linguistic imperialism is a short-lived candidate, however, for one critical reason: the issue at hand here is not the dominance or hegemony of any one specific language—be it English, French, or otherwise. This was not a case of a dominant language being imposed upon a subject population, as witnessed in certain colonial language policies during the modern period.

The terms *Western imperialism* and *Eurocentrism* are also not quite right. In a counterfactual universe, after all, had the IOC chosen Cairo, Yerevan, Bangkok, or Yangon to host the 2008 games, the pseudo-universalism of IOC regulations would have been compatible, at least linguistically, with any such hypothetical situation. Arabic and Armenian are alphabetic scripts. Thai and Burmese are alphasyllabaries, or abugidas. The Parade of Nations could have unfolded according to existing regulations, and the universalist myth could have lived to mystify another day.

The hegemony at play, then, is not primarily a matter of Occident and Orient, West and East, Roman and Exotic, or even Europe and Asia. It is not reducible to any such crude binaries. Instead, the divide is one that pits *all alphabets and syllabaries against the one major world script that is neither: character-based Chinese writing*. It is a new hierarchy of script that tells us: while some alphabets and syllabaries are more compatible with modernity than others, all alphabets and syllabaries can take pride in their superiority over Chinese. There is a missing term in our discussion, it would seem: one that would enable us to pay keen attention to the historic origins of this hegemonic system in Euro-American imperialism, while at the same time recognizing how this hegemony enrolls a wide diversity of scripts, both Western and non-Western, into its configurations of power. The true fault line at play here is not the West and the rest, but *pleremic* and *cenemic*. So long as a script is *cenemic*—a writing system in which graphemes represent meaningless, phonetic elements, based on the Greek term *kenos*, meaning *empty*—then the IOC's claim to universality stands, as will those made by the likes of Remington, Underwood, Olivetti, Mergenthaler, IBM, Adobe, and more. It is only in the case of *pleremic* script—writing systems like Chinese in which graphemes represent meaningful segments of language, based on the Greek *plērēs*, meaning *full*—that this universalism breaks down, as it did on August 8, 2008. Thus, while the origins of this hegemony are undoubtedly connected to the history of modern imperialism, nevertheless its expression takes the form of a different kind of binary altogether, one that divides the diverse multitude of cenemic scripts on one side from a singular pleremic script of immense scope and historical span on the other: Chinese.

TO BE OR NOT TO BE, THAT IS NOT THE QUESTION

China has undergone profound changes over the past five hundred years. At the midpoint of the last millennium, Ming dynasty China was one of the engines of the world economy, one of its largest population centers, and a sphere of unparalleled cultural, literary, and artistic production. Over the ensuing centuries, China witnessed a transformative conquest by a non-Chinese dynasty from the northern steppe; a doubling of the empire's size, as the consequence of immense Eurasian military campaigns into present-day Mongolia, Xinjiang, and beyond; a period of unprecedented economic and demographic growth during the eighteenth century; the emergence of ecological and demographic crises that spawned the largest and most destructive civil war in human history; colonial incursions by multiple Western nations that rewired the circuitry of global power; the demise of an imperial system over two millennia old; and a period of widespread political and social experimentation and uncertainty.

During the anxiety-ridden nineteenth and twentieth centuries in particular, Chinese reformers of multiple political persuasions engaged in thoroughgoing critical reevaluations of Chinese civilization in an attempt to diagnose the cause of China's woes, and to identify which aspects of Chinese culture would need to be transformed to ensure their country's transition into a new global order intact. Targets of criticism included Confucianism, government institutions, and the patriarchal family unit, among many others.

For a small but vocal group of Chinese modernizers, some of the most impassioned criticism was trained on the Chinese language. Chen Duxiu (1879–1942), a founding member of the Chinese Communist Party, famously called for a "literary revolution" to overthrow the "ornate, sycophantic literature of the aristocracy," and to promote the "plain, expressive literature of the people!"[14] "In order to abolish Confucian thought," the linguist Qian Xuantong (1887–1939) wrote, "first we must abolish Chinese characters. And if we wish to get rid of the average person's childish, naive, and barbaric ways of thinking, the need to abolish characters becomes even greater."[15] The celebrated writer Lu Xun (1881–1936) was yet another member of this anti-character chorus. "Chinese characters," he argued, "constitute a tubercle on the body of China's poor and laboring masses, inside of which the bacteria collect. If one does not clear them out, then one will die. If Chinese characters are not exterminated, there can be no doubt that China will perish."[16] For these reformers, abolishing characters would constitute a foundational act of Chinese modernity, unmooring China from its immense and anchoring past.

To abolish character-based writing invited serious peril, however. What would become of China's vast corpus of philosophy, literature, poetry, and history, all written in Chinese characters? Might not this inestimable heritage be lost to all but the epigraphers and specialists of tomorrow? Were China to abandon characters, moreover, what would become of the country's pronounced linguistic diversity? Cantonese, Hokkienese, and other so-called "dialects" of Chinese are as mutually distinct as Portuguese and French. Indeed, many have argued that the coherence and persistence of the Chinese polity, civilization, and culture have in no small part been predicated upon the unifying influence of a common character-based script. Were China to go the route of phonetic writing, would not these linguistic differences in the oral realm be made more insurmountable, and politically charged, once formalized in writing? Might the elimination of character-based writing precipitate the breakup of the country along fault lines of language? Might China cease to be one country, and instead become a continent of countries, like Europe?

The puzzle of Chinese linguistic modernity would appear, then, to be a perfectly irresolvable one. Characters held China together, but they also held China back. Characters maintained China's connection with its past, but so too did they isolate China from the Hegelian sense of historical progress. How then was China to make this seemingly impossible transition?

Returning to the twenty-first century, where the passages of Lu Xun and Chen Duxiu continue to bejewel the syllabi of countless undergraduate lecture courses on modern Chinese history (and scholarly writings alike), a world presents itself that hardly anyone could have anticipated at the dawn of the twentieth. Chinese characters were *not* exterminated, and yet China did not perish. Not only are Chinese characters still with us, clearly, but they form the linguistic substrate of a more vibrant world of Chinese information technology than even its most avid defenders could have dreamt of: a spectacularly large and growing presence in electronic media, widespread literacy, an ever-increasing network of Confucius Institutes and early education immersion programs propelled by foreign interest in the acquisition of Chinese as a second language, and continued popular fascination with Chinese characters that has manifested itself in not a few regrettable tattoos. More than ever before, Chinese is a world script. Throughout most of the past century, most assumed that such an outcome was conceivable *only* if China abandoned character-based writing, and underwent thoroughgoing alphabetization—which it did not. This outcome was not supposed to have been possible, and yet here we are. What happened? What did we miss?

The answer to this question is complex. At the outset, however, one key element can be outlined plainly: In sharp contrast to the popular trope that *winners write history*, in the case of modern Chinese language reform it is the *losers* of history who have managed to command the greatest attention of scholars—the Chen Duxius, Lu Xuns, and Qian Xuantongs. Collectively, we have been enamored of this vocal minority's brand of *easy iconoclasm*: incandescent, seductively quotable, yet ultimately naive calls to abolish characters, or to replace Chinese writing wholesale with English, French, Esperanto, or one of a variety of competing Romanization schemes. Meanwhile, we know virtually nothing about those who made possible China's contemporary information environment: iconoclasts who were no less passionate, and yet whose work was grindingly technical, dogged by intractable challenges, but ultimately of unparalleled success and significance. Unlike their celebrated and well-known abolitionist counterparts, the builders and users of the modern-day Chinese information infrastructure never appear on course syllabi, nor are their writings canonized within source compendia on the history of modern China. Indeed, they were often anonymous even in their own times, leaving behind fragmentary sources about their work, and in all but a very few cases never achieving any celebrity.

For these language reformers, the question of Chinese linguistic modernity was never the stark binary advocated by Lu Xun and Chen Duxiu: *In the modern age, are Chinese characters to be, or not to be?* Theirs was a far vaster, more open-ended, and thus more complex question: In the modern era, and particularly in the modern information age, *what will Chinese characters be, and how will the "information age" itself be transformed in the process?* Seductive though it may be, *To be or not to be?* was never the primary question of Chinese linguistic modernity. The question was: *To be, but how?*

When we move away from the simplistic iconoclasm of character abolitionists, an entirely new history of the Chinese language comes into focus. No longer in the realm of Confucian ethics or Daoist metaphysics, where Chinese characters were criticized by some as the very repositories of antimodern thought—as the "the very nests and lairs in which poisonous and corrupt thoughts reside," to pull another gem from Chen Duxiu's abolitionist jewel box[17]—we find ourselves in the admittedly less sensational yet decidedly more vital realm of Chinese library card catalogs, phone books, dictionaries, telegraph code books, stenograph machines, font cases, typewriters, and more—the infrastructural subbasement of Chinese script whose systems of inscription, retrieval, duplication, categorization, encoding, and transmission make it possible for the aboveground "Chinese canon" to function. We find ourselves in the plumbing and the electrical grid of Chinese.

Throughout the early twentieth century, just as some language reformers were critiquing the Confucian classics, many publishers and educators decried the average time required to find Chinese characters in the leading dictionaries of the era; library scientists lamented how long it took to navigate a Chinese card catalog; and state authorities bemoaned the inefficiency of retrieving names or demographic information within China's immense and growing population. "Everyone knows that characters are hard to recognize, hard to remember, and hard to write," one critic wrote in 1925. "But there is a fourth difficulty in addition to these three: they are hard to find."[18] These were problems, moreover, that could not be solved through mass literacy, the simplification of characters, vernacularization, or any of the other lines of action so often treated as synonymous with "language reform." And if these problems proved insoluble—if it proved impossible to build a telegraphic infrastructure for Chinese, or a Chinese typewriter, or a Chinese computer—then arguably even the most well-meaning efforts at mass literacy and vernacularization would not be enough to realize the ultimate goal: to usher China into the modern era.

CONTINUITY IS STRANGE

One of the few celebrated explorations of Chinese characters that dwells completely within the space of *To be, but how?* comes not from the world of scholarship, but from conceptual art. In 1988, the artist Xu Bing unveiled *Book from the Sky* (*Tianshu*), a work composed of four thousand *fake* Chinese characters. Despite their uncanny resemblance to actual Chinese, the graphs that formed the *Book from the Sky* resist all attempts to be read, offering the viewer no knowable sound (*yin*), meaning (*yi*), or shape (*xing*).[19]

This triad—*yin-yi-xing*—is of very old provenance in China, and many would say constitutes the three fundamental dimensions through which to define and understand Chinese writing in all its structural, stylistic, phonemic, and heuristic qualities. For the paleographer and the calligrapher, the most significant part of this triad is the *xing* or *shape*, the axis along which it becomes possible to engage with different historical forms of the Chinese character—such as seal script of the first millennium BCE, or clerical script of the Qin and Han dynasties—or with different calligraphic styles, such as "running script" or "grass script." For the poet and the philologist, by contrast, it is perhaps the *yin* or *sound* that stands paramount, being the ontological axis along which one can think about and engage with archaic pronunciations of Chinese words, or craft passages of great lyrical elegance. For the journalist and essayist, meanwhile, *yi* or *meaning* is at the core of one's concern, being the axis along which it becomes possible to find *le mot juste* or perhaps invent "new terms for new ideas."[20] These axes coexist and cooperate, no doubt: the poet cares for the *yi* and *xing* as well; and the essayist concerns herself arguably as much with *yin* as with *yi*. Important for us is not to distinguish these three axes, but to note that—in popular conceptions—they seem to exhaust all of the myriad possible understandings of what Chinese writing is, and thus can be.

Xu Bing's *Book from the Sky* explodes this idea. Exiting the three-dimensional, *yin-yi-xing* universe altogether, Xu carves a moat of infinite depth between his faux Chinese graphs and anyone—be they poet, essayist, calligrapher, philologist, or mere reader—who would seek to commune with them by means of sound, meaning, or shape. Fundamentally, they should not be "Chinese" at all. Something is wrong, however. If *Book from the Sky* constitutes a total rupture of the *yin-yi-xing* triad, and if this triad constitutes the entirety of what makes Chinese *Chinese*, how then are we still able to recognize *Book from the Sky* as somehow unmistakably Chinese?

The answer is that the *yin-yi-xing*/sound-meaning-shape triad does not, in fact, exhaust the entirety of what makes the Chinese character. This three-dimensional space, while accounting for many of the aspects of Chinese script that have occupied our attention for the greater part of history, is nested within further dimensions of writing that are largely invisible, inaudible, and unconcerned with meaning. In this book, I will refer to these dimensions collectively as the *technolinguistic*.

To inaugurate our discussion of the technolinguistic realm, I draw inspiration from typographer and type historian Harry Carter, who once reminded a somnolent world of a basic fact:

Type is something that you can pick up and hold in your hand. Bibliographers mostly belong to a class of people for whom it is an abstraction: an unseen thing that leaves its mark on paper. For their convenience it has long been the practice to talk about a typeface, meaning, not the top surface of a piece of type, nor even of many pieces of assembled type, but the mark made by that surface inked and pressed into paper.[21]

In describing the process of creating *Book from the Sky*, one as compelling as the finished work, Xu Bing helps us gain a clearer sense of what these dimensions entail. "My requirement of these characters," he begins by explaining, "was that they resemble Chinese characters to the greatest extent possible, while still not being Chinese." To achieve this objective, Xu set out on a painstaking analysis of real Chinese characters in order to extract those qualities he would imbue in his fake font. To begin, the particular number of fake characters he chose to produce—four thousand, as compared to one hundred or perhaps one million—came not at random but because it mimicked the statistical realities of character frequency and common usage within actual Chinese. "I decided to make just over four thousand fake characters," Xu explained, "because average reading material is also made up of around four thousand different characters." "When you've learned four thousand or more characters, that is to say, you can read and thus you're an intellectual." Structurally, Xu explained, the creation of his fake characters "required me to observe the internal structural principles of characters." To that end, Xu carefully examined the *Kangxi Dictionary* to determine the average stroke count of real Chinese characters, as well as their distribution across a curve from lower- to higher-stroke-count characters—all of which would then inform his production process. Stylistically and aesthetically, meanwhile, Xu Bing did not develop a "fake" font for *Book from the Sky*, but instead patterned his character forms after one of the most conventional Chinese fonts of all: the *Song-style character* or *songti*, widely used in printed Chinese matter well into the present day. "As for the font," he explained, "I thought to use Song-style characters. Song-style is also called 'court-style.' Often being used in important documents and in serious matters, Song-style is the font with the least individual flavor to it—the most standard font."[22] Xu Bing even pushed his exercise in verisimilitude into the realm of *taxonomy.* He created his own system for organizing his movable type blocks, so that he could retrieve them during the printing process—just like a real compositor.

When viewed through the *yin-yi-xing* framework, then, *Book from the Sky* might strike us as an exercise in rupture and discontinuity. When viewed within the technolinguistic realms of taxonomy, instrumentality, statistics, and materiality, however, we see that *Book from the Sky* was just the opposite: an exercise in *continuity*, or more accurately, an exploration of just how far one could push technolinguistic continuity and at the same time produce a Chinese script that violated everything that a Chinese script was supposed to be within the confines of the classic *yin-yi-xing* triad.

Rather than pursuing the history of Chinese script through the conventional fiction of the *yin-yi-xing* triad, a fiction in which script "comes loose from the metal" (to draw again from Carter), our examination will dwell primarily in a technolinguistic realm that makes the sound-meaning-shape triad possible.[23] We will climb into the manholes, crawlspaces, and airshafts of Chinese, exploring all of the complex and fascinating meaninglessness that makes meaning tick.

In concrete terms, what happens when we shift our attention to the technolinguistic? What happens when, following Carter's observation, we do *not* let the face come loose from the metal? First, we find ourselves less prepared. Traditionally, scholars of China are exceedingly well-trained at excavating meaning from the sound-meaning-shape triad, and attuned to detecting shifts therein. Indeed, as soon as the words "Chinese language reform" are broached, the thoughts of the historian turn almost instinctively to familiar subject matter: the deluge of newly coined Chinese words derived from other languages; the vernacularization movement of the 1910s and its call for greater congruence between written and spoken Chinese; the widespread development of specialized and professionalized Chinese-language discourses in fields as diverse as paleontology, aesthetics, law, constitutional reform, ethnology, feminism, and fascism; and efforts to forge a "national language" (*guoyu*) out of the welter of mutually unintelligible dialects. Other common invocations include calls for the Romanization of Chinese and the simplification of Chinese characters for the purposes of mass literacy.

If scholars of China are conditioned to think critically about changes such as these, transformations in the technolinguistic realm find us far less prepared. Changes in the organization of a Chinese phone book, the application of Western-style punctuation marks to Chinese texts, the reorientation of Chinese texts from vertical to horizontal alignment, the employment of numerical coding schemes to transmit Chinese characters over filaments of wire, the statistical analysis of character frequency in order to refine the parameters by which computers retrieve Chinese characters from memory—in short, the humble yet immense information infrastructure that enables the Chinese script to "work"—all seem to constitute "nondestructive" edits within the history of the language. They are changes, to be sure, but not ones that transform Chinese characters in any essential sense, whether in terms of structural makeup, phonetic value, or semantic meaning. What does it matter that a Chinese text is adjusted to read left-to-right where once it was read top-to-bottom? What does it matter that Western-style punctuation marks have been added, or an index, or page numbers, or a bar code? What does it matter that a once paperbound Chinese text is now found in proprietary digital formats like PDF? As long as the structures of the characters, their phonemic values, and their meanings remain the same, has not Chinese stayed the same—and with it the still unbroken, 5000-year hitting streak of the world's "oldest continuous civilization"?[24]

The answer is *no*. The technolinguistic domain does not stand apart from the better known *yin-yi-xing* triad; in fact, the historical transformations that take place within it—and especially those that throw it into crisis—are arguably more critical than those of sound, meaning, and shape. Consider, for example, what happens when we take a technolinguistic view of three of the subjects that China historians treat as synonymous with the subject of "Chinese language reform": the simplification of Chinese characters, the vernacularization of Chinese, and the pursuit of mass literacy. When viewed through cognitivist and sociocultural approaches, these three initiatives constitute the very core of language reform, pursuits around which the very question of language crisis seems to pivot. What happens, however, when we are faced with historical actors who were no less passionate about language reform, but whose goals included the creation of a Chinese telegraph code, a Chinese typewriter, Chinese braille, Chinese stenography, Chinese word processing, Chinese optical character recognition, Chinese computing, Chinese dot matrix printing, and more? For these reformers, many of the conventional topics of language reform actually made the problem of Chinese linguistic modernity even *harder* to solve—or at best had no impact on their pursuits one way or another.

Simplified characters are a case in point. While these were no doubt pertinent to questions of literacy and language pedagogy, a character reduced from sixteen strokes to a mere five—as with "dragon" (*long*), in its transformation from the graph 龍 to 龙—was no easier to transmit across telegraphic wires, set in movable type, or type on a Chinese typewriter than its "traditional" counterpart. "Simplification" here simplified *nothing*.

Vernacular Chinese, or *baihua*, actually made things palpably *worse*. Vernacular Chinese texts are invariably lengthier than equivalent messages in literary or "Classical" Chinese, and so this movement in the early twentieth century quite literally multiplied the challenges of transmission, inscription, and retrieval. To send a message in vernacular Chinese meant having to send *longer* messages, which in turn exacerbated the fundamental question even further: namely, of how to transmit, type, save, or retrieve a Chinese character—*any Chinese character*—in the first place.

Most counterintuitively, *mass literacy*—that preeminent concern of language reform—exacerbated the problem of Chinese information technology most of all. No longer able to rely upon certain kinds of assumed literate subjects—the literati and Civil Service examinees of old, for example—developers of the modern Chinese information infrastructure not only had to build all of these new and challenging technolinguistic systems, but they had to do so with only vague, imperfect, and constantly shifting understandings of what the millions of new Chinese *users* of their systems would be like. How literate would these users be, and in what ways? What dialects would they speak? What would their professions and educational backgrounds be? Would they be men or women, girls or boys? In what kinds of technical environments would they be operating? These questions would impinge upon the systems being built, and yet at no point in this history would anyone have a stable answer.

And *who* precisely had the authority to make such decisions? As party to this vexed transition from Qing imperial subject to informed (and informational) citizen of the new Republic, Chinese intellectual and political elites were undergoing a tortuous transition of their own. Over the course of the late nineteenth century, and most notably in 1905 with the abolition of the Civil Service Examination system, controls that the state and establishment intellectuals enjoyed over the Chinese language steadily crumbled, eliciting acute anxieties over when and how a new regime of language would take shape, what it would look like when it did, and who would be positioned at the apex of this new hierarchy. The uncertainty of this period was further compounded by the rise of an entrepreneurial cultural class, one that rushed into the vacuum left by disintegrating state power in hopes of establishing private and profitable cultural enterprises. Just as concerns over modern information management were reaching fever pitch, then, the combined absence of state control and the rise of the "business of culture" made this an ever giddier and more unsettled time.[25]

Above all, by turning our attention to the technolinguistic we begin to appreciate just how strange continuity truly is—a point that returns us to Xu Bing and his *Book from the Sky*. Continuity is strange because, despite commonsense understandings, it is in no way synonymous with conservatism. To *continue* something—in this case, to continue character-based Chinese script—can be avant-garde, iconoclastic, radical, and even destructive. Phrased differently, while it is virtually a cliché to speak of the "destruction" often entailed in acts of creativity, rarely do we pause to reflect upon the destruction central to acts of continuation. What is more, continuity and discontinuity are not antithetical concepts, as we see with Xu Bing and his fake characters. The question is not *to continue or not to continue*. The question is *what* to continue, and which kinds of discontinuity will be essential for achieving that goal. If the diverse actors in our history could be said to share one worldview in common, it could be summarized through recourse to a famous passage from the twentieth-century Italian novel *Il Gattopardo* by Giuseppe Tomasi di Lampedusa. Uttered by Tancredi Falconeri, a young prince in a wealthy aristocratic family in Sicily, and nephew to the book's protagonist, the brief passage contemplates how his family's position within society might ever hope to survive the tumultuous and unificationist impulses of the Risorgimento as it swept through their Sicilian homeland on its way northward:

Se vogliamo che tutto rimanga com'è, bisogna che tutto cambi.
In order for everything to stay the same, everything must change.

Giuseppe Tomasi di Lampedusa was no turn-of-the-century Chinese language reformer, and yet this passage elegantly captures a belief and impulse that propelled many of the protagonists we will meet in our history. Like Tancredi, they too believed that *in order for everything to stay the same, everything must change*. "Everything"—or *tutto*—is the operative term here, of course, which in its repetition corresponds to two separate referents. The first *tutto*—the one we want to stay the same—is the very *yin-yi-xing*/sound-meaning-shape triad discussed above: that part of script that lies above the surface, and through which the immense Chinese-language corpus is inscribed, read, appreciated, and more. This is the part of the script that naive iconoclasts would have done away with, whose utterly impractical schemes would have seen it replaced with Esperanto, or French, or one or another alphabetic scheme. By contrast, the second *tutto* corresponds to something else entirely, something which through its wholesale transformation might enable the salvation of the *yin-yi-xing* triad. This *tutto* corresponds to our *technolinguistic*: the infrastructure of language that, in its modest immensity, enables the language to work in the first place. If this *tutto* could be ripped apart, broken down, and reconstituted—the way that Chinese characters were categorized, retrieved, transmitted, materialized, ontologized, and indeed conceptualized—then perhaps the Chinese script could survive and even thrive despite our age of alphabetic hegemony.

FIELD NOTES FROM THE ABYSS

In this book, we will focus on one of the most important and illustrative domains of Chinese technolinguistic innovation in the nineteenth and twentieth centuries: the Chinese typewriter. In addition to being one of the most significant and misunderstood inventions in the history of modern information technology, so too is this machine—both as object and as metaphor—a historical lens of remarkable clarity through which to examine the social construction of technology, the technological construction of the social, and the fraught relationship between Chinese writing and global modernity.

If only it were peppered generously with moments of success and triumph, the story of the Chinese typewriter, and modern Chinese information technology more broadly, would be simpler and more joyous for the telling. If the Western typewriter marked a "revolution in production," as one historian phrased it, and one that "greatly increased the speed and reduced the cost with which a written document could be produced,"[26] our hope would be to make similar claims for the Chinese typewriter—to establish it as the unrecognized equal of its better-known Western counterpart. Another strategy might be to follow the path carved out by wildly popular "object histories," a veritable cottage industry in which "authors attribute to their chosen commodities an exaggerated, mysterious, almost godlike power," as Bruce Robbins has phrased it.[27] If tulips, codfish, sugar, and coffee have all changed the world, it stands to reason that perhaps the Chinese typewriter did, too.

No such triumphal story awaits the reader. While it might be tempting to identify one or another of our historical figures as the "Chinese

Charles Babbage," the "Chinese Grace Hopper," or perhaps the "Chinese Steve Jobs," this would amount to little more than a distracting parlor trick. While the Chinese typewriter *did* find its way into major Chinese corporations, as we will see, as well as metropolitan and provincial governments across the country, the Chinese typewriter did *not* transform the modern Chinese corporation or the functioning of Chinese government. For better or worse, no history of the Chinese typewriter can stake its claim upon the idea of "impact."

With all of this in mind, then, a fair criticism at this point might be: Did China need the typewriter at all, and if not, why do we need a history of it? Would it not be more accurate to say that China "skipped" the typewriter, moving directly to the computing age in the same way that certain parts of the world "leapfrogged" landline telecommunication, and entered directly into the world of cellular phones?[28]

In one respect, the answer is *yes*. To the extent that we demand certain things of a technology prior to admitting it into the category of "typewriter"—that it will have transformed the history of modern business communications and record-keeping, that it will have become a fundamental part of the feminization of the clerical workforce, that it will have become a cultural icon whose impact in popular culture extended far beyond its role as a business appliance—then the Chinese typewriter we are about to examine will not seem like much of a "typewriter" at all. Would it not be best simply to come clean, then, and admit that Chinese script is unsuited to the technology of typewriting, and that between the alphabetic and the nonalphabetic lies what has been called a "technological abyss"?[29]

In another more important way, the answer is resoundingly *no*. Chinese typewriting may not have approached the scale or centrality of its counterparts elsewhere in the world, and yet in many ways China experienced and engaged with the age of typewriting—and indeed telegraphy and computing as well—much *more* intensively than did the alphabetic world. As early as the 1870s, the novel inscription technology was known in China, and looked upon with admiration. In his account of the 1876 Philadelphia Centennial Exposition, Chinese Customs official Li Gui wrote of the "ingenious" device:

It was set up on a small square table and was only about a [Chinese] foot high and eight inches wide and made of iron. In the middle of it was a clever device embedded with ink, and an iron plate was set up under which were arranged all the letters of the foreign alphabet, twenty-six, like chess pieces, operated by a woman on the staff [i.e., a type-writer]. Paper is placed on the iron plate then, using a technique similar to that of playing the piano in foreign countries, it prints certain letters by means of her hands pressing certain alphabetical keys, while inside the machine an impression of each letter is struck. These are connected together to form words very nimbly and quickly. Offices all buy one, since its uses are many and its cost is only in the range of a little over one hundred dollars. Unfortunately, however, it does not print Chinese characters.[30]

To build a Chinese typewriter—a sentiment and desire just beneath the surface of Li Gui's closing remark—would be no small feat. To bring a nonalphabetic script into a technological domain that was built with alphabets in mind, engineers, linguists, entrepreneurs, and everyday users had no choice but to bring both script and technology into a shared critical space, posing questions that today might sound like irresolvable Zen kōans, but which in their original contexts were deeply practical ones: What is Morse code *without letters*? What is a typewriter *without keys*? What is a computer where *what you type is not what you get*? The Chinese typewriter was not a new type of mining drill, nor a new type of artillery, nor anything like most of the technologies being imported from abroad during the modern period—technologies that, while they undoubtedly demanded their own forms of intangible cultural, political, and economic practices and worldviews, could at the very least be "switched on" the moment they reached Chinese soil. As linguistically embedded and mediated technologies, Chinese telegraphy, typewriting, and computing explode conventional narratives of "technology transfer" and "diffusion" that have long guided our understanding of how industrial, military, and other apparatuses and practices circulated from Western loci of invention to non-Western loci of adoption.[31] Technolinguistic systems such as typewriting, telegraphy, stenography, and computing came with much steeper requirements. Because these systems were conceptualized and invented in direct association with alphabetic script, even the most basic functionality of a Chinese typewriter or a Chinese telegraph code required inventors, manufacturers, and operators to subject both the Chinese script and the technologies themselves to unprecedented forms of analysis and reconceptualization—to scrutinize *both* Chinese *and* type-writing, telegraphy, computing, and more. In order for everything about Chinese characters to stay the same, that is, everything about Chinese characters *and* modern information technology would have to change.

And clearly something did change—radically. In today's world, China is not only the largest IT market on the planet, but also home to a script that is among the fastest and most successful within our era of electronic writing—despite being nonalphabetic. Even if we accept, then, that from the nineteenth century onward the "technological abyss" between alphabetic and nonalphabetic was real, the fact remains that something *happened* in this abyss that has entirely escaped our attention. Indeed, if this book can be said to have one primary argument, it is that we must venture into the technological abyss to recover something of great importance that took shape here while the world was not paying attention—something that cannot be captured through conventional, celebratory, impact-focused histories of technology. This expedition, however, requires us to dispense with the easy iconoclasm of the character abolitionists, and equally so with any implicit desire for all histories of technology to be histories of triumph. Our story will be composed of what can only be called a long cascade of short-lived experiments, prototypes, and failures, where even the most successful devices lived only brief lives before disappearing into obscurity. Indeed, many of our Chinese telegraph codes, character retrieval systems, and typewriters were little more than speculations, wild ideas about how Chinese script *might* survive and function in the age of alphabetic hegemony. Counterintuitively, however, it is precisely within these speculations, short-lived successes, and outright failures that we witness the intensity of China's engagement with questions of technolinguistic modernity most clearly, and where both the material and semiotic foundations of the modern Chinese-language information infrastructure were slowly and unconsciously being laid. The history of modern Chinese information technology is not one that derives its importance and relevance from the magnitude of its immediate *effect*, but from the intensity and endurance of its *engagement*.

CAN WE HEAR THE CHINESE TYPEWRITER?

As we stand at the rim of the abyss, making final arrangements and provisions for our expedition down into its depths, a troubling question remains: When we finally encounter the myriad elements that populate this abyss—bizarre codes and speculative machines—will we be able to engage with them seriously, as anything other than pale imitations of their "real" counterparts elsewhere in the world? When we learn how many characters per minute a typical Chinese typist could manage in the 1930s, will our minds not immediately juxtapose this against the speeds we know to have been achieved by operators of Remingtons and Underwoods? When we see the chassis of a Chinese typewriter, will our aesthetic sensibilities not instinctively contrast it with the elegant, downright sensual design of the Lettera 22 by Olivetti? And when we first begin to listen to Chinese typewriters, will we be able to hear them through anything other than a soundtrack of modernity that, in our minds, is synonymous with the *rat-a-tat* cadence of the QWERTY keyboard? Our question is not *Can the Chinese typewriter speak?* but rather *When it speaks, will we be able to hear it?*

In 1950, American modernist composer Leroy Anderson (1908–1975) debuted a frenetic piece called "The Typewriter" in which he transformed this Western business appliance into a musical instrument. A soloist—the symphony percussionist, most likely—took his place at the most downstage position, in front of the orchestra, seated before a mechanical typewriter. Nested in an accompanying melody, the typist-percussionist set off on a blazing, nearly unbroken run of staccato thirty-second notes, punctuated by artfully placed rests and, to great comic effect, strikes of the typewriter bell indicating that the end of the line had nearly been reached. The piece was played "allegro vivace," or "bright tempo," with its 160 beats per minute evoking Rimsky-Korsakoff's "The Flight of the Bumblebee." Slightly less known than his "Syncopated Clock," Anderson's 1950 work found its way into popular consciousness, and has remained an infrequent yet always warmly received part of the cultural repertoire (it was recently performed in Ludwigshafen, Germany, by the Strauss Festival Orchestra Vienna, as well as at the Melbourne Fringe Festival). Perhaps its single greatest promoter came from outside the world of symphony: comedian Jerry Lewis, who performed a mimed rendition in his 1963 film *Who's Minding the Store?*

Anderson's mid-century piece is revealing, for it alerts us to the fuller spectrum of the alphabetic typewriter as an icon of twentieth-century modernity. The typewriter's day job might have been as an inscription machine and a business appliance, but it also moonlighted as one of the auralities of mass modernity: a sixteenth- and thirty-second-note soundscape in which we have lived now for over a century, continuing to absorb its cadences in the age of computing to the point where it is a taken-for-granted feature of our world. This soundscape was a long time in the making, moreover. In 1928, two decades before Anderson, someone attempted to capture in language the awesome and awful sound of the King Arms Thompson machine gun. While some called it the "Tommy Gun," riffing on its namesake, others dubbed it the "Chicago Typewriter"—the rat-a-tat of the typewriter likened to the rat-a-tat of bullets fired from a gun. This nickname inadvertently closed a historical circle, it bears noting, with the first mass-manufactured typewriters coming off the assembly line of the one-time Civil War–era weapons-maker Remington—a fact that prompted Friedrich Kittler to draw his famous analogy of the typewriter as a "discursive machine gun." By the 1930s, it was no longer the typewriter that took its name from the machine gun, however, but the machine gun that took its name from the typewriter.[32]

Aurality is only one part of the typewriter's iconography. Within the history of film, the typewriter was promoted long ago from mere set piece to all-but-dues-paying cast member. Whether in *His Girl Friday, The 400 Blows, The Shining, All the President's Men, Jagged Edge, Barton Fink, Naked Lunch, Misery, Schindler's List, The Lives of Others,* or any number of other examples, the typewriter has become an agent of narrative, sometimes even the fulcrum around which entire scenes and stories revolve. One of the machine's most audacious appearances was in the 1970 *Bombay Talkie,* in which one scene finds actors dancing on a gigantic typewriter as part of the film's culminating musical number. Referring to the typewriter as a "fate machine," the film expands upon this dramatic moniker by explaining that "typewriter keys represent the keys of life and we human beings dance on them. And then when we dance, as we press down the keys of the machine, the story that is written is the story of our fate." The film's famed Bollywood number, "Typewriter Tip Tip Tip," captures this same sentiment with evocative onomatopoeia:

A type writer goes *tip tip tip tip*
It writes every story of life.[33]

The Chinese typewriter we will meet in this book sounded nothing like Anderson's virtuoso, and it did not go *tip tip tip*. Neither did it leave its mark on a single famous Chinese writer. You will find no Chinese coffee table books featuring the likes of Lu Xun, Zhang Ailing, or Mao Dun at their faithful Chinese machines, cigarettes drooping James Dean–style from their lower lips. Likewise, there are no museums dedicated to the Chinese typewriter (yet), and with a few exceptions, nothing on par with the global network of collectors and nostalgics that has formed around its alphabetic counterpart. In more ways than one, then, the Chinese typewriter might not strike us as much of a *typewriter* at all.

As we set out to examine and understand this machine, and the broader history of modern Chinese information technology, the question that must constantly be invoked is thus: are we capable of doing so? To return to the metaphor of sound: if the Chinese typewriter cannot be heard except through Anderson's score, the Tommy Gun, and the *tip tip tip* of Bollywood, is it in fact possible to *hear* this machine at all? This is the principal methodological challenge of this book.

Depending upon the reader's disposition, this book puts forward answers to this question that might seem either naively optimistic or crushingly pessimistic. I *do* think it is possible to write a history of the Chinese typewriter, and with it the broader history of Chinese techno-linguistic modernity, but only to the extent that we abandon any and all fantasies of hearing this machine "on its own terms." No such aural space exists or has ever existed: no autonomous, unspoilt soundstage waiting to be reconstructed by the historian that, when rediscovered, will redeem the Chinese typewriter by returning it to its rightful place. The aurality of the Chinese typewriter was and has always been a compromised space, at all times somehow related to, completely ensphered within, and yet distinct from the global soundscape of the "real" typewriter of the West. In listening to the Chinese machine, we can never hope to isolate ourselves in the peace and quiet of an anechoic chamber, dwelling upon the fine texture of its sounds through high-fidelity speakers. The analytical space we occupy is more like a crowded café, music blaring throughout, in which we are straining to hear faint sounds. There exists no such thing as a "China-centered" history of the Chinese typewriter—nor of Chinese modernity.[34]

Methodologically, the posture I adopt in this book can best be described as *agonistic*: a posture in which our ultimate goal is not to arrive at a singular, harmonic, conflict-free, and final description of the history in question, but one that makes ample space for, and even embraces, dissonances, contradictions, and even impossibilities that are understood to be productive, positive, and ultimately more faithful to the way human history actually takes shape. To hear the Chinese typewriter, I argue, it is necessary *both* to interrogate and deconstruct our own longstanding assumptions about technolinguistic modernity—a practice that by now comes naturally to the historian—*and* to eschew all expectations that the act of critical reflexivity has the power to liberate us from these assumptions. No matter how intently I have listened to the Chinese typewriter over the past decade, and no matter how intently I have sought to denaturalize the cadences of the Remington machine and the QWERTY keyboard that play on perpetual loop in the recesses of my mind, there has never been a point during that time when I could hear the Chinese typewriter all by itself.

Chinese typewriters make sound, of course. They even have their own onomatopoetic counterpart to the *tip tip tip* of *Bombay Talkie*, and yet this counterpart cannot be found so easily, and is certainly not celebrated in popular culture. The sound of Chinese typewriters, as it was experienced by those who worked and lived with them, is instead buried in archival documents, where I found that the distinct rhythm and tonality of the Chinese machine was sometimes captured through the clip-clopping term *gada gada gada* (嘎哒嘎哒嘎哒). In this formulation, the *ga* refers to an initial sequence of movements entailing the depression of the typing lever, the resulting insertion of the metal slug into the type chamber, and the striking of the platen; while the *da* refers to a second sequence, entailing the type chamber falling back to its original position, and the ejection of the metal slug back into its original location on the tray bed matrix.

Sound is not the same thing as aurality, however. Even when I heard the machine for myself, with its *gada gada gada* rhythm, it was Anderson's typewriter that continued to set the sonic backdrop against which this sound was playing. *Gada gada gada* has its own rhythm, to be sure, but as for its *velocity*, my mind could not help but hear it as a half- or whole-note accompaniment to the thirty-second-note-clip of the *real* typewriter with its *tip tip tip* and *rat-a-tat-tat*.

What I steadily came to realize over the course of this research was that the Andersonian score was not something that my historical actors or I had a "view of" or "feelings about"—expressions that suggest some kind of critical distance between the object and the person. More accurately, the technolinguistic consciousness of the modern age *is* Remington, and to engage the Chinese typewriter is at all times to do so *inside* the Remington field. To undertake our exploration critically and productively, then, requires engaging with the question of agonism introduced above, beginning with a basic observation: Within and of itself, the deconstruction of our assumptions and categories does not rid us of these assumptions or categories. To historicize and deconstruct something is merely to destabilize it momentarily, to open up brief and fleeting windows of time in which something—anything—might happen that would be impossible if a given concept were allowed to reside in numb, dumb slumber. But the act of deconstruction does not endure. At most, it serves to contribute a tiny pulse of energy to a collective and sometimes exhausting struggle to keep a given concept or configuration "in play" just one moment longer. In this act we perpetually exert ourselves to drag concepts back mere inches from the precipice toward which everything slides inexorably: the precipice that separates the realm of critical thought from the vast wasteland of *the given*. Pessimistic as this may sound, I consider this invigorating struggle to be what gives critical thought its primary meaning—and to be one of the most pointed answers I can give when, particularly in the present day, humanistic thought is placed beneath the interrogator's lamp and demanded to justify its existence in our technophilic, antiintellectual age. Furthermore, I would argue that to eschew or retreat from this agonistic process is to deplete historicism and deconstruction of their only real power. For the scholar who demonstrates the constructedness of a thing, and in any way pretends to have transcended or made unreal the thing thus deconstructed; the scholar who proclaims the decentered-ness of his or her approach, and in any way pretends to have smudged this center from our maps; the scholar who dislodges a "master narrative" by multiplying or pluralizing it (modernity/modernities, enlightenment/ enlightenments), and imagines that anything has happened other than a consolidation of the master narrative by other means; such acts amount to exiting the field of intellectual struggle altogether, abandoning one's post, and leaving one's comrades immersed in a struggle now made incre-mentally more difficult by being "one intellect down." As we move forth to understand the history of the Chinese typewriter and Chinese infor-mation technology, we must adopt a critical relationship with our own "Remington selves," to be sure, but at all times remind ourselves that the mere act of critical self-awareness cannot by itself free us from this heuristic and experiential framework. We are not Remington, and we are.

A WORD ON SOURCES

This book is based on an eclectic body of sources, compiled over the course of ten years. It encompasses oral histories, material objects, family histories, and archival texts from more than fifty archives, museums, pri-vate collections, and special collections in nearly twenty countries. The global scale and diversity of this archive merits attention in at least two ways. First, it speaks to the challenges and inequalities built into writ-ing the history of modern Chinese information technology, especially in constituting the archives one requires to build our histories in the first place. While the history of the information age in the West enjoys countless museums and archival collections dedicated to the subject, nothing comparable is enjoyed by the historian of the information age in China, and arguably in the non-Western world more broadly. For that reason, I had no choice but to build an archive from the ground up, working in collections in China, Taiwan, Japan, the United States, Italy, Germany, France, Denmark, Sweden, Switzerland, the United Kingdom, and elsewhere. The history of the modern Chinese information infra-structure has to be pieced together from across a diverse, transnational, and all-but-overlooked cast of characters who fashioned an immense and complex repertoire of coordinated technolinguistic systems that by now govern the Chinese-character information environment in the form of indexes, lists, catalogs, dictionaries, braille, telegraphy, stenography, typesetting, typewriting, and computing.

Second, the scale and diversity of this archive speak to the fundamentally transnational nature of the history in question. Although we will be discussing the "Chinese typewriter," the term *Chinese* should not be taken as an adjective describing the nationality, mother tongue, or ethnicity of all our protagonists. Defying simple categorization, the actors in our story comprise a diverse and unusual cast of characters who hailed from all over the world, and yet endeavored to solve the puzzle of Chinese writing in the modern age. To tell the story of the Chinese typewriter requires us to travel, not only to Shanghai, Beijing, Tongzhou, and elsewhere in China, but also to Bangkok, Cairo, New York, Tokyo, Paris, Porthcurno, Philadelphia, and Silicon Valley. As it turns out, by writing the history of the Chinese typewriter, one necessarily sets out upon a global history of the information age.

To begin our exploration of Chinese technolinguistic modernity, in fact, we first head to San Francisco, where we will examine the invention of a Chinese typewriter that will enjoy worldwide fame and transform popular ideas about modern Chinese information technology—all despite the fact that this particular Chinese typewriter will never actually exist.

1

INCOMPATIBLE WITH MODERNITY

It makes the mind dizzy to think what a Chinese typewriter must be.
—*The Far Eastern Republic*, 1920

To handle a Chinese typewriter is no joke, meaning it is.
—Anthony Burgess, author of *A Clockwork Orange*, 1991

If a standard Western typewriter keyboard were expanded to take in every Chinese ideograph it would have to be about fifteen feet long and five feet wide—about the size of two Ping-Pong tables pushed together.
—Bill Bryson, 1999

The first mass-produced Chinese typewriter was a figment of popular imagination. It was first sighted in January 1900, when the *San Francisco Examiner* spread word of a strange new contraption housed in the city's Chinatown neighborhood, in the back room of a newspaper office on Dupont Street. The machine boasted a *twelve-foot keyboard* complete with 5,000 keys. "Two rooms knocked into one apartment afford shelter for this remarkable contrivance," the author explained, describing a machine so large that the "typist" was something akin to a general commanding forces over a vast terrain (figure 1.1). The piece was accompanied by a cartoon in which the caricatured inventor sat atop a stool and shouted Cantonese-esque gibberish at "four muscular key-thumpers through a

A CHINESE TYPEWRITER

1.1   Cartoon in the *San Francisco Examiner* (1900)

large tin megaphone."[1] *Lock shat hoo-la ma sho gong um hom tak ti-wak yet gee sam see baa gow!![2]*

One year later, some 1,700 miles to the east, the *St. Louis Globe-Democrat* featured strikingly similar imagery. In form, the Chinese machine bore a resemblance to Remington typewriters growing in popularity at the time, but in size it was monumental—complete with two stairways patterned after those one might find in the Forbidden City in Beijing (figure 1.2).[3] Here, the Chinese "typist" literally climbed up and down stairways of keys, haplessly searching for his desired character.

In 1903, a name was at last given to the imaginary inventor of this apocryphal machine. Photographer and columnist Louis John Stellman christened the inventor *Tap-Key*, a deft pun that played upon faux Cantonese and onomatopoeia.[4] "I see by one of the papers that a Chinaman has invented a typewriter which writes in the Celestial language," Stellman wrote, his description augmented by a drawing of yet another absurdly large contrivance (figure 1.3). No fewer than five Chinese

A CHINESE TYPEWRITER AT WORK.

1.2 Cartoon in the *St. Louis Globe-Democrat* (1901)

operators clacked away simultaneously at this immense keyboard, while five more fed immense sheets of paper through a platen of industrial proportions. Evidently, the number of personnel needed to operate a Chinese typewriter had doubled since the machine first debuted three years earlier.

Tap-Key and his monstrous machine never existed in the flesh—only in the imagination of foreigners. In another sense, however, this imaginary machine constituted the first "mass-produced" Chinese typewriter in history, one that circulated across space and time more widely than many of the real machines we will encounter. From its first appearance in 1900, this colossal Chinese contraption would make regular cameos in

1.3 Cartoon of Chinese typewriting (1903)

popular culture, whether in print, music, film, or television, demonstrating with each appearance the technological absurdity of character-based Chinese writing. These fantasies and their discomfiting portrayals of both the Chinese language and Chinese people are not vestiges of an unsavory past, moreover, but have lived on well into the present day. One of the more peculiar appearances of this imagined object took place in 1979, in the made-for-television movie *The Chinese Typewriter*, featuring Tom Selleck as a womanizing weapons-expert-turned-private-detective.[5] The plot centered around Selleck's attempt, in the role of Tom Boston, to recapture a passenger jet stolen by one Donald Devlin (played by William Daniels), a high-powered executive discovered to have embezzled millions of dollars from his company and fled to South America. Knowing of Devlin's avaricious tendencies, Boston and his partner Jim Kilbride (played by James Whitmore, Jr.) develop a plan to lure this well-connected and cautious criminal out of hiding with the prospect of a wild new business venture: a functioning Chinese typewriter.

KILBRIDE: Donald Devlin, his game is Money, right? So if you could figure out a way to make him richer, he'd probably shimmy up a cactus plant to get to it, right? But it would have to be something ... something so big, so exotic, something that appeals to the imagination. Something that hasn't been invented yet, I mean a real industrial bonanza of some kind. Like a ... a product for foreign export.

Camera cuts to the QWERTY typewriter on his desk, then back to Kilbride, who has begun to laugh.

KILBRIDE: Chinese typewriter.

BOSTON: Chinese typewriter?

KILBRIDE: Yes, yes. Yes, a Chinese typewriter.

The scene cuts to Kilbride's office, a think tank something along the lines of the Silicon Valley IDEO corporation *avant la lettre* in which free-range geniuses are seen tinkering on all manner of complex models, blueprints, and equations. Kilbride continues his explanation:

KILBRIDE: You see in China there's no such thing as a typewriter. They got a hundred different dialects, three thousand characters in the Chinese alphabet. So when a guy wants to type a letter he has to go to another guy, stand in front of this huge rack taking out each character one by one. It takes half a day to type a paragraph.

BOSTON: So what?

KILBRIDE: Well, so for years they've been trying to come up with a cheap computerized Chinese typewriter, one that can be sold and manufactured for 50 to 100 dollars per unit. The damn computer gotta be so large, the cheapest version they can come up with costs several thousand dollars. Too expensive to mass-produce. [Kilbride finds what he is looking for: the blueprints to a Chinese typewriter.] And these plans are useless 'cause they don't work.

Credit for the most memorable and impressive invocation of the Chinese typewriter, though, goes to Oakland-born rapper Stanley Burrell, better known by his stage name MC Hammer. In the music video to his 1990 multiplatinum hit "U Can't Touch This," Hammer debuted a bit of footwork that would go on to become one of the most well-known dances of the decade. Known as the "Chinese Typewriter"—a name that appears to have been coined not by the artist himself, but popularly and emergently—the dance featured Hammer side-stepping in rapid, frenetic movements. This step, someone apparently decided, mimicked the alien virtuosity of a Chinese typist as he navigated an absurdly massive keyboard crowded with tens of thousands of characters. Not unlike Tap-Key racing up and down stairways, Hammer's imaginary Chinese typist traversed great distances at breakneck speeds, and yet was the very embodiment of hopeless inefficiency, one whose very life force was gulped down by a lumbering behemoth that produced hardly anything in return.

If Hammer and Selleck have invoked the imagined Chinese typewriter in the arena of popular culture, still others have brought it into the arena of popular and academic scholarship. In 1999, celebrated writer Bill Bryson assured readers of his popular study of the English language that "Chinese typewriters are enormous, and most trained typists cannot manage more than about ten words a minute."[6] Drawing upon a vibrant imagination, an alphabet-centered understanding of information processing, and a total lack of familiarity with the technology of which he spoke, Bryson invoked the image of a hulking piece of equipment measuring some seventy-five square feet—"two Ping-Pong tables pushed together"—on which even a trained operator could not help but limp along at a comically slow pace. Walter Ong agreed. "There can be no doubt," he emphasized in his landmark *Orality and Literacy*, "that the characters will be replaced by the Roman alphabet as soon as all the people in the People's Republic of China master the same Chinese language ('dialect'), the Mandarin now being taught everywhere." "The loss to literature will be enormous," Ong continued, "but not so enormous as a Chinese typewriter using over 40,000 characters."[7]

Returning to Tap-Key and his monstrous machine, we are immediately struck by the dehumanizing and exoticizing caricatures. What concerns us primarily here, however, is not the charged racism of this imagery, but another aspect that easily escapes notice. In each of these many portrayals of the mythological Chinese machine, one invariably encounters massive *keyboards* with thousands upon thousands of *keys*. The question we will ask in this chapter is, quite simply: Why *keys*? Why did Stellman call his fictional protagonist *Tap-Key*? Why, when Bill Bryson imagined a Chinese typewriter, did his mind turn to a fifteen-by-five-foot *keyboard*? Why is it that, for us in the present day, as for those in 1900, simply to hear the words "Chinese typewriter" brings to mind a monstrous Rube Goldberg contraption featuring an immense keyboard upon which each of the language's tens of thousands of characters is assigned its own dedicated key? If it "makes the mind dizzy to think what a Chinese typewriter must be," where does our mental dizziness come from exactly?

A tempting reaction to this question would be to invoke "common sense": typewriters are, by definition, machines with keys and keyboards, making it perfectly natural that our minds should turn to such metaphors when imagining a Chinese "version" of this device. We could take this logic further, in fact, and contemplate all of the many subtle properties that we associate with the typewriter, often without realizing it. Visualizing a mechanical English-language machine in our minds, we might depress the key marked "A" and watch as the machine impresses the corresponding letter on the page, in lowercase. Automatically, the carriage advances one space, horizontally and to the left. If we depress the key marked "L," the carriage advances again, exactly the same distance, despite the difference in width of the letters "l" and "a." Our hands and fingers, poised above the keyboard, are also worthy of note. The very form of the machine enforces a visceral distinction between the different "strengths" of the digits of the hand: the pinky is *weak*, and the forefinger *strong*. We press the carriage return, the platen rotates a set distance, and the machine sails back across your line of sight, once again horizontally, but this time to the right. Such is the "essence" of the typewriter.

However commonsensical all of these qualities might sound to us now, none of them were predestined to become part of our taken-for-granted understanding. Were we to conduct the same thought experiment circa 1880, at a time when typewriting was a novel arena of practice still very much in formation, many more images would have come to mind—most of which have since left our collective memory. In the early years of American and European typewriting, as we will see, there were many different *types* of typewriters that did not necessarily contain the features we now consider part of the typewriter's inherent essence. Some machines were designed to be operated using only one hand, as with the Malling-Hansen Writing Ball, designed by Danish inventor Rasmus Malling-Hansen (1835–1890) and famously owned by Friedrich Nietzsche during the 1880s to compose letters during a time of rapidly declining health. Others arrayed the letters of the alphabet around a swiveling circular plate, as in the Lambert typewriter of 1904. Still others had *no keys or keyboards at all*, as with the American Visible Typewriter of 1891. Indeed, only one form of typewriter embodied all of the features we now consider the sine qua non of typewriting.[8] This was the keyboard-based, single-shift machine. The Remington, the Underwood, the Olivetti.

Returning to Tap-Key and the imaginary Chinese typewriter, then, this chapter mounts a counterintuitive argument: when we view denigrating cartoons of monstrous Chinese machines, or seemingly neutral statements about Chinese technolinguistic "inefficiency," we are in fact staring at the death mask of our own once vibrant technolinguistic imagination—the collapse of a once rich ecology of both machines and *ways of thinking about machines* that has since disappeared into the monoculture of the Remington world. In the wake of this collapse, and in the context of the Remington monoculture, it has become increasingly difficult to imagine anything *other* than keys and keyboards—and thus to imagine anything *but* monstrous Chinese absurdities equipped with thousands of keys. The Chinese monster in our minds is not a static image, that is to say—a photograph in an album that we retrieve and contemplate from time to time. It is the outcome of a kind of mental program that is occasionally jolted from dormancy and allowed to run its course. The program runs as follows:

A typewriter is an object with keys.
Each of these keys corresponds to one letter in the alphabet.
Chinese possesses no alphabet, but rather entities called "characters."
There are tens of thousands of characters in Chinese.
A Chinese typewriter must be an enormous device with many thousands
    of keys.

Every time it is initiated, this conceptual algorithm guides its thinker to the same invariant conclusion, all while producing the belief that he or she has arrived at this conclusion spontaneously and autonomously. The immensity of the Chinese typewriter is not something that must be ruminated upon. Rather, it feels true because it simply *insists*. It is this conceptual algorithm, and not the Chinese language, that is a cause of our dizziness.

The "Chinese typewriter" as monstrous Other was, in this way, the byproduct of a collapsing technolinguistic imagination in twentieth-century America and Western Europe. It derived from popular notions of Chinese exoticness and alterity, to be sure, but much more importantly from emergent and often unconscious beliefs about what constituted the "normal" relationship between language and machines in the alphabetic world. To understand our imagined Chinese typewriter, then, we must pay *less* attention to the "Chinese" part of this dyad, I argue, and far more to early Western conceptions of the "typewriter" itself—for it was through machines like the typewriter that many in the Euro-American world came to form deep-seated opinions about their own languages, as well as non-Western, non-Latin, and especially nonalphabetic scripts. We must dig deep into the history by which "keyboards" and "keys" became inseparable from our understanding of "typewriter."

Once we understand where this idea of the Chinese typewriter-as-monstrosity comes from, we are in a position to appreciate the potent ideological work it has performed over the course of its long career. Stipulating that Louis Stellman, Bill Bryson, Tom Selleck, and MC Hammer are not high on the list of individuals we often turn to for insight into the history of China, or the global history of modern information technology, nevertheless I argue that there is profound analytical value in understanding the process by which diverse groups of individuals can, when presented with the words "Chinese typewriter," consistently arrive at more or less the same monstrous and absurd outcome. This mental algorithm is a site of acute importance, for it is within this algorithm that we find insight into a central question of our larger story: the historical process by which longstanding nineteenth-century critiques of Chinese writing managed to survive the decline of those evolutionist and social Darwinist arguments on which they had, for more than a century, been based. The image of the absurd Chinese keyboard is thus neither frivolous nor innocuous. It is the successor to a discourse that in the previous century had been rooted squarely in notions of racial hierarchy and evolutionism. More than successor, in fact, this technological monster rehabilitated and rejuvenated Orientalist discourses, insofar as calls for the abolition of Chinese characters in the twentieth century and beyond have no longer needed to traffic in gauche, bloodstained references to Western cultural superiority or the evolutionary unfitness of Chinese script. Now the same arguments could be made more forcefully through the sanitized, neutral, and supposedly objective language of comparative *technological fitness*. After all, if a Chinese typewriter is really the size of two Ping-Pong tables put together, need anything more be said about the deficiencies of the Chinese language?

Before we move on to examine real Chinese information technologies beginning in the next chapter, then, it is vital for us to examine the history of the illusory ones, for in this history there emerged a pervasive and powerful interpretive framework from which real Chinese information technologies—real Chinese typewriters, in particular—never escaped over the course of their own histories. This history of our collapsing technolinguistic imaginary took place across four phases: an initial period of plurality and fluidity in the West in the late 1800s, in which there existed a diverse assortment of machines through which engineers, inventors, and everyday individuals could imagine the very technology of typewriting, as well as its potential expansion to non-English and non-Latin writing systems; second, a period of collapsing possibility around the turn of the century in which a specific typewriter form—the shift-keyboard typewriter—achieved unparalleled dominance, erasing prior alternatives first from the market and then from the imagination; next, a period of rapid globalization from the 1900s onward in which the technolinguistic monoculture of shift-keyboard typewriting achieved global proportions, becoming the technological benchmark against which was measured the "efficiency" and thus modernity of an ever-increasing number of world scripts; and, finally, the machine's encounter with the one world script that remained frustratingly outside its otherwise universal embrace: Chinese. Across this history, we will see how the rise of Remington in particular transformed the material, conceptual, and financial departure points for all subsequent thinking about typewriting for the world's languages. When Remington conquered the world, it was not "the typewriter" in any abstract sense that made its way into practically every corner of the globe—it was specifically the single-shift keyboard that achieved global saturation. This particular *type* of typewriter became the machine against which every writing system in the world would be measured, with profound implications for every one of them it absorbed—even more so for the one writing system it could not.

ASIA BEFORE REMINGTON

Our history of Tap-Key and the imaginary Chinese typewriter begins, not in China nor in the United States, but in Siam. Here, in the year 1892, the first Siamese typewriter was invented by Edwin Hunter McFarland, second of four children of Samuel Gamble and Jane Hays McFarland.[9] Before their children's birth, the McFarlands had put down roots in Siam and established the family as missionaries, doctors, educators, and philanthropists with access to the highest rungs of elite society.[10] Edwin graduated from Washington and Jefferson College in 1884, and returned to Siam to serve as the private secretary of Prince Damrong Rajanubhab, son to Mongkut King Rama IV, and half-brother of Chulalongkorn King Rama V.[11] In 1891, Prince Damrong dispatched Edwin to the United States with a very particular charge: to develop a typewriter for the Siamese script, just one of the court's many reform and modernization initiatives.[12]

Edwin enjoyed considerable resources in accomplishing this task. He had trained with his father in the art of printing, and could draw upon his father's work on the first printed dictionary of the Siamese language.[13] Even more importantly, Edwin had at his disposal a much wider array of approaches to the question of typewriting than would be true only a few decades later. At a time before Western typewriters had settled into the form we now take for granted, Edwin had before him different types of typewriters to choose from, each offering a different starting point from which to engage with this exotic, non-Latin script.

As he contemplated the written Siamese language, with its forty-four consonants, thirty-two vowels, five tones, ten numerals, and eight punctuation marks, Edwin would have encountered three typewriter paradigms, each presenting different affordances and limitations. One option was the index typewriter, a form of typewriter that did not have keys or a keyboard, but instead employed a flat or circular plate upon which the letters of the alphabet were etched. Using a pointer, the typist operated the machine by moving the pointer to the desired character, and then depressing a type mechanism.[14] The earliest known index machines were the Hughes Typewriter for the Blind (1850), the Circular Index (c. 1860, maker unknown), and the Hall Typewriter, developed in 1881 by the American inventor and entrepreneur Thomas Hall. One of the advantageous features of index machines was the interchangeability of types, fonts, and thus languages— a feature that inventors and entrepreneurs celebrated and promoted to potential customers. Like his contemporaries, Hall had global ambitions for his invention, setting out to internationalize the machine practically as soon as the first model was released in Salem, Massachusetts. As early as 1886, Hall began to promote his interchangeable typewriter plates for Armenian, Dutch, French, German, Greek, Italian, Norwegian, Portuguese, Russian, Spanish, and Swedish.

For Edwin's purposes, Hall's machine had its limitations, however. Like other American typewriter inventors, Hall thought almost exclusively in terms of Western European writing systems—whether Latin, Greek, or Cyrillic—formatting each of his interchangeable metal plates with the same eight-by-nine matrix format. With a total of seventy-two possible symbols, Hall's machine served such languages as Italian and Russian quite well, but fell just short of the number required for Siamese.[15]

A second option was the single-shift keyboard typewriter, exemplified in the manufactures of the Remington Typewriter Company. Founded by Eliphalet Remington in 1816, the company began life as a Civil War–era weapons manufacturer based in Ilion, New York. As the cataclysmic war came to an end, and as the United States entered the postbellum period, Remington set out to reallocate its efforts, collaborating with the typewriter companies Yost and Densmore, and the inventors Christopher Sholes, Carlos Glidden, and Samuel Lewis. In 1873, Remington debuted the Sholes and Glidden Type-Writer. This single-keyboard system featured an interface upon which each key corresponded to both the lower- and uppercase versions of each letter. The operator could toggle between cases using the now familiar "shift" key.

The limitations of the Remington device would also have been apparent to Edwin, however. In English, there is a sharp distinction in terms of frequency between lowercase and uppercase letters, one that made it eminently reasonable to sequester uppercase letters to the harder-to-reach "shift" level. In English, capital letters constitute between 2 and 5 percent of all printed matter, with lowercase letters accounting for most of the balance. Out of the 2,641,527 letters that constitute Jane Austen's *Pride and Prejudice*, for example, only 14,177 or 2.56 percent are capital letters. Melville's *Moby-Dick* exhibits only a slightly higher percentage of 2.91 percent. *Ulysses* by James Joyce and Shakespeare's *Hamlet* fall toward the outer bound of capitalization, ranking at 4.58 percent and 5.61 percent, respectively.[16] By offloading these little-used uppercase letters to secondary "shift" keys, typewriters could be reduced in size, without affecting ease of use or output.

The same could not be said of Siamese, whose alphabet does not distinguish between lower- and uppercase forms. As such the "shift" function of the single-keyboard machine would have required Edwin to relegate half of the Siamese alphabet to the more cumbersome, two-stroke operation. This was certainly possible, but far from opportune.

A third option, and the one that Edwin ultimately chose, was the double-keyboard machine designed by the Smith Premier Typewriter Company. Alexander T. Brown, an inventor from Cortland, New York, had founded the company in 1880 when, like his counterparts at Remington, he teamed up with the weapons manufacturer Lyman C. Smith. In a machine shop in Syracuse, they worked on the typewriter, and eventually made it a primary focus of their business. Indeed, thanks to the immense plant they constructed at 700 East Water Street, Syracuse came to be known by many as "Typewriter City." The company's flagship model at the time was the Number 4, with which Smith Premier established itself as the leader in double-keyboard or "complete keyboard" machines, as the manufacturers themselves referred to their design.[17] With 84 keys on the keyboard, the double-keyboard machine "provides a key for every character," the company's advertisements boasted, enabling greater accuracy than its shift-keyboard counterpart, saving time for the operator, and prolonging the life of the apparatus (insofar as there was no "shift" key that would shoulder the burden of heavy usage, and wear down or break

1.4 The Smith Premier double-keyboard typewriter

in the process). "In a shift-key machine," the Smith Premier company explained, "there is danger of error, the operator taking the hand out of its natural field to depress the shift-key."[18]

When it came time for Edwin to finalize a manufacturing agreement, it was in Syracuse, and not at the factories of Hall or Remington, that he found the best fit for the needs of Siam's modernization efforts (figure 1.4). Siam was to be Smith Premier country.

Having chosen his technolinguistic starting point, Edwin now needed to work with engineers to revisit some of the integral design principles of the machine to bring it into compliance with the specifications of

Siamese writing. One of the primary requirements was to retrofit the machine with a greater number of so-called "dead keys," a technical term referring to keys that do not advance the platen after an impression is made. Equipped with such dead keys, Edwin's retrofitted Smith Premier would thus be capable of handling Siamese accents, first registering an accent and then superimposing the letter.[19] Having built up the letter, the carriage then advanced to type the next one.

The Siamese script would need to change as well, alerting us to the fact that this technolinguistic negotiation was by no means a lossless one. Even with its ample 84 keys, Edwin's brother George later recalled, the Smith Premier machine "lacked two of the number needful to write the complete Siamese alphabet. Do what [Edwin] would, he could not get the whole alphabet and tone marks on the machine [figure 1.5]. So he did a very bold thing; he scrapped two letters of the Siamese alphabet." "To this day," he continued, "they are absolutely obsolete."[20]

In 1895, loss befell the Siamese royal family and the McFarland clan alike. The designated successor to the Siamese throne, Crown Prince Maha Vajirunhis, died of typhoid, whereupon Maha Vajiravudh acceded as the eldest son of King Rama V.[21] In the same year, Edwin died an early

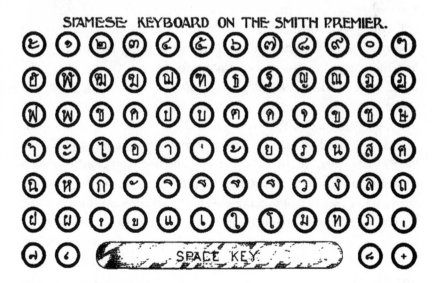

1.5 Keyboard of Siamese typewriter

death as well, exacting an emotional toll on the family, and leaving to his younger brother George the McFarland Siamese Typewriter. "From 1895, the typewriter became a part of the fabric of my life," George recalled. "On Ed's death it devolved on me to introduce the use of the Siamese typewriter. He had made it but it was not yet appreciated and wanted."

George McFarland was not an inventor. He was a dentist. Deeply embedded in Siamese society like others in his family, he managed the Siriraj Hospital of Siam and established the first private dentistry office in Bangkok circa 1891.[22] Here in his office, George placed his late brother's typewriter on display for his patients as a kind of deeply personal museum exhibit. Perhaps inspired by the curiosity it generated, or by the memory of Edwin, George soon took a much bolder step two years later: he opened a Smith Premier store of his own in Bangkok, thereby continuing the unlikely sisterhood between Syracuse and the Siamese capital.[23] "During the next few years," George recalled, "thousands of these machines were imported and the day came when no Government office felt it could do business without a Smith Premier."[24]

OUR COLLAPSING TECHNOLINGUISTIC IMAGINATION

The year 1915 marked the second, abrupt turn in the history of Siamese typewriting, and one that would in the end push the McFarland clan out of the typewriter business entirely. This change came, not from dynamics within Siam, but because of corporate maneuvers taking place half a world away in the United States. Having joined the Union Typewriter Company in 1893, a trust corporation that encompassed Caligraph, Densmore, Remington, and Yost typewriters, the Smith Brothers found their profits threatened by the new innovation of "visible typing" pioneered by the Underwood company. Up to this point, leading typewriter models were structured in such a way that the printing surface of the page was not viewable by the typist, being oriented inward toward the machine's type bars. To examine the text one was typing, a typist needed to lift up the chassis to view the typewritten text. Underwood's new model boasted a fully visible alternative that met with widespread approval and consumer demand.

Prevented by trust regulations from making sweeping structural changes to their typewriter, however, the Smith brothers sold all shares in the Smith Premier Typewriter Company, departed from the trust, and reorganized as L. C. Smith & Brothers Typewriters Inc. Their first model—the "Standard"—incorporated visible typing design, and in the process also abandoned their original double-keyboard format, moving to the increasingly dominant single-keyboard, shift-key typewriter form. As a consequence, the global supply of double-keyboard machines dried up—a shift that mattered perhaps little in the English-language market, but one that took out of circulation the device that formed the basis of Siamese typewriting to that point.

This change was cemented in 1915, when Remington purchased Smith Premier. As George recalled, it was "decreed that no more non-shift type-writers were to be manufactured."[25] This transition was captured in a pair of photographs from the period, the first showing George's storefront before the acquisition, and the second showing the newest outpost of Remington's worldwide network (figure 1.6). "It was a particularly dark day for Siam," George lamented, "because the Smith Premier had been so admirably suited to a language possessing so many characters."[26] As for the new model, "no one wanted it"—"no one knew how to use a shift machine: everyone cried for the old No. 4's and 5's. I was at my wits' end. I did not know what to do."[27]

George had little choice but to convert to the cause of shift typewriting or abandon the business altogether. While on furlough, he assisted Remington in developing its first portable Siamese machine. "The little machine was so attractive and convenient," he later confessed, "that people were induced to try to use a shift machine."[28] The Remington keyboard was adopted eventually by all manufacturers of Siamese type-writers, and at the same time, the models and styles of such machines proliferated along the same lines as those for other languages. Remington soon began marketing a Siamese Portable, a Siamese Standard, and a Siamese Accounting Machine, as well as a network of Siamese typewrit-ing schools centering on the Touch Method—at least one of which was founded and overseen by McFarland himself (figure 1.7).[29] The future of Siamese typewriting belonged to the typewriter form made famous by Remington.

1.6 The old McFarland store and its Remington Company replacement

1.7 Siamese typewriter by Remington, manufactured c. 1925 (USA), Peter Mitterhofer Schreibmaschinenmuseum/Museo delle Macchine da Scrivere, Partschins (Parcines), Italy

With the transition to a single-keyboard design, elements of Siamese writing once considered compatible with the technology of typewriting were suddenly flagged as "problems." The "characters are so numerous," remarked Abel Joseph Constant Cousin (1890–1974), inventor and French priest who worked with Remington's competitor, the Underwood corporation, to develop a new model of Siamese machine.[30] This was true not only for Siamese, he felt, but more broadly for "foreign languages of the Asiatic groups." The "problem was left unsolved," Cousin claimed in his patent application, "of adapting the Siamese language to a standard typewriter keyboard, which has only 42 keys." What was required was to "reconcile the discrepancy which exists in the requirements for 94 characters to be typed upon a 42-key-machine, since the latter is capable of operating only two characters for each key, or 84 characters in all." Certain limitations would need to be observed, however. "The cost of making a small number of enlarged machines sufficient to supply the market," he explained, "each having many extra types and keys, would be prohibitive, as it would involve practical redesigning of the machine thoroughly and would make it necessary to incur prohibitive outlay for newly designed manufacturing dies, patterns and equipment." Only by further pruning the Siamese script would it be possible to "bring the typing of Siamese substantially on par with that of modern European languages."

This passage by Cousin is revealing for three reasons. First, we see in real time how Siamese script became a "problem," and how this problem took shape as part of the relationship between Siamese orthography and the shifting grounds of technolinguistic imagination in the world of type-writing. Second, we notice the peculiar way Cousin went about assigning blame for this new "problem." Namely, for Cousin it was not that the Underwood was incompatible with Siamese, but Siamese that was incompatible with the Underwood. Finally, we notice the broader category Cousin deployed in making his claim about Siamese: the Siamese problem was not restricted to Siamese, in fact, but was instead an instantiation of a broader problem of the "Asiatic"—where "Asiatic" in this usage was effectively synonymous with scripts that exhibited an abundance of orthographic modules exceeding the capacity of his machine.

When George McFarland published his memoir in 1938, much had changed.[31] A small batch of photographs bear witness to this transformation, scarcely larger than postage stamps, printed in vivid black and white, and now housed at the Bancroft Library in Berkeley as part of McFarland's papers. Two young girls appear in the photographs, kneeling upon the ground and holding up a sign. The sign reads *Remington,* and flanking it are two other young girls, this pair wearing neckties, holding a winged Remington typewriter ensconced by a wreath, the machine posed as if flying through the air (figure 1.8).

1.8 Remington Siamese typewriter event photos

The wider scene behind the girls encompasses more than twenty-five school-age children, all encircling this homage to the typewriter, in the background standing the equestrian statue of King Rama V and the Ananta Samakhom Throne Hall.[32]

Siam was now part of the Remington empire.

"REMINGTON AROUND THE WORLD"

Remington's acquisition of the McFarland shop on the corner of Burapha and Charoen Krung roads was but one part of a much larger global effort, decades in the making. Remington first presented its new product to the world in 1876 at the Centennial Exposition, although with little fanfare; it was outshone there by Alexander Graham Bell's telephone, which captured worldwide attention. It was not until the 1880s and early 1890s that the company measurably increased its reach into markets both at home and overseas. In 1881, the company sold no more than 1,200 machines in all. In 1882, however, the sales agency Wyckoff, Seamans, and Benedict took over as the company's sales agency, bringing the device to markets worldwide.[33] Direct sales representatives were soon stationed in Germany in 1883, France in 1884, Russia in 1885, the United Kingdom in 1886, Belgium in 1888, Italy in 1889, Holland in 1890, Denmark in 1893, and Greece in 1896. As early as 1897, the company boasted of branches in Paris, Bordeaux, Marseilles, Lille, Lyons, Nantes, Antwerp, Brussels, Lisbon, Oporto, Madrid, Barcelona, Amsterdam, Rotterdam, and The Hague, among other European cities; and representatives throughout the Americas, Asia, Africa, and the Middle East (in locations including Algiers, Tunis, Oran, Alexandria, Cairo, Cape Town, Durban, East London, Johannesburg, Beirut, Bombay, Calcutta, Madras, Simla, Colombo, Singapore, Rangoon, Manila, Osaka, Hong Kong, Canton, Fuzhou, Macao, Hankou, Tianjin, Beijing, Jiaozhou, Saigon, and Haiphong).[34]

In 1897, Remington also began to herald its Number 7 model as the company's omnilingual flagship, encompassing "every language which uses the Roman characters" as well as Russian, Greek, Armenian, Arabic, and "a complete line of polyglot keyboards."[35] Ten years later, in 1907, the company would release its first front-strike, visible typewriting machine, the Number 10, and in 1915 it was selected as the Official Typewriter for the Panama-Pacific International Exposition in San Francisco (which would feature an illustrious Remington Pavilion, and for which all typed official communications were to use Remington machines).[36]

Part and parcel of Remington's ascendancy was the decline and disappearance of those alternate typewriter forms that Edwin McFarland had contemplated not long ago. The diverse ecology of approaches that had once characterized early typewriting steadily thinned out, replaced by a technolinguistic monoculture populated exclusively by varieties of the single-shift design. Double-keyboard machines like those from Edwin and George's past disappeared entirely from the market, while non-keyboard index machines all but vanished.[37] As McFarland's generation gave way to the next, moreover, new cohorts of inventors chose almost without exception to use the single-shift machine as their mechanical starting point when contemplating the design of foreign-language machines. The single-shift keyboard typewriter became a magnetic core, attracting ever-increasing numbers of patent applications, with Remington and other companies becoming the hubs of global sales, marketing, and distribution networks.

The globalization of typewriters was a source of immense pride and prestige for these companies. "When travelling, take a portable in your trunk," one pithy Remington ad read, portraying a caravan of elephant-riding Arab traders tracing a path across an unnamed desert, transporting a small rope-bound load of unmarked wooden crates (figure 1.9).[38] "Not everybody knows that there is an organized government in Mongolia," a *Wall Street Journal* article read in 1930, "but Remington Rand has filled an order for 500 Remington Typewriters for that government."[39] From its headquarters in Ivrea, the Italian manufacturing firm Olivetti (founded in 1908) shared in this broader discourse of global typewriting. In the pages of *Rivista Olivetti*, readers learned of the company's penetration of markets in Vietnam, Cambodia, and Laos. As the company's report explained, these societies remained charmingly ancient, and yet had also "adapted to modern life." "It is not without pride," the story remarked, "that Olivetti contributes to their forward march with its supply of typewriters."[40] Olivetti lauded its creation of an Arabic typewriter as well, assigning to it a practically civilization-shifting impact upon the Arab world. "Yes, for Arabs too have their typewriter," one *Rivista* article read,

1.9 Remington around the world: an advertisement for Remington typewriters

1.10 Advertisement for Olivetti Arabic typewriter

as if in dialogue with an incredulous reader, "and they owe it to the type-writer, which is now daily used by them, if they have been able to free themselves from the last practical difference which they still had with regard to the Europeans" (figure 1.10).[41]

The globalization of this technology, coupled with the rise of a techno-logical monoculture, exerted a profound effect on the cultural imaginary of script, technology, and modernity. A typewriter in Cairo would now look, feel, and sound *exactly* the same as one in Bangkok, New York, or Calcutta—all except for the symbols on each machine's keyboard. This unified, *rat-a-tat* cadence of the single-shift typewriter—soon referred to simply as *the typewriter*—would become part of the soundtrack of a new global modernity.

The apotheosis of the single-shift machine is best appreciated viscer-ally by perusing the holdings of some of the world's most extensive type-writer museums and private collections. Whether at the Peter Mitterhofer

Schreibmaschinenmuseum in Partschins, the Musée de la Machine à Écrire in Lausanne, or the Museo della Macchina da Scrivere in Milan, one must practically press his face up to the glass to discern the languages of the machines on display. Hebrew, Russian, Hindi, Japanese kana, Siamese, Javanese, and more are all practically indistinguishable, feeding into a peculiar effect in which language itself seems to be merely a *feature* or *amenity* of the machine.[42] What emerges is the effect of an omnilingual, omnicompetent, reified *ur*-typewriter that "comes in" Burmese, Korean, Arabic, Georgian, or Cherokee, the same way that its high-gloss exterior might "come in" black, gray, red, or green.

To achieve this effect was no small feat, it bears noting. Indeed, the globalization of the single-shift keyboard required nothing short of engineering brilliance. Even as *Remington News* and *Notizie Olivetti* blurred the lines between typewriters for Arabic, Hebrew, Russian, French, and Italian, presenting them as identical in every way except for their keyboards, engineers knew better that keyboards alone did not an Arabic, Hebrew, or Russian typewriter make. Keys and keyboards were but the most visible—one might reasonably say *superficial*—manifestations of a highly technical process of translating the materiality of the English-language machine into forms that could handle other languages and scripts. It was only inside the machine, amid its orchestra of subcomponents, that a typewriter could be said to "have" or "be" language—carriage advance mechanisms, spacing mechanisms, the selective use of dead keys. In fact, for engineers and manufacturers, it was not even in the devices themselves that language lived, but in the casts, dies, molds, presses, lathes, and assembly processes of the factory. More than the Remington typewriter, it was the Remington typewriter *factory* that constituted English. To translate the English-language Remington machine into Arabic, Khmer, Russian, or Hebrew was, in actuality, to translate the Remington factory itself.

Just like Edwin McFarland, inventors who set their sights globally confronted different challenges and complexities along the way. Each "problem," and indeed each trivial adjustment and nonissue, was party to a dialectic between their target script and their technolinguistic starting point, emerging not from any fundamental properties of either the writing system or the machine, but out of the tensions and fortuitous compatibilities that emerged between them.

In the era of Remington, scripts were measured not against the "English language" or the "Latin alphabet" in any abstract sense, but against the concrete, technolinguistic configurations of the single-shift keyboard machine as it had been built for the English language: a limited set of keys, limited capacities of character superimposition, with an equidistant, leftward-advancing carriage. All of these qualities, which perhaps seemed "invisible" and "natural" within the context of English typewriting, became either useful or obstructive properties that had to be reimagined and redesigned, one by one.

Engagements with exotic writing systems were not simple binaries, then, pitting Self against Other, or alphabetic against nonalphabetic. They involved a complex spectrum along which each of the alphabetic and syllabic scripts of the world were ranked on a scale of greater and lesser compatibility with the modern. At one end was the taken-for-grantedness of the English language, neighbored by those scripts that required of the English typewriter little more than cosmetic adjustments to the keyboard and the key surfaces. In the cases of French, Spanish, and Italian, for example, the fortuitous overlap of alphabets demanded at most a new layout of letters on the keyboard, so as to approximate better their relative frequency in different languages. Russian required only a somewhat more complicated transformation, in this case by outfitting the machine with the Cyrillic alphabet, which numbered a mere 33 letters.

At the other end of the spectrum, however, were arrayed those scripts whose technolinguistic performance demanded much more challenging engagements. The plasticity and universality of the typewriter form was tested to a far greater extent by Hebrew and Arabic, for example. These scripts demanded all of the more facile changes considered thus far—new letter frequency analyses, font-creation, the resurfacing of the keyboard—but also more complex shape-shifting. In Hebrew, the operative difference that concerned engineers was not the difference of alphabet, but the right-to-left directionality of the script. In a machinic sense, Hebrew was English *backward*, with engineers concentrating their attention upon what they saw as the salient part of the English-language machine in need of modification: the carriage-advancing mechanism. In 1909, Samuel A. Harrison filed a patent for an "Oriental Type-writer," which would be based upon the American-built Yost machine. Harrison's patent focused

exclusively on adjusting the Yost typewriter such that "by a certain adjustment the same operating device will cause the relative advancement ...
to be reversed whereby the feed of the member that is advanced is in the
opposite direction."[43] "By this means," Harrison explained, "the typebar
or type-carrying mechanism can carry two or more different alphabets,
one of which such as the English is printed from left to right on the
paper" while for the other "the printing will read in the reverse direction from right to left; as for instance the Hebrew language."[44] In 1913,
London-based inventor Richard A. Spurgin accomplished the same, this
time as assignor to the Hammond Typewriter company. Beginning with
the Hammond machine, he focused his efforts on creating a "reversible
carriage" for "Hebrew and those languages that require the operation of
the machine in a manner reverse to that of our language."[45]

In creating these slightly revised machines, Western designers and
manufacturers reopened certain "black boxes" within the structure and
behavior of the English-language device—in this case, the otherwise
taken-for-granted mechanism of leftward carriage advance. The typewriter form here had to grow, as it were, to encompass its mirror image,
wherein the depression of keys triggered a rightward rather than leftward movement, and the "return" key initiated the opposite. In the legal
realm, this required the filing of new patents and the drafting of appro-
priately worded explanations of the mechanism. In the realm of manufacturing, it required the adjustment of dies and molds, those negative
spaces that would be used to manufacture the new Hebrew "version" of
the typewriter form. Through the course of these adjustments, however,
engineers had to take great care. The starting "typewriter-self" could be
stretched and twisted, certainly, but engineers had to take pains not to
stretch or twist it so far as to "cut" or "tear" it—that is, to violate the starting condition in some fundamental way. The basic substance of the nor-
mative, English-language typewriter had to remain consistent: Hebrew
could not require a full-scale reimagination of the typewriter, only a varied type of performance.

Fortuitously for engineers, the solution to the Hebrew problem
brought them halfway through the Arabic problem as well, Arabic being
a script also written from right to left. The Arabic problem required still
another adjustment of the typewriter form, however, this one to address

the cursive ligatures that govern the flow of Arabic letters. Although type-writer engineers were pleased by the relative "economy" of Arabic let-ters, in terms of overall number, many Arabic letters are inscribed in one of four different ways, depending upon their relative positions within a word. Letters can appear at the beginning of a word (initials), betwixt letters (medials), at the close of a word (finals), or by themselves (iso-lated), challenging engineers to "fit" each of these graphemic variants on a device incapable of handling them all.

In 1899, a self-identified artist from Cairo, Selim Haddad, patented one of the earliest designs for an Arabic typewriter.[46] Arabic possessed only twenty-nine letters, he explained in the patent, but their different shapes and connections had "swelled the number of characters or type to the enormous number of six hundred and thirty-eight."[47] Haddad proposed an ingenious solution: for each Arabic letter, he would use only two vari-ants, rather than four, one variant to handle all initials and medials, and the other to handle all finals and isolates. Constructing "my new letters without a connection-bar solely on their right sides," he explained, "and constructing the middle and beginning letters with a connection-bar on their left I have derived advantages of great importance."[48] "It allows me to use one and the same shape of the letter for both beginning and middle positions and one and the same shape for both the end and isolated posi-tions."[49] With this innovation, he explained, he would reduce the overall number of variant forms from over six hundred to a mere fifty-eight—well within the compass of the single-keyboard apparatus.[50]

Inventors did not always agree on the best way to transmute the single-keyboard typewriter form to achieve such technolinguistic perfor-mances. The problem of Arabic orthographic variation was later revisited by Baron Paul Tcherkassov of St. Petersburg, Russia, and Robert Erwin Hill of Chicago.[51] Calling their machine a "Universal Eastern alphabet type-writer," oriented toward what they collectively described as scripts "such as the Arabic, Turkish, Persian, and Hindustani," Tcherkassov and Hill contended that the "Arabic problem" was to be resolved using a group of specially crafted, semantically meaningless graphemes that could be combined with real Arabic letters to produce all necessary ligatures. Sim-ply put, their Arabic typewriter would produce certain letters using con-ventional, single-key acts of inscription, while others would need to be "built up" using multiple keystrokes (some being real Arabic letters, and others being meaningless "connectors").[52]

No matter the disagreements, however, typewriter inventors in the twentieth century all subscribed to one powerful orthodoxy: never should the encounter with exotic scripts throw the single-keyboard typewriter form itself into question in any fundamental regard. "It is highly desirable in building a special machine of this sort," one inventor noted, capturing this orthodoxy succinctly, "that as much of the machine as possible may be of the ordinary standard form and of such a character that the special machines can run as far as possible through the ordinary course of manufacture for which the factory is organized and for which the tools are adapted."[53] Such motivations are readily understandable when we return to the discussion above regarding the factory as the site of language. Companies like Remington, Underwood, Olivetti, Olympia, and others had built factory floors fine-tuned to the stamping and assembly of metal parts, compositing them into precision devices to be shipped all over the world as part of a lucrative market. While a powerful economic motivation was to produce as many different types of foreign-language machines as possible, nevertheless the mantra of the period was, quite reasonably, that of *minimal modification*.[54]

By the midpoint of the century, the single-shift keyboard had conquered practically the entire world, with traces of its own historical particularity all but erased. Encounters with Siamese, Hebrew, and Arabic may have challenged the typewriter form, demanding it to stretch beyond its English-language and even Latin alphabetic origins, and prompting engineers to reopen such "black boxes" as leftward carriage advance, dead keys, and more, and yet never did such modulations threaten the machine's core mechanical principles. The fundamental blueprints of the starting point remained identical in every instance, as did the machine's underlying processes of casting and assembly. The single-keyboard machine had not only conquered the global typewriter market. It would seem to have conquered script itself.

The globalization of the single-shift keyboard typewriter had profound effects on the writing systems it absorbed into its expanding family. The most profound impact was reserved, however, for the one world script that it failed to absorb: Chinese.

TAP-KEY AND THE CHINESE MONSTER

Chinese characters eluded Remington, conspicuously and frustratingly absent from the company's growing roster. Although thousands of Western-style keyboard machines were indeed sold on the Chinese market, this was exclusively in support of the expatriate and Western colonial offices in China's treaty ports and missionary outposts. Although typewriter companies made widespread claims about the universality of their machines—that their machines could handle all languages—such claims quietly excluded a vast subset of the human population. The typewriter's "universality" was anything but.

When we reflect on the approach of engineers and inventors, the reasons for this absence are not difficult to surmise. If Hebrew had challenged engineers to make their machines bidirectional, vertical Chinese writing challenged them to imagine a machine that moved along a different axis altogether. If keyboard designers strained in their statistical analyses of Siamese, Russian, Arabic, and Hebrew, Chinese confronted them with an entirely nonalphabetic script. Albeit unintentionally, Chinese writing served as vigilant witness to the pseudo-universalism of the typewriter form, this false pretender to transcendence. That Chinese played this role, it bears reminding, was by no means preordained. Had no solution presented itself for the "Arabic typewriter problem" or the "Siamese typewriter problem," one or more of these scripts might have stood beyond the outer bounds of the typewriter's plastic embrace. One or more of these scripts might themselves have achieved the status, not simply of "an other," but as "the Other": alterity so radical that the Western typewriter form could not become it, except through a metamorphosis so intense as to annihilate this typewriter-self in the process. A solution was found for each of these puzzles, however, sometimes elegantly and at other times awkwardly: Hebrew became English "backward," Arabic became English "in cursive," Russian became English "with different letters," Siamese became English "with too many letters," French became English "with accents," and so forth. While different in many ways, Arabic, Hebrew, and Siamese were, in some fundamental sense, commensurable with the typewriter—and therefore, commensurable with the technolinguistic modernity it represented.

For predictable financial reasons, typewriter developers and manu-facturers never entertained the notion of abandoning their pseudo-universal typewriter form in the face of a recalcitrant Chinese script. What ensued was precisely the opposite. They dispensed with all of the romantic notions of civilizational possibility that characterized their engagements with other languages. They abandoned their seemingly boundless willingness to interrogate and reimagine many of the typewriter form's most taken-for-granted features. Instead, they marshaled all of the material and symbolic resources at their disposal to set out on what would become an unrelenting, multifront character assassination of the Chinese script itself—a kind of technolinguistic Chinese exclusion act. From this moment forward, it would be the Chinese script, and not any limitations of the single-keyboard typewriter form, that would bear the full weight of responsibility for the "impossibility" of Chinese typewriting—if Chinese was technolinguistically "poor," the script alone was responsible for its poverty. Phrased differently, the single-keyboard typewriter form would finally realize its universality by excommunicating from this universe one of the world's oldest and most widely used writing systems. In the Kristevan sense, the Chinese script was marked as the "abject form": an object or condition existentially intolerable to a given system or state of affairs, and as such one that had to be banished from ontology itself.

Here, then, we return to Tap-Key and the comical monstrosity of the imagined Chinese typewriter to explore the second question raised at the outset of this chapter: What ideological work is being performed by images and ideas such as these? What does it mean that the Chinese typewriter has been derided and denigrated by the strange bedfellows of MC Hammer, Walter Ong, Bill Bryson, *The Simpsons*, Qian Xuantong, Anthony Burgess, Tom Selleck, the *Far Eastern Republic*, the *St. Louis Globe-Democrat*, the *San Francisco Examiner*, the *Chicago Daily Tribune*, Louis John Stellman, and countless others?

To answer this question, we must momentarily return to a time before the typewriter, when critiques of Chinese writing were not grounded in terms of technology so much as those of race, cognition, and evolution. In *The Philosophy of History*, Georg Wilhelm Friedrich Hegel posited that the very nature of Chinese writing "is at the outset a great hindrance to the development of the sciences."[55] Arguing that the very structure of Chinese grammar rendered certain habits and dispositions of modern thought unavailable—ineffable and perhaps even unimaginable—he found that those who thought and spoke in Chinese were inhibited by the very language they used from ever taking the stage of progressive History with a capital H. All human societies were by language possessed, in other words, yet Chinese people had the misfortune of being possessed by one that was incompatible with modern thought.

Within the larger history of the anti-Chinese discourse, Hegel's role was one of transmitter and popularizer but not innovator. As many scholars have argued, the nineteenth century witnessed the formation of a powerful strain of social Darwinist thought which, like its parent theory, organized the totality of human language into a hierarchy of progress and backwardness.[56] The organizing principle, again reflecting its epistemic heritage, valorized the Indo-European language family, and deemed as developmentally retarded those languages that lacked such properties as declension, conjugation, and, above all, alphabetic script. As linguist, missionary, and sinologist Samuel Wells Williams (1812–1884) observed, "Chinese, Mexican, and Egyptian were alike morphographic; sometimes called ideographic." Among these, "Mexican" was barbarously destroyed by Western invaders, and Egyptian ultimately yielded to phoneticization. China alone tenaciously held on to this dying system of writing, "upheld by its literature; strengthened by its isolation; and honored by its people and their neighbors who had no written language."[57] What ensued was a "mental isolation caused by the language": "it has attached them to their literature, developed their conceit, given them self reliance, induced contempt of other nations, hindered their progress."[58] Such languages were understood as being stuck in a state of arrested development that, in turn, froze in time those who spoke and thought in, with, and through these languages.

Chinese was long a preferred target of social Darwinism. Comparativists dwelled on its "ideographic" script, tones, and lack of conjugation, declension, gender, and plurality. Chinese was, for many, the quintessential antipode, a conviction so powerful that even apologies for the Chinese language could be drafted into the service of its critique. In 1838, Peter S. Du Ponceau (1760–1844) mounted a painstaking argument in which he refuted the long-held idea of Chinese as an ideographic language, demonstrating that the majority of characters were in fact composed of both categorical and phonetic components.[59] Faced with this seemingly destabilizing proposition, one that might have closed the space of alterity between Chinese and non-Chinese languages, reviewers of Du Ponceau's work seized upon this idea of semi-phoneticization to recast the language as an evolutionary half-breed—a written language on its way to, but which never arrived at, full alphabetization. Du Ponceau, one review read, "has successfully combated the old and general opinion, that the Chinese system of writing is ideographic; showing that the characters do not represent ideas but words, which recall ideas."[60] At the same time, the reviewer continued, Du Ponceau had also demonstrated that the Chinese were linguistically inferior to even "the savage tribes of the New World." The latter, "though destitute of all literature and even of written language," the review continued, "are found to be in possession of highly complex and artificial forms of speech ... while in the Old World, the ingenious Chinese, who were civilized and had a national literature even before the glorious days of Greece and Rome, have for four thousand years had an extremely simple, not to say rude and inartificial language, that, according to the common theories, seems to be the infancy of human speech."[61] The lowest of the New World, it would seem, surpassed the highest of the Old.

Fetishization of the alphabet operated as a powerful trope within many disciplines, tending to show up wherever Western scholars engaged in the comparison of linguistic scripts and ruminated on the relative strengths and weaknesses thereof. In 1853, Henry Noel Humphrey wrote in his *The Origin and Progress of the Art of Writing* that the Chinese "never carried the art of writing to its legitimate development in the creation of a perfect phonetic alphabet."[62] "The Chinese language," a 1912 tract reported, "is the most horrible that any sane man can be called upon to acquire."[63] "The Chinese language must go."[64] "Phonetic characters in-the-making, like the Chinese," W.A. Mason echoed in his 1920 tract *The History of the Art of Writing*, "long since arrested in the development of its written characters at an early stage."[65] "Away with the old ideography," Bernhard Karlgren proclaimed cavalierly in his classic 1926 study *Philology and Ancient China*, "and replace it with phonetic writing."[66] A

1932 report phrased it even more bluntly: "The writing of Chinese in the Chinese manner is, as a proposition, simply 'too bad.'"[67]

Over the course of the twentieth century, however, voices both inside and outside the social sciences began to question the social Darwinist program, and with it, notions of Chinese as evolutionarily "unfit." In 1936, American sinologist Herrlee Glessner Creel (1905–1994) published "On the Nature of Chinese Ideography," an essay in which he mounted a painstaking critique of the widely shared belief that Chinese script constituted an orthographic half-breed caught between the presumed origins of all written language—pictography—and their presumed destiny of full phoneticization. Creel mounted a critique not only of anti-Chinese discourse but also of the West's broader preoccupation with cenemic script. "We Occidentals have come, by long habitude," he argued, "to think that any method of writing which consists merely of graphic representations of thought, but which is not primarily a system for the graphic notation of sounds, in some way falls short of what writing was foreordained to be, is not indeed writing in the full sense of the word."[68] Creel took aim directly at authors who believed in the supremacy of the alphabet, and the related idea that the grammar of Chinese rendered certain forms of thought—particularly those forms deemed critical to modernity—ineffable.

Creel's argument hinged upon a pivotal critique being leveled at the time against broader notions of comparative civilization and race science—a critique exemplified in the work of Franz Boas (1858–1942). While the impact of Boas's work could be felt in other disciplines, Creel explained, it was to be lamented that no parallel shift had yet taken place in writings on nonalphabetic languages such as Chinese. "In philosophy, in the study of society, in biology, we have at last abandoned the theory of unilinear evolution," Creel explained. He continued:

We no longer suppose that we can range all living creatures in a single line, from the protozoan to man. We recognize that phenomena are various and intractable, not fitting easily into our preconceived schemes. We have learned that theories must be cut to fit the facts, not facts to fit the theories. But in this matter of writing the old idea lingers on. If Chinese does not fit into the predetermined top of the scale, then it follows that Chinese is primitive.[69]

Culminating Creel's argument was a profound and deceptively simple pronouncement: "It is as natural for the Chinese to write ideographically as it is for us to write phonetically."[70]

Over time, evolutionist arguments against Chinese have fallen steadily into dubious standing. In his 1985 work *Writing Systems*, Geoffrey Sampson dedicated an extended meditation to refuting the notion of Chinese insufficiency.[71] Meanwhile, those who had once lent authority to such arguments began to backpedal. Referring to his and others' work in the influential volume *Literacy in Traditional Societies*, Jack Goody later stated that "We certainly gave greater weight than we should to the 'uniqueness of the West' in terms of communication, a failing in which we were not alone."[72] Goody began to tread more cautiously around the question of Chinese, and retreated from earlier claims of Western exceptionalism. "The logographic script inhibited the development of a democratic literate culture," Goody added, in line with his earlier claims, but "it did not prevent the use of writing for achieving remarkable ends in the spheres of science, learning and literature."[73] Goody steadily distanced himself from one-time fellow travelers, and from overreaching and Eurocentric scholarship that confidently presented the alphabet as the catalytic agent of the "Greek Miracle."[74] Although Eric Havelock saw fit to posit that "the Chinese script is a historical irrelevance," and although Robert Logan blamed Chinese script for the absence of a Chinese scientific revolution, Goody now went so far as to hint at the possibility of Chinese *advantages* and Western *disadvantages*.[75] "With a vastly reduced number of components," Goody wrote in 2000, "it becomes initially more difficult, but in the end easier to learn. Logographs, such as Chinese characters, can be learned one by one. Everyone, even without schooling and language learning, can be partially literate. In Japan, I have only to recognize the sign, not the word, for *entrance* or *men*, to be able to use the parking lot or the toilet; I do not have to understand a whole system, as with the alphabet."[76] "Of these there are some 8,000 in current use, although basic Chinese for popular literature needs a range of only 1000–1500 characters," Goody continued. "In these respects it is the most conservative of contemporary writing systems."[77]

Concepts of alphabetic supremacy and Chinese linguistic unfitness were not so easily dispelled, and yet those who continue to champion such views have found themselves increasingly marginalized. In the late 1970s and early 1980s, linguist and psychologist Alfred Bloom briefly took up the torch of the Chinese-as-nonmodern camp. The lack of a subjunctive mood in the Chinese language, he argued in the course of a 1979 article, rendered it impossible for Chinese thinkers to conceive counterfactually, thereby limiting their capacity to conceive of or generate the sort of hypothetical propositions that were so vital to the development of science and innovation.[78] This same view also informed the work of sinologist Derk Bodde who, in describing a China he saw as "linguistically handicapped," argued that "written Chinese has, in a variety of ways, hindered more than it has helped the development of scientific ways of thinking in China."[79] Inheriting and elaborating upon the long heritage of Chinese antimodernity, William Hannas has more recently attempted to resurrect many of the same arguments, contending that Chinese, Japanese, and Korean orthography "curbs creativity" and helps explain Asia's inability to compete in the world of technology and innovation.[80]

As William Boltz calmly noted, however, in reference to the work of Bloom, and by extension all those who mount arguments couched in ideas of cognitive limitation: "No serious linguist who knows Chinese has had any difficulty refuting it." Consonant with Creel's "principle of effability," Boltz underscored a claim that was fast becoming accepted fact—that "languages in their capacity to express human thought are all equal, at least in the sense that every language has the capacity or potential to express what its speakers want to express."[81]

With the decline of race science and the rise of cultural relativism, the story of the twentieth century would seem to be one of steadily growing cross-cultural engagement and understanding. Notions of Chinese linguistic "unfitness" have largely disappeared, or at the very least have become decidedly quieter and less self-certain. Those who carry the torch of previous generations have come to seem archaically Eurocentric and gauche—if not the dross of airport bookshop paperbacks, unworthy of serious intellectual engagement.

In reality, however, the concept of Chinese linguistic unfitness not only survived the decline of evolutionism and race science, but flourished in the new century. This rejuvenation and fortification owed its second life to technology, wherein questions of Chinese linguistic fitness henceforth moved out of the politically untenable realm of race, and into the sanitized realm of technological devices like the typewriter. Principal among the denigrators were technologists themselves. "From Ancient to Olivetti," the Lettera 22 campaign of the early 1950s read, this slogan displayed prominently above a pair of contrasting images: the Olivetti typewriter, which stood as the emblem of sleek and functional modernity, and a potpourri of Chinese characters of the kind found on oracle bones from the Shang dynasty (1600–1046 BCE)—used as the token placeholders for antiquity (figure 1.11).

By the second half of the twentieth century, there existed a global echo chamber in which tropes of Chinese technolinguistic absurdity and irrelevance resonated and repeated, unchecked and uncritically. In 1958 Olivetti proclaimed that their machines "write in all languages" (figure 1.12).[82] Like Remington and Underwood, Olivetti could enjoy such a statement only to the extent that the company barred from view the script that stood frustratingly outside its embrace: Chinese.

Meanwhile, the rest of the world's love affair with the keyboard typewriter grew ever more passionate. The typewriter was an inscription machine, first and foremost, but far beyond that, it had become a thick symbolic ecology made up of imagery, aesthetics, iconography, and nostalgia. In service to the cult of authorship, the typewriter became a legitimating mark of the artist. Any writer of repute (real or imagined) had to be photographed at some point before his or her favorite model, immortalized in the smoky act of creation. By mid-century, the cult of the typewriter was so strong that Allen Ginsberg, in the famous closing passage of *Howl*, could even use this inscription device to announce the sacredness of the device itself. In more ways than Ginsberg himself may have realized, the typewriter was indeed *Holy!*[83]

Meanwhile, more than any other symbol, the "Chinese typewriter" as imagined object became the single most widespread and damning piece of evidence in the rejuvenated trial against Chinese characters—one in which Chinese script was again found to be incompatible with modernity and deserving of abolition. What at first ran parallel to, and helped to illustrate its more dominant evolutionist counterpart, soon inherited the throne as the only acceptable mode by which concepts of Chinese unfitness could be deployed. By conjuring up farcical and absurdist images

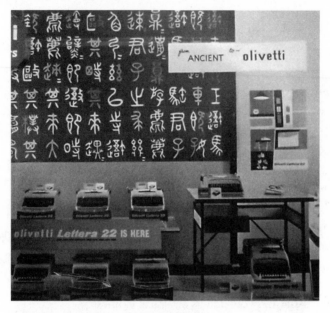

1.11 Advertisement for Olivetti Lettera 22

of monstrous Chinese typewriters, the criticism of Chinese inoculated itself against claims of unsavory evolutionism, and recast itself in the sanitized and supposedly objective language of *technological fitness*. While it became déclassé for one to concur with Bodde, Havelock, Bloom, and others on evolutionist grounds, or with their more distant Hegelian antecedents, the ongoing trial against the modernity of the Chinese language was renewed in the twentieth century with redoubled vigor in the seemingly neutral realm of *technolinguistics*. Perhaps Chinese speakers were able to express themselves as completely as those of Western languages in a cognitive sense, and so Hegel was wrong. Yet *technologically*, speakers and writers of Chinese were demonstrably hindered by their onerous script, one that obstructed literacy and the adoption of modern information technologies such as telegraphy, typewriting, stenography, punched-card computing, and more—and so Hegel was *right*. This modern technological critique of Chinese—one framed in the clean, plastic-and-metal world of carriage advance and platens, rather than blood-temperature terms such as cognition, culture, race, social Darwinism, and evolutionism—would

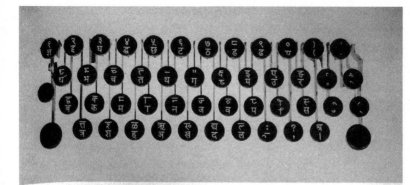

Le macchine Olivetti
scrivono in tutte le lingue

Le nostre fabbriche producono per tutti i mercati
macchine per scrivere con 170 diverse tastiere

« Illustre A;ico, Le ;audia;o un elenco di L wone... ». La perfetta dattilografa, la campionessa delle 500 battute al minuto alle gare di dattilografia, la segretaria modello che scrive — secondo i rigidi canoni dei libri didattici — senza degnare di uno sguardo la tastiera ed il foglio di scrittura, facendo due errori di battuta su trenta lettere quotidiane, rimane esterrefatta davanti a tale ignominia, da lei commessa. « Che mai succede? Capogiro? Reazione del subconscio? ».
Si tranquillizzi, signorina, lei sta benissimo. Non stia a disturbare Freud. Suo unico difetto è stato quello di non controllare, prima di iniziare la lettera, che la macchina su cui scrive fosse veramente la sua, quella che adopera abitualmente. Qualcuno, infatti, forse per scherzo, ha sostituito la sua macchina, con tastiera italiana, con una inglese, che ha alcuni tasti disposti differentemente. Le battute erano precise, e se fossero state eseguite sulla solita macchina la frase sarebbe risultata esatta: « Illustre Amico, le mandiamo un elenco di 5 zone... ».
Questo episodio, probabilmente non accaduto, ma che

potrebbe benissimo succedere, serve ad introdurci nelle non semplici vicende delle tastiere delle macchine per scrivere, assillante preoccupazione di inventori e costruttori.
Già l'avvocato Giuseppe Ravizza di Novara, sfortunato predecessore dei fabbricanti di macchine per scrivere (i suoi « cembali scrivani », realizzati artigianalmente dal 1855 al 1881, erano strumenti rudimentali che anticipavano i principi base delle moderne macchine per scrivere), intuì che il problema della tastiera rappresentava un elemento di primaria importanza nella sua invenzione.
« Decisamente — egli annotava — il maneggio del " cembalo scrivano " è ben diverso da quello del pianoforte. In questo la mano scorre e salta continuamente ed ha bisogno di un certo agio, nel mio la mano deve stare ferma o quasi raccolta, e le sole dita lavorano in lunghezza ed in larghezza quanto più si può concentrata e ristretta ed i tasti avere quella sola larghezza che comporta la dimensione delle dita e non più ».

1

1.12 Olivetti article from 1958

in the last will and testament of its aged and estranged forebear deftly and quietly be named the inheritor of its entire discursive fortune.

Having now prepared ourselves to meet real Chinese typewriters—to view them with our own eyes, and listen to them with our own ears—we will need to remain conscious of the fact that, at all times, the interpretive frameworks through which we grasp and understand such sights and sounds will never stop being shaped by and refracted through the imaginary Chinese machine we have just come to know. Our eyes and ears are not our own private possessions, that is to say, but products of the very history examined in this chapter. Rather than disdain *Tap-Key* or feign liberation from him and his descendants—efforts that would not only be disingenuous but also unproductive—our posture should instead be one of uncomfortable embrace. The typewriter form globalized by Remington, Underwood, Olivetti, Olympia, and others is not an object with which we have a "relationship," in the sense that it sits at a distance from us, separated by Cartesian emptiness. The typewriter form that emerged during the twentieth century, and which spilled out into a broader iconography, is an object we think *with* and *through*, not an object we think *about*. By accident of history, our consciousness at this particular moment in time *is* Remington.

We travel now to Ningbo, in southeast China, but to a time prior to the emergence of the typewriter. As we will see, the "puzzle" of Chinese typewriting—that is, of fitting a nonalphabetic script containing thousands of characters within a novel information technology—is one that first emerged in the 1800s among foreigners who contemplated the relationship between the Chinese language and two earlier technolinguistic systems: movable type and telegraphy. It was here in the 1800s, before the advent of the typewriter, that the puzzle of Chinese typewriting first began to take shape.

2

PUZZLING CHINESE

Of all the languages in the known world, the most difficult to represent by movable type is, without controversy, the Chinese; having hitherto baffled the most skillful European typographers.
—Marcellin Legrand, 1838

That the mosquito-wing-thin pages of William Gamble's notebook survived those four tedious years is a marvel. It would have accompanied the Irish-born American printer to China in 1858, where he was dispatched to oversee operations at the Presbyterian Mission Press in Ningbo, one hundred miles south of Shanghai. This was Gamble's workbook for practicing Chinese characters.

Each page of the journal was divided into a fifteen-by-fifteen grid, enabling Gamble (1830–1886) to practice just over two hundred Chinese characters before moving on to the next sheet. Within each cell, a smaller four-by-four matrix served as guidelines to help the printer work toward composing characters with structural balance and elegant dimension. Preserved now at the Library of Congress in Washington, DC, it is one of the more intimate parts of the Gamble collection.[1]

Gamble did not use this particular journal to practice his penmanship, however. Instead, he used it as a kind of accounting ledger with which he and two Chinese assistants tallied data about the relative frequency of Chinese characters within an immense corpus of Chinese texts. For four painstaking years, Gamble and his Chinese colleagues—referred to by Gamble as "Mr. Tsiang" and "Mr. Cü"—examined roughly 1,300,000 Chinese characters in total, spread out over four thousand pages.[2] Working line-by-line, they made a record of every one of these characters, counting every time each character appeared, and organizing the data into handwritten tables.

If only we enjoyed a time lapse film of these three men's labors, the obsessive quality of their venture would become clear: bodies gyrating in a mechanical *anti-reading* of such texts as the *Great Learning* and the *Daode-jing*, dismembering and dissolving their substance into elemental units, and then ranking these elements into classes of frequency.[3] The *Zhuangzi* as living text might confound and delight—*Now I do not know whether I was then a man dreaming I was a butterfly, or whether I am now a butter-fly, dreaming I am a man*—but this was not Gamble's principal concern. For Gamble, what mattered over the course of those four years was to determine the raw composition of the *Zhuangzi*—to determine, as it were, that the *Zhuangzi* was nothing more or less than a composite made up of approximately 8 percent *zhi* (之), 5 percent *er* (而), 5 percent *bu* (不), 4 percent *ye* (也), and only a small handful of "butterflies" (*hu* 蝴 and *die* 蝶).[4]

Faced with the algorithmic obsession of William Gamble, a question that instinctively comes to mind might be: *What did he discover? Which Chinese characters were the most frequent, and least? What might this mean for the question of Chinese technolinguistic modernity we are pursuing? Was this the answer?* While this chapter addresses these important questions, our primary concern will be to dig beneath Gamble's work and ask: *How did he come to settle upon this puzzle in particular, and what were the consequences of this particular puzzling?* How did he come to decide that, among all the possible questions one might raise about Chinese script, this was the one he would let dominate four years of his labor and his obsession? As we set out to examine the "puzzle" of Chinese writing, this is to say, the first questions we will ask are not about the solutions offered to this puzzle, but about the puzzles themselves. How and why did the Chinese script come to be viewed as a puzzle in need of a solution in the first place, let alone one so addictive that it could command someone's concentration for four painstaking years? Moreover, why did Gamble choose to turn Chinese into this *kind* of puzzle—one whose solution was premised on counting and statistics?

It would be tempting to assume that the Chinese language, owing to its alien immensity, simply demands this kind of obsessive research. The Chinese lexicon numbers in the tens of thousands, after all, and has undergone steady expansion over the course of history (figure 2.1). The *Shuowen Jiezi*, an early character dictionary compiled by Xu Shen (ca. 58–ca. 147) in the Eastern Han dynasty, included 9,353 characters and 1,163 variants, forming the foundation for character compilation work thereafter.[5] The *Da Song Chongxiu Guangyun*, completed by Chen Pengnian in 1011, contained more than twice that number, with a total of over twenty-six thousand. In 1716, with the completion of the *Kangxi Dictionary*, the number of characters totaled some forty-seven thousand, escalating further in the twentieth century with three immense compilation projects—the *Da Kan-Wa jiten*, the *Hanyu dazidian*, and the *Zhonghua zihai*, tallying 49,964, 54,678, and 85,568 characters, respectively. It seems only natural then that practically anyone—be they printer, scholar, or student—would instinctively arrive at the same riddle: how can one possibly "contain" this overwhelming Chinese abundance, whether within the confines of human memory, a printer's rack, a telegraph code, or a typewriter? The language itself seems to be all the evidence one needs as to its puzzling inscrutability.

Common sense fails us here, however, as it so often does. As this chapter will argue, no matter how natural or inevitable a particular "Chinese puzzle" might seem to us in retrospect, all "Chinese puzzles" are in fact historically constructed and variable. Phrased differently, before Chinese can be said to be "puzzling," it must first be "puzzled" by a particular group of people, in a particular time, and within a particular technolinguistic framework. There has never been an inherent or a priori "Chinese puzzle"—one simply willed into existence by the intrinsic complexity or strangeness of Chinese writing—only *specific* puzzles, formulated within specific historical and technolinguistic contexts, some of which succeed, others of which do not. For puzzlings that succeed, they go on to be reinscribed and remembered as stable, natural, and a priori; while those that fail, fail because they could not for whatever reason attract would-be puzzle solvers into the ludic labor of their solution. These are the puzzlings that fail to become puzzles, and are quietly forgotten.

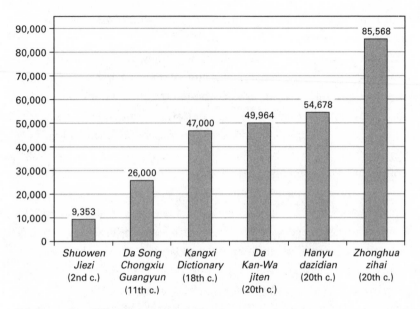

2.1 Expansion of Chinese lexicon over time

The choice of words here—*puzzling Chinese* rather than *Chinese puzzles*—might seem peculiar at first, and yet is motivated by two core observations. First, while it might be tempting to think of Chinese script as inherently puzzling owing to its lack of an alphabetic writing system, history teaches us that China has for the greater part of its history been marked by a spectrum of civilizational achievement and catastrophe as awesome as what one finds among any of its alphabetic neighbors. At the midpoint of the last millennium, Ming dynasty China was one of the engines of the world economy, one of its largest population centers, and a sphere of unparalleled cultural, literary, and artistic production—all *without an alphabet*. To have visited the China of the sixteenth and seventeenth centuries would have been to find a society undergoing accelerated urbanization, a demographic explosion, and with a vibrant print culture, not to mention the rise of fantastically wealthy families who made their fortunes as part of a growing interregional trade, and through an empire-wide banking and financial system ingesting ever greater quantities of silver from New World mines in Potosí—all *without an alphabet*. Thinking ahead to 1911, six years before the great Russian Revolution, Chinese revolutionaries overthrew a 2,000-plus-year-old imperial system *without an alphabet*. From the 1940s onward, the largest Communist state in human history was formed, a Great Leap Forward launched to cataclysmic ends, a Cultural Revolution waged, a post-Mao Reform Era inaugurated, and a new economic superpower established—all *without an alphabet*. To imagine that Chinese is intrinsically puzzling—and thus that it has always been so—is pure fallacy. Second, even as we move into the nineteenth century, and into a time when the emergence of powerful alphabet-centered information technologies placed Chinese script at an objective disadvantage, even here the Chinese script never constituted a "puzzle" in any singular or stable way. To the contrary, the "puzzling" nature of Chinese was always in the eye of its beholders, and within particular technolinguistic contexts—never due simply to any inherent properties of the Chinese script itself.

In this chapter, we will examine three remarkably different "puzzlings" of Chinese, each of which emerged during the nineteenth century. We will call these three puzzlings *common usage, combinatorialism,* and *surrogacy*. As exemplified in the work of William Gamble above, the first mode—common usage—was premised upon an assumption about Chinese that is so widespread and taken for granted that it might strike us as unnecessary even to articulate: namely, that the elemental unit of Chinese writing is the "character," and that there are tens of thousands of these characters in Chinese. Setting forth from this point of departure, the common usage approach to Chinese technolinguistic modernity was one geared toward reducing the Chinese lexicon to its most essential units, which in turn required the kind of painstaking statistical work undertaken by Gamble and his associates. The technolinguistic objective of this common usage approach was that of building inscription technologies that contained only the most frequently used characters within the overall language. The common usage approach to Chinese technolinguistic modernity will be the focus of the first part of this chapter.

Precisely at the same time that Gamble was feverishly counting up each of the language's tens of thousands of characters, however, others were problematizing Chinese in a dramatically different way, based upon far different assumptions about the script. The second mode—combinatorialism—was premised upon reimagining Chinese writing as a form of quasi-alphabetic script in which one could decompose Chinese characters into a set of modular shapes that the operator would then use to build, or "spell," characters piece by piece. Chinese characters could not be "spelled" in the sense of phonetic letters, yet perhaps they could be "spelled" using a set of recurring, modular *shapes*. In this approach, what was called for was not a reduction of the Chinese lexicon, but a critical reimagining of the very essence of Chinese writing itself, transposing the idea of "letters" and "spelling" atop Chinese writing and reimagining the structural components of Chinese characters—often referred to as "radicals"—as the equivalent of letters in the Latin alphabet. In forming the puzzle this way, combinatorialism thus denaturalized one of the most seemingly self-evident and "obvious" characteristics of Chinese—namely, the idea that Chinese was made up of an overwhelming number of elemental units called "characters." If the Chinese character was *not* the irreducible atom of Chinese writing, as maintained in the common usage approach—if there existed something still more fundamental and repetitive into which characters could be further divided—then the real puzzle of Chinese technolinguistic modernity was waiting to be solved here at this level. The method that grew out of this puzzle would come to be known as "divisible type," and will be the focus of the second part of this chapter.[6]

Still others puzzled and problematized Chinese in yet another way—a third mode we will refer to as *surrogacy*. In contrast to both common usage and combinatorialism, this third mode was premised neither on counting and ranking Chinese characters nor on cutting them up into pieces, but on symbolic systems that could be used to stand in for, or refer to, Chinese characters—particularly in the arena of the emerging technology of telegraphy. In this mode, Chinese characters remained the fundamental unit of Chinese, but were not to be trafficked in directly. Rather, they were to be sequestered to separate spaces—whether a character code book, a database or, most abstractly, to the realm of human memory and the "mind's eye"—and thereafter "retrieved" according to specific protocols. In this way of puzzling Chinese, the primary task of the puzzle solver was neither the statistical conquest of Chinese lexical abundance, as in common usage, nor the dismemberment of Chinese characters into their component elements, as in combinatorialism, but the development of ever more efficient techniques of reference, address, databasing, search, and retrieval. Surrogacy will occupy our attention in the final part of this chapter.

Instead of treating the "Chinese puzzle" as a taken-for-granted starting point for our history, then, this chapter excavates the underlying and often invisible assumptions that made each of these distinct types of puzzles imaginable, meaningful, solvable, and worth puzzling over in the first place. As we will see in subsequent chapters, moreover, it will be these logics and assumptions that persist and continue to shape the pursuit of Chinese technolinguistic modernity long after Gamble and his contemporaries have been forgotten. Each of these three logics will reappear, that is, in the age of typewriting to come.

SETTLING THE NOMADIC TYPESETTER: MOVABLE TYPE, COMMON USAGE, AND THE DESIRE TO SURROUND LANGUAGE

Returning to the question raised at the outset of this chapter—what propelled William Gamble in his algorithmic, statistical labors—we find important clues from the printer's own accounts. As Gamble explained in the preface to his 1861 work, *Two Lists of Selected Characters Containing All in the Bible and Twenty Seven Other Books*, "not only have the type themselves taken up much room," referring to Chinese fonts, "but the compositor in going from case to case for each type has unavoidably consumed so much time, as thereby to render composition both expensive and tedious."[7] Decades before linguists, engineers, and language reformers would attempt to resolve the "incompatibility" they saw between Chinese and typewriting, then, Gamble sought to resolve a still earlier "incompatibility": between the Chinese script and movable type.

During the mid-nineteenth century, it became commonplace for foreign printers to criticize Chinese for the challenges it presented to movable type printing—a curious criticism when we consider that the technology of movable type printing had been invented in China four centuries prior to Johannes Gutenberg in Mainz. In the eleventh century, Bi Sheng

(990–1051) developed earthenware movable type, which was arranged into an iron frame lined with paste to secure the fonts during the printing process. In the fourteenth century, records by Wang Zhen (fl. 1290–1333) cite the usage of wooden movable type, with bronze and other metals being used in the process by the late fifteenth century.[8] While Chinese printing was dominated by xylographic or woodblock printing methods throughout this period, nevertheless the technical prowess of the system continued to be refined. An illustrative case in this regard comes from the early Qing period. In 1773, work began on a massive compilation project authorized by the Qianlong emperor (r. 1736–1795). Its objective was the collection, publication, and circulation of 126 rare Chinese works selected from the larger *Four Treasuries* (*Siku quanshu*) collection.[9] In his first memorial to the emperor, supervisor of the Imperial Printing Office (*Wuying dian*) Jin Jian requested authorization to cut a total of 150,000 movable wooden type, and further stipulated the need to cut between 10 and 100 duplicates of approximately 6,000 of the most commonly used characters within the font.[10] In contrast to xylography and the hand carving of master copies, movable type printing required operators to be acutely aware of the total number of *distinct* characters needed to print a given text and, relatedly, the number of duplicate sorts that would be needed for each of these characters—that is to say, their relative frequency.

Jin Jian partitioned the Chinese lexicon, and with it the printing office itself, into two broad categories in accordance with the energetic output of the typesetter's body—with the lexical categories of high and lower frequency characters being translated into somatic categories of *near* and *far*, *vicinity* and *distance*. "There are some rare characters that are seldom used," Jin wrote, "and for which few type will have been prepared. Arrange small separate cabinets for them according to the twelve divisions mentioned above, and place them on top of each type case from where they may be seen at a glance." It was only then, after this primary partition, that Jin organized his characters secondarily according to the "radical-stroke" system, a taxonomic system dating back to the late Ming dynasty (1368–1644) and then canonized in the early-eighteenth-century Qing dynasty (1644–1911) reference, the *Kangxi Dictionary*.[11] In this dictionary, named in honor of the Kangxi emperor (r. 1661–1722) who ordered its compilation, upward of 40,000 characters are organized into 214 categories based on the primary character component (or radical [*bushou*]) out of which each character is composed. Characters such as *ta* (他 "he") and *zuo* (作 "to do/make") are categorized along with others built using the "person" radical (亻), while characters such as *hong* (洪 "flood") and *hu* (湖 "lake") are assigned to the group of characters listed under the "water" radical (氵). A second-order taxonomic rule places each of these 214 categories in a sequence based on the number of strokes in the component radical. For example, insofar as the person radical is composed using two brushstrokes and the water radical using three strokes, the characters *ta* and *zuo* above precede *hong* and *hu* in the dictionary.[12] Within each of these 214 classes, a third-level taxonomic rule further organizes characters according to the number of residual strokes required to complete the character, thereby placing the five-stroke character *ta* prior to the seven-stroke character *zuo*.[13] Well into the twentieth century, this was the dominant method of organizing Chinese-language dictionaries in China.

When Jin Jian and his typesetters walked from case to case, and between cases and the printing press, then, they were quite literally *walking through a physical model of the Chinese language itself*.[14] Rather than surrounding the Chinese script, the way their counterparts in the Western world surrounded the alphabets that lay before them in the type case, it was the Chinese script that surrounded them. This distinct spatial quality of Chinese movable type vexed Gamble and a number of his contemporaries, whose increasing criticism of Chinese movable type was premised on what we might term an almost imperceptible "drift" in the operative definition of "movable type" itself. While aware that this printing technology had been invented in China and for Chinese characters, Gamble and others nevertheless expressed their belief that a time still awaited when Chinese would finally be brought "within the compass of European typography," as one contemporary phrased it.[15] "European typography" in this formulation carried with it a particular meaning, namely the idea of Chinese movable type as a lesser form thereof, lesser precisely because of the nomadic, peripatetic practices of Jin Jian and others—this process of *walking through* the Chinese script. For Gamble, true movable type was clearly tied to an ideal of *sedentary mastery*—the way in which a Western typesetter could stand fixed before his type rack and *surround* script. Operating beneath all these assumptions was a shifting definition of the term "movable type": no longer limited to its strict technical meaning, signifying a technique and technology of printing in which "type" were cut and cast as modular, "movable" pieces (a criterion Jin Jian's method undoubtedly fulfilled), a new paradigm had emerged in which "movable type" was only truly *movable* to the extent that the typesetter was *stationary* (something the Chinese typesetter was most certainly not).[16]

The first of our three "puzzlings" of Chinese took shape in this context—one in which algorithmic reading, frequency analysis, and ever-increasing Chinese corpora would quickly become the mainstays of a new epistemic framework. A foundational moment for this new "Chinese puzzle" came in 1810, when the English-speaking world was introduced for the first time to the *Great Qing Legal Code*, as translated by George Staunton. Practically from the moment of its publication, scholars and statesmen alike crowded about the window it opened up into the once mysterious legal logic of the Celestial Empire—a "landmark in Western knowledge of not just Chinese law but also Chinese civilization in general," as historian Li Chen has argued.[17] Called Britain's first sinologist by some, Staunton enjoyed a personal and professional pedigree that established him as an ideal translator and editor for such a work. The son of diplomat and Orientalist George Leonard Staunton, and page boy to George Macartney during the 1793 embassy to emperor Qianlong's court, Staunton commenced his training in the Chinese language at a precocious age, later going on to serve as a member of British Parliament and as a senior official in the East India Company.

Staunton's translation of the *Great Qing Legal Code* was foundational for the formation of modern sinology in a second, less apparent way. Upon its publication, Orientalists, printers, educators, and publishers became fascinated by a realization that Staunton had stumbled upon in the course of his translation work—specifically, a *number* that Staunton arrived at when he tallied up how many distinct Chinese characters he encountered along the way. For all its size and complexity, Staunton revealed, the *Great Qing Legal Code* was composed of roughly two thousand different Chinese characters in all—but a small fraction of the approximately forty-seven thousand found in the *Kangxi Dictionary*, the leading Chinese dictionary of the day. This remarkable number—2,000, more or less—implanted a thrilling possibility in the imagination of many: if a mere five percent of all Chinese characters were needed to print and read a legal text of such importance, what might also be true of the Chinese-language canon more broadly? What might this mean for foreign printers and students of the Chinese language, who had long believed it impossible to conquer its "tens of thousands" of characters? Far more than providing one potential solution to the "Chinese puzzle," Staunton's work offered up something even more powerful: a means by which to transform Chinese *into* a puzzle, or in other words, the puzzle itself.

Staunton's "discovery" of Chinese common usage sent a shock wave through the transnational community of sinologists that could be felt for decades. For foreign printers of Chinese, his observations raised the possibility of dramatically scaled-down Chinese fonts—fonts containing perhaps five thousand sorts, rather than fifty thousand—that would nevertheless be sufficient for their needs. As suggested in the preliminary findings of Staunton, the objective of surrounding the Chinese language—and thus sedentarizing the nomadic Chinese typesetter—might be achieved by erecting fences to divide up the rolling terrain of the Chinese language. This puzzling of Chinese script was thus primarily one of boundary formation: determining through rigorous and painstaking analysis which characters in the Chinese lexicon were truly "essential" and which were frivolous or incidental. For foreign learners of Chinese, the notion that their energies could be focused on mastering a core subset of the language—rather than dispersing their energies across "useless" characters—was particularly attractive in an era when no formalized pedagogical system was in place for acquiring the Chinese language. Printers and students alike quickly took up the task of establishing boundaries between essential and inessential characters, dissolving Chinese texts in the acid bath of reason with the goal of scientifically determining where to concentrate their mental energies and financial resources. The age of Chinese "distant reading" had begun.

In quick succession, scholars began to extend Staunton's initial observations far past the Qing legal code. Bengal-based missionary Joshua

Marshman announced to his readers that "all the works of Confucius contain scarcely three thousand characters."[18] The epic novel *Romance of the Three Kingdoms* was composed of a mere 3,342 distinct characters, the *Missionary Herald* reported. William Gamble, the missionary printer we met at the start of this chapter, determined more precisely that the Four Books of Confucianism required only 2,328 unique characters to print; the Five Classics only 2,426. In the aggregate, the Thirteen Classics—embodying *The Book of Poetry* (*Shijing*), *The Book of Documents* (*Shujing*), *The Rites of Zhou* (*Zhouli*), *The Ceremonies and Rites* (*Yili*), *The Record of Rites* (*Liji*), *The Book of Changes* (*Yijing*), *The Commentary of Zuo* (*Zuo zhuan*), *The Commentary of Gongyang* (*Gongyang zhuan*), *The Commentary of Guliang* (*Guliang zhuan*), *The Analects* (*Lunyu*), *Luxuriant and Refined Words* (*Erya*), *The Classic of Filial Piety* (*Xiaojing*), and *The Mencius* (*Mengzi*)—contained only 6,544 characters.[19] With each passing year, it seemed, foreigners were steadily solving the puzzle of common usage Chinese.

In puzzling Chinese so, however, the "common usage" framework gave rise to a profound and arguably irresolvable tension—a politics that would haunt this approach to Chinese technolinguistic modernity, perhaps forever. In contrast to our analogy of nomadic and sedentary, plains and fences, the terrain of Chinese language itself was one that had always morphed and changed, and always would. As historians know well, the Chinese lexicon changed and expanded dramatically during the closing decades of the nineteenth century and the opening decades of the twentieth, with thousands of new Chinese-language terms pouring in from neighboring Japan, and still more developed by Chinese translators of foreign-language texts. In the context of common usage Chinese movable type, neologisms and other lexical changes exerted an influence unknown in Western, alphabetic movable type. If we imagine typesetters in the Western world having, at certain points in history, to confront neologisms such as "hegemony," "colonialism," and others, such terms posed no steeper challenge for the alphabetic typesetter than to set the three comparably common words "my," "he," and "gone," or "is," "on," "oil," and "calm," respectively. No matter the novelty or obscurity of a word in German, English, French, or Italian, nevertheless its composition in alphabetic movable type required the same letters as its more common counterparts—save perhaps in the printing of foreign loanwords that required special letters not present in one's script. In the Chinese case, by contrast, "common usage" was a zero-sum game, in the sense that the *inclusion* of any new character necessitated the *exclusion* of another—or phrased differently, it required constantly reestablishing the boundary between "frequent" and "infrequent."

What is more, for the very foreign missionaries and printers who became attracted by the "common usage" puzzle, they in fact *wanted* to change the lexical ground beneath China's feet: to introduce new terms and concepts, whether those connected to modernity, Christian salvation, or something else. To that end, if foreign printers had determined with great precision how many distinct Chinese characters they would need to *reproduce* the Chinese-language corpus as they found it in the nineteenth century—to reprint the Confucian classics, for example, in what we might think of as the reproductive or *descriptive* imperative within the common usage puzzle—they also made sure to calculate the number of characters they would need to *intervene* in this corpus—to print works in Chinese for the first time in what we might term the *prescriptive* imperative of common usage. In Chinese, the Old Testament contained 503,663 characters, Gamble discovered, and yet required no more than 3,946 unique characters to produce. The New Testament contained 173,164 characters, comprising only 2,713 distinct ones. For both Old and New, a total of 4,141 characters sufficed. In 1861, Gamble synthesized his findings in a work entitled *Two Lists of Selected Characters Containing All in the Bible*, a text geared toward missionary printers and pedagogues engaged in the work of proselytizing. Gamble declared with not a small measure of pride, "It has been from a knowledge of these facts therefore that we have been enabled to arrange a large fount of Chinese metallic type in so compact a manner that the compositor can reach any type he wants without moving more than a step in any direction; and by having placed say five hundred of the most numerous characters together, he has more than three-fourths of all he uses just under his hand, almost as conveniently as an fount of roman type are arranged in an English printing office."[20]

Common usage constituted the first and most widespread way in which foreigners set out to "puzzle Chinese" during the nineteenth century. When supplied with sufficient time, and one or another subset of the Chinese-language corpus, it now became possible for any foreign reader of Chinese to contribute to this great dissolution of the Chinese language into its constituent parts. To dissolve Chinese in the acid bath of statistical reason—to help one's Western brethren demarcate clearly between what must be known and what can be ignored—became a new logic and lens through which these foreigners could make sense of Chinese as a nonalphabetic script. Thanks to "common usage," Chinese script was no longer insurmountable in its abundance. Now it was a puzzle, with a solution.

As we will see in the next chapter, this particular way of puzzling Chinese soon became dominant in China itself, this time among Chinese intellectuals in the first half of the twentieth century. The number and scale of "common usage Chinese" analyses exploded, with Chinese scholars and their assistants subjecting ever-larger swaths of the Chinese-language corpus to algorithmic readings. If Staunton, Gamble, and others had decomposed the Confucian classics, the legal code, and the Bible, Chinese scholars would soon decompose newspapers, textbooks, literature, and more. The "common usage" approach to Chinese technolinguistic modernity would also be extended beyond movable type and character primers to a promising new information technology: that of the typewriter.

To puzzle Chinese in this way would come with a price, however. As we will also examine, Chinese linguistic modernity premised upon the theory and practice of "common usage" could never strive for either comprehensiveness or stability. For any technology premised on common usage—be it within movable type, pedagogy, typewriting, or computing—at no time could the Chinese language as a whole be permitted to take part. Only a small subset of the language could, at any given time, take part in Chinese technolinguistic modernity so conceived. What is more, the boundaries that segregated "included" versus "excluded" characters could never hope to remain set or stable for long—they would constantly shift with the dictates of the age. The power and authority to determine these boundaries, moreover, would forever be fought over by competing social and political factions. This would be the price of any technolinguistic modernity premised upon this mode of puzzling Chinese.

HOW DO YOU SPELL CHINESE? DIVISIBLE TYPE AND THE REIMAGINATION OF CHINESE SCRIPT

The early eighteenth century was a golden age in the history of translation, witnessing not only the first English-language version of the *Great Qing Legal Code* by Staunton, but also a bevy of Western-language translations focused upon the great philosophical and religious traditions of Asia. In 1838, Orientalist Jean-Pierre Guillaume Pauthier added to this growing corpus, setting his sights not on the legal structure of the Qing empire, but on one of the philosophical pillars of Chinese civilization: Daoism. In that year, he published the first French translation of the *Daodejing*.

As a member of the Société Asiatique de Paris, Pauthier was one of the leading Orientalists of his day, already responsible for the French translation of *A Record of Buddhist Kingdoms* (*Foguoji*) and the *Great Learning* (*Daxue*).[21] With the publication of *Le Tao-Te-King, ou Le Livre Révéré de la Raison Suprême et de la Vertu*, Pauthier's celebrity only grew, cementing his reputation as one of the leading scholars of China of his generation.[22] As in the case of Staunton's translation of the Qing legal code, what concerns us here is not the content of Pauthier's translation but the condition of its production—or more accurately, the "extracurricular" work Pauthier carried out during the translation process.

When Pauthier began work on *Le Tao-Te-King*, he could have easily followed the path carved out by his Orientalist predecessors and colleagues. He could have employed one of the numerous conventional Chinese fonts available at the time, for example, or he could have undertaken a statistical analysis of his own. Likely aware of George Staunton and the nascent "common usage Chinese" movement, he could have subjected the *Daodejing* to the same kind of algorithmic analysis being applied to other subsets of the Chinese canon, thereby arriving at a custom-tailored font for his printing needs—just as Gamble would do for the Old and New Testaments.

The common usage approach to puzzling Chinese did not puzzle Pauthier, apparently, at least not enough to elicit from him the same commitment of time and obsessive labor as Staunton before him, or Gamble to follow. This is not to say that Pauthier was satisfied with the status quo, however. Propelled by the same desire to bring Chinese "within the compass of European typography," Pauthier instead set out to puzzle Chinese in his own way. Specifically, he set out to design a new Chinese font that featured not only characters, but also *pieces* or *fragments* of character that he would use to build characters combinatorially. By harnessing "radicals and primitives" as the ontological foundations of Chinese script—rather than Chinese characters themselves—such a font offered the possibility of a forty-fold reduction in the size of a complete Chinese font, from tens of thousands of sorts to somewhere in the order of two thousand. Treating Chinese characters as quasi-"words" and radicals as quasi-"letters," Pauthier set out to puzzle Chinese by posing a Zen-kōan-like question: *How does one spell Chinese?*

Divisible type was simple, but deceptively so. To print the character *ming* (明 "brightness"), Pauthier reasoned, one could combine two smaller metal sorts to create a composite: one metal sort featuring the sun, *ri* (日), and one featuring the moon, *yue* (月). To print the character *shi* (時 "time"), meanwhile, one could reuse "sun," but this time pair it alongside another metal sort containing *si* (寺)—a character that by itself signified "temple," but in the character *shi* would be used solely for its phonetic value. By breaking characters up into their component parts, Pauthier imagined, individual sorts could be allocated in a variety of different contexts, not unlike how Latin letters recombine to form different French words.

At first glance, the story of divisible type would appear to be one of technological triumph: the Western mind, with its *esprit d'analyse* in full bloom, prevailing over the immensity of the Chinese language through the dogged deconstruction of reality into fundamental elements. Unhindered by convention and custom, the rational French mind had opened up a pathway by which Chinese writing could be ushered into the modern era.

Divisible type was not as simple as it seemed, however. For the system to work, Pauthier realized, he could not rely on the 214 conventional

Chinese radicals of the *Kangxi Dictionary* in their isolated forms, since radicals possess an enormous variety of sizes and placements depending upon the characters in which they appear. In the examples of *ming* and *shi* above, for example, the sun component is located on the left side of the character, composed so as to occupy slightly less than half of the character's horizontal span, and slightly less than all of the character's vertical. By comparison, the sun component operates quite differently within the characters *dan* (旦 "dawn"), *han* (旱 "arid"), and *xi* (昔 "the past"), among many other possible examples. Within *dan*, the sun component is positioned atop the character, occupying nearly the entire space of the character body. Within *han*, the sun is also positioned atop the character, but here occupying less than half of the overall space. Within *xi*, meanwhile, it is positioned along the bottom half of the character. Still another form is, of course, the character *ri* itself (日)—the Chinese word "sun," in which the form occupies the space in its entirety.

Here, then, an entirely different puzzling of Chinese script was underway, one that entailed its own kinds of joyous and obsessive tedium. In order to harness the productive, combinatorial possibilities of Chinese radicals, Pauthier would first need to set out upon a full-scale analysis of radicals in all their many orthographic and positional variations. This labor, although every bit as engrossing as that of Gamble and his algorithm, and although focused on the same objectives, was nevertheless conceptualized in entirely different ways.

To pursue this experimental method in Chinese type design, Pauthier sought the assistance of Marcellin Legrand, an esteemed member of the nineteenth-century engraving and type design establishment. Decades earlier at the Imprimerie Royale, Legrand had been commissioned to develop new faces meant to rejuvenate the press, contracted in 1825 to engrave fifteen roman and italic bodies collectively known as "Types de Charles X." On the strength of his work, Legrand was later named the official engraver of the press.[23] He was also one of the era's foremost designers of "Oriental fonts" and "exotic type." Legrand was responsible for the carving of a Middle Persian font under the direction of French-German scholar Jules Mohl; an "Éthiopien" font in 1831 under the direction of the Irish-Basque geographer and explorer Antoine Thomson d'Abbadie; a font for Gujarati, carved in 1838 under the direction of French Orientalist Eugène Burnouf; and a host of others.[24]

Legrand was quickly enthralled by Pauthier's puzzle. Upon learning of his idea, Pauthier later recounted, Legrand "for the interest of science, was willing."[25] "Of all the languages in the known world," Legrand himself soon pronounced, evidently enthused, "the most difficult to represent by movable type is, without controversy, the Chinese; having hitherto baffled the most skillful European typographers."[26] The results of this research took the form of an at-a-glance brochure published in 1845, entitled the "Table of 214 Keys and Their Variants" (*Tableau des 214 clefs et leurs variants*).[27] The brochure included its own makeshift symbolic notation, including cartouches, hollow circles, asterisks, and carets, and used them as graphic placeholders to demonstrate all of the possible permutations of each radical, and thus which metal sorts would be needed in the font (figure 2.2).

Although it required months if not years of work to accomplish, Legrand and Pauthier's analysis of Chinese radicals was but the first, and arguably the simplest, of the challenges that confronted divisible type printing. There was a second problem as well, one that emerged not because of the structure of Chinese characters but because of the aesthetic ideologies and commitments of Pauthier and Legrand themselves. Operating at the apex of French Orientalism, with translations focused upon canonical texts like the *Daodejing*, Pauthier and his contemporaries were enduringly committed to capturing Oriental "essences." We should recall that, when aesthetes of this age sought out words to praise the various non-Latin "exotic types" being designed by contemporary Western type designers, the words they used centered upon *authenticity* and *fidelity*, not iconoclasm or innovation. The same was no less true in the case of Chinese. "The complete Chinese air they assume," as one contemporary Chinese font would be praised, "so as not to be distinguishable from the best style of native artists, together with the durability of the letter, would recommend them to universal adoption."[28] As a member of the rational, modern state of France, then, Pauthier may have aspired to revolutionize Chinese printing methods; but as a connoisseur of vaunted exotic pasts, he most certainly did not wish to disrupt Chinese orthographic aesthetics.

2.2 Table of 214 keys by Marcellin Legrand

Legrand shared his client's aesthetic commitments, and captured the challenge of divisible type Chinese printing most succinctly. The core puzzle, as Legrand phrased it, was "to solve the problem of representing the figurative language of China, with the fewest possible elements, without, however, altering the composition of the symbols."[29] This objective, while simple to articulate, would not be simple to achieve. For while their objective was a printing method capable of producing Chinese characters of impeccable aesthetic composition, the very principle upon which Pauthier and Legrand's technique of divisible type was based—of cutting up characters into pieces, and then using these pieces to build up characters on the printed page—diverged completely from all longstanding and mainstream approaches within Chinese textual practice. In divisible type, the central conceit involved drawing an equivalence between letters of the Latin alphabet, on the one hand, and structural components of Chinese characters, or "radicals," on the other. If successful, the homologization of letters and "radicals" would enable one to *spell* a Chinese character in ways analogous to the spelling of a French *mot*.

There was a problem with this plan, however. When we compare the compositional principles of divisible type alongside those of the *longue-durée* history of Chinese calligraphic aesthetics, it had always been the *stroke* (*bihua*) that was understood as the primary compositional element of characters—never the "radical." Strokes, not radicals, were the fundamental components of Chinese composition that students of calligraphy were expected to master, and whose mastery was understood as part of a broader practical and aesthetic education. Beginning in the Han dynasty, but accelerating during the fifth and sixth centuries, China witnessed a proliferation of treatises in which Chinese script forms were carefully categorized.[30] Jin dynasty calligrapher Wang Xizhi (303–361), the "sage of calligraphy," put forth the theory of the "eight fundamental strokes of the character *yong*" (the calligraphic model we examined in the introductory chapter, and which in part formed the organizational basis of the 2008 Olympics Parade of Nations). Wang Xizhi's eightfold classification was later elaborated upon by Li Fuguang, who expanded upon Wang's taxonomy and extended it to thirty-two strokes in all.[31] The celebrated calligrapher of the Eastern Jin Wei Shuo (272–349), known as Lady Wei, created a further fine-grained examination of character structure, and arrived at seventy-two fundamental strokes.[32]

In practicing and mastering the fundamental strokes of Chinese characters, one goal had always been that of composing characters that *held together*. As outlined in disquisitions by calligraphic masters across the centuries, characters whose component parts sagged about the page were considered unrefined and "lazy," exhibiting a rattling laxity instead of embodied wholeness. "In the writing of those who are skillful in giving strength to their strokes," as Lady Wei explained:

The characters are "bony"; in the writing of those who are not thus skillful, the characters are "fleshy." Writing that has a great deal of bone and very little meat is called "sinewy"; and writing that is full of flesh and has weak bones is called "piggy." Powerful and sinewy writing is divine; writing that has neither power nor sinews is like an invalid.[33]

The quality of the calligrapher was thought to be captured in his or her brushwork: *Writing is like the person (zi ru qiren)*, as the well-known adage phrased it. "When one is pleased," renowned Ming dynasty calligrapher Zhu Yunming (1461–1527) explained, "then the spirit is harmonious and the characters are expansive / When one is angry, the spirit is coarse and the characters are blocked / When one is sad, the spirit is pent up and the characters are held back / When one is joyous, the spirit is peaceful and the characters are beautiful."[34]

If longstanding Chinese calligraphic practice privileged the "stroke" above all, Pauthier and Legrand's divisible type sought to usurp that position on behalf of the "radical." The very concept of the "radical" as these Frenchmen imagined it, however, was a fictional one, invented by foreign observers of China in their attempts to draw mental equivalents between familiar linguistic concepts in Indo-European languages and the language they encountered in China. In the Chinese context, the *bushou*—what foreigners often translated as "radical"—was fundamentally a *taxonomic* concept more faithfully translated as "classifier" or "chapter heading," referring to the sectional divisions of Chinese lexicons and dictionaries. While *bushou* themselves did correspond structurally to the characters so classified, nevertheless *bushou* were never regarded as the generative "roots" or "radices" that grew into such characters. Their place was a taxonomic or perhaps etymological one, used to organize and subsequently locate characters within textual sources, but not a compositional or typographic one. Delving into manuals and guides to Chinese calligraphy, one would be hard-pressed to find any descriptions of Chinese composition that portrayed it as an act of assembling radicals, the way that French and English handwriting manuals might explain the act of composing words as the sequential inscription of their constituent letters. In divisible type printing, however, this was *precisely* how the radical would be made to work: as the productive *root* or *radix* of the Chinese character, one that *grew* and produced semantic varieties and variations in the same way as in inflectional or agglutinative languages. As a productive force, then, they could be harnessed: they could be made to produce, not haphazardly but rationally and efficiently, as a kind of scientific forestry of language.

Divisible type departed from calligraphic practices in a second, vitally important way, as well. By setting their "radicals" on fixed metal pieces, and then building characters out of these metal pieces, Pauthier and Legrand inverted the classic relationship between *part* and *whole* in Chinese orthography. In all other forms of Chinese inscription, whether manuscript, movable type, woodblock, or otherwise, it was the compositional coherence, integrity, and beauty of the character as a whole that governed the ways in which any given component of that character was to be articulated, assessed, or valued. For Legrand and Pauthier, however, the material conditions of divisible type depended upon precisely the opposite relationship between part and whole. Fixed on metal sorts, the components of divisible type were numb and unresponsive, stubbornly maintaining fixed postures no matter their shifting contexts. Exhibiting none of the versatility of the hand-drawn stroke, Legrand and Pauthier's modules inverted the dynamics of Chinese script, subordinating the logic of the character to that of its elemental units.

From the outset, then, Legrand and Pauthier had set for themselves a challenging, and arguably impossible, task: to produce characters that upheld Chinese calligraphic standards, while at the same time producing such characters using principles that themselves violated, or at the very least dramatically departed from, those very aesthetic standards. For Pauthier and Legrand, this tension between modular rationality and aesthetic composition manifested itself in the Chinese font they ended up creating. Most basically, Pauthier and Legrand did not cut up or divide all of the Chinese characters they could have, and in principle *should* have. Instead, they categorized their characters into two broad rubrics: what one observer at the time termed "typographically divisible" versus "typographically indivisible" characters—characters that could be cut up, and those that could not.[35] Through their research, Legrand and Pauthier settled upon a font composed of approximately 3,000 elements in all, part of which was dedicated to divisible type components, but with another sizable portion dedicated to conventional, full-body Chinese characters.

Why such mercy and restraint? When confronted with Chinese characters that lay motionless on the operating table, what force or presence stayed our surgeons' hands? When a character was said to be "divisible" or "indivisible," who or what was making such claims? As we see with the characters *liu* (流), *hai* (海), and *dang* (蕩), three criteria determined whether a given character would be dissolved into component parts— as in the case of *liu* and *hai* (figures 2.3 and 2.4)—or left intact and treated as a full-body font—as with *dang* (figure 2.5). First, the shape of a divisible character had to be susceptible to a clean, two-part split, along either a straight horizontal or vertical seam. Building up a character out of three or more components, while in theory possible, was avoided by Pauthier and Legrand, presumably because of the added complexity such a procedure would have entailed. In the case of the first two characters, such a division was possible, with the water radical (氵) being separable from *liu* (㐬) and *mei* (每), respectively. By contrast, the character *dang* was not amenable to such a clean division, not at least with respect to the water radical. To have factored out the water radical in *dang* would have required the creation of either a cumbersome tripartite division of the character into a water radical, a grass radical (艹), and the residual *yi* (易), or an even more awkward L-shape containing the grass radical and *yi* together. It was deemed more economical simply to leave *dang* intact.

The second criterion pertained to the relative frequency of each module. In our earlier example of *dang*, it is clear that, while a vertical separation of the water radical would have been too cumbersome, a horizontal separation of the uppermost grass radical was feasible. This would have divided the character into two parts: the grass radical and the residual

2.3 Character *liu* in divisible type

2.4 Character *hai* in divisible type

2.5 Character *dang* in full-body type

character *tang* (湯 "soup"). It can be surmised that the reason such a division was not made was because there was no advantage to doing so. Unlike the modules *liu* and *mei* above, each of which appears in many other characters in precisely that position and scale, a *tang*/"soup" module would have been relatively useless. A module *tang* of this scale and proportion would not be usable in practically any other character. Once again, it was simply more advantageous to leave the character *dang* intact.

The most important rule—the one that Pauthier and Legrand clearly felt could never be violated—was that the "bones" of a character could never be broken. It was impermissible to cut through a stroke and set its fragments on separate metal sorts. While a small conceptual leap might have led these Frenchmen away from the idea of combining "radicals" to that of combining asemantic graphemes—for example, forming the character *yi* (一 "one") through the combination of two half-length dash-shaped marks (--)—this was a step they were clearly unwilling to make. Not unlike Cook Ding and his ox, speaking to King Wen via the *Zhuangzi*, Pauthier and Legrand strived to follow the "natural makeup" of the character body, to "strike in the big hollows, guide the knife through the big openings, and follow things as they are." As with the butcher's blade, when passing his scalpel through the spaces between the strokes, "even places where tendons attach to bones give no resistance, never mind the larger bones!"[36]

Having partitioned their font into these two broad categories of divisible and indivisible, Legrand and Pauthier composed their partial-body and full-body fonts in markedly different ways, as exemplified in the examples of *liu*, *hai*, and *dang*. In the character *liu* (figure 2.3), we discover a wide space separating the water radical on the left from the remainder of the character body. Although small in terms of absolute size, this micrometric, interstitial void was sufficient to hermetically seal the water radical within a completely autonomous zone. The same hermetic seal is found in the character *hai*. Each of these characters was clearly produced using divisible type, each of them splayed out owing to an artificial injection of space. In the full-body character *dang*, however, we see that the third stroke of the water radical sweeps boldly up and into the adjoining region, creating the type of coherent interpenetration so valued within calligraphic composition. In the divisible type characters *liu* and *hai*, by contrast, the constituent parts again do not venture into each other's zones. They are completely divided along an imaginary *y*-axis, creating a character that, in calligraphic parlance, might be considered "lazy" or "loose."

The same discrepancy can be detected in a comparison between the characters *ran* (然) (figure 2.6) and *wu* (無) (figure 2.7)—both composed using the fire radical (灬), but with *ran* being composed using divisible type, and *wu* composed as a full-body character.

Ran splayed open, its upper and lower halves separated by a distance that, while small in absolute terms, was altogether cavernous. The fire within *ran* was decidedly bolder than that of *wu*, precisely because, once again, the method of divisible type depended upon the absolute bound-edness of its modules. Space was not an inherent enemy of the calligra-pher, of course: calligraphers had always had at their disposal the playful enlargement of space, and even distortions of the character body. For Legrand, however, space was not an option but a necessity: it alone made possible the rationalized modularity that set their printing method apart from all others. A further survey of the font reveals the same pattern. Whenever Pauthier and Legrand created full-body fonts, they took pains to forge characters *à la chinoise*. When creating partial-body fonts, how-ever, the imperatives of their system dictated that modules be kept abso-lutely separate. At its inception, combinatorialism was characterized by a paradox in which the tiny spaces between radicals were something the system depended upon and *depended upon negating*.

Here, then, was the second mode by which foreigners in the nine-teenth century reconceptualized Chinese script as a "puzzle," a mode that put forth a strikingly different vision of Chinese writing than the common usage approach of Staunton, Gamble, and others. By harnessing "radicals and primitives" as the quasi-alphabetic, ontological elements of Chinese writing, it promised a phenomenal, forty-fold reduction in the size of a complete Chinese font, from tens of thousands of sorts to somewhere in the order of two thousand. What is more, the combina-torial approach offered the promise of a unified technology of Chinese inscription. Unlike the common usage system, in which infrequently used Chinese characters would be ostracized from the communitas of the machine, divisible type printing offered the possibility of welcoming

2.6 Character *ran* in divisible type

2.7 Character *wu* in full-body type

all characters within the collective embrace of a single system of textual production—provided that some underwent dismemberment.

Gaining a measure of popularity and momentum over the course of the nineteenth century, divisible type prompted publishers and printers in Europe, the United States, and in missionary-cum-colonial outposts to set out on their own scientific dissections and dismemberments of Chinese characters. In 1834, Samuel Dyer was reported to be carrying out a similar divisible type project at the Anglo-Chinese College in Malacca.[37] In 1844, the Chinese Mission of the Board of Foreign Missions of the Presbyterian Church published a specimen list of 3,041 type based on the principle of divisible type.[38] Divisible type printing was later picked up and developed further by Auguste Beyerhaus of Germany. Praised as "one of the most remarkable typographical displays" at an exhibition of industry and applied sciences, Beyerhaus's "Berlin Font" contained 4,130 Chinese type pieces in all: 2,711 complete characters, 1,290 2/3-sized partial characters, and 109 1/3-sized partial characters. Meanwhile, just as the puzzle of common usage gave rise to new Chinese fonts, and just as these common usage fonts found their way into new Chinese-language publications, so too was the puzzle of divisible type enrolled into the world of printing. In 1834, Julius Heinrich Klaproth's edited translation of the *Nipon o daï itsi ran* or *Annales des empereurs du Japon* was produced using Legrand's divisible type method, as was Klaproth's 1836 republication of *Foe Koue Ki* or *Relation des royaumes bouddhiques*.[39]

As we will examine in the following chapter, the combinatorial approach to Chinese technolinguistics persisted into the twentieth century, forming a potent counterdiscourse to the more dominant puzzle of common usage. Just as inventors and engineers would go on to develop experimental Chinese typewriters based on the common usage approach to puzzling Chinese, so too would other engineers attempt to develop Chinese typewriters premised on the divisible type concept of "spelling" Chinese characters.

As with common usage, however, the puzzle of combinatorialism would also be haunted by its own internal tensions. In London, Legrand was awarded a prize for the "general excellence of his types" for his Chinese font, praised for rendering character-based Chinese script compatible and commensurable with European typography.[40] Offering a measure of criticism, however, one reviewer commented: "The form of some of the characters is a little stiff, and disproportionate, owing partly to inexperience, and partly to the attempt which the French have made, to split and combine the elements of various characters, so as to prevent the necessity of cutting a new punch for each separate symbol; but on the whole they are exceedingly neat and handsome."[41] If common usage was beset by an irresolvable politics of presence and absence—of determining over and over again what parts of the Chinese language were to be included or excluded—the combinatorial approach would be beset by a politics of aesthetics. In their attempt to revolutionize Chinese movable type, but without upsetting the delicate compositional balance of those Chinese characters produced using this revolutionary method, practitioners of divisible type had made themselves the servant of two masters, neither of whom could be served loyally.

THE POLITICS OF PLAINTEXT: SURROGACY, SEMIOTIC SOVEREIGNTY, AND THE CHINESE TELEGRAPH CODE

During the early 1860s, the appeal of divisible type extended outside the worlds of printing and typography, and into the novel arena of telegraphy. In particular, yet another eccentric Parisian in our story was drawn to the work of his compatriots, seeing in Pauthier and Legrand's method an untapped potential that had eluded even its inventors. While his fellow Frenchmen were bedeviled by the tiny yet evident empty spaces upon which the system of divisible type printing relied, this man realized that the method could be exploited within the framework of telegraphy *without* inheriting this vexed politics of beauty.

In an 1862 essay, *De la transmission télégraphique des caractères chinois*, Pierre Henri Stanislas d'Escayrac de Lauture proposed a system of Chinese telegraphic transmission based on the principles of divisible type.[42] Transporting Pauthier and Legrand's method to the world of wire, electricity, and code, Escayrac de Lauture explained:

The question is simpler for telegraphic transmission than for typography. In consequence of the diversity of forms, which, according to its position, the same character radical or phonetic may receive, Marcellin Legrand must have engraved 4,220 different types. Telegraphy, considering the elements, without taking into account the appearance assigned to them in writing on account of their association, need only occupy itself with about 1,400 characters. By means of these 1,400 characters, the whole Chinese language may be transmitted.[43]

To transmit the character *shuo* (說, signifying "to speak"), for example, Escayrac de Lauture's method involved sending a series of ciphers that indicated, not the character itself, but the components out of which the character was built: in this case, the modules *yan* (言, signifying "language") and *dui* (兌, signifying "exchange") (figure 2.8). Upon deciphering these modules, the recipient would have the necessary criteria by which to retrieve the intended character: composed of the modules *yan* and *dui*, the one possible candidate was *shuo*. The same operation could be carried out for other characters, such as the transmission of *hai* (海 "sea") via the transmission of the water radical (氵) and the character radical *mei* (每). There would be no need to assemble or combine these radicals in a material or physical way, as in divisible type printing—the reconstitution of these Chinese characters would happen in the mind of the recipient.

Here we encounter the third mode of puzzling Chinese—what we term "surrogacy." Although inspired by divisible type, and although harnessing the very same Chinese character modules that Pauthier and Legrand had engraved and fashioned in metal, Escayrac de Lauture no longer concerned himself with metal pieces or their physical reassembly. This may have been the puzzle of combinatorialism, but not of surrogacy. Chinese characters and radicals remained the fundamental substrate of the language for Escayrac de Lauture, as in common usage and divisible type, but they were not to be manipulated directly. Instead, they were to be sequestered to "off-site" locations—such as a telegraph code book—and then "retrieved" by means of agreed-upon transmission protocols. It was these codes and protocols that constituted the primary concern for Escayrac de Lauture and the jigsaw pieces in the surrogacy puzzle—not Chinese characters themselves. The puzzle of surrogacy was the puzzle of metadata.

Escayrac de Lauture was born in Paris in 1826. An explorer, scholar, and writer, he studied at the college of the Oratoriens at Juilly, where his capacity for language became evident. In 1844, he entered the Ministry of Foreign Affairs as an attaché, serving the state in the punitive expedition against Madagascar, coordinated with the British. His career and

2.8 Transmission strategy using the Escayrac de Lauture system

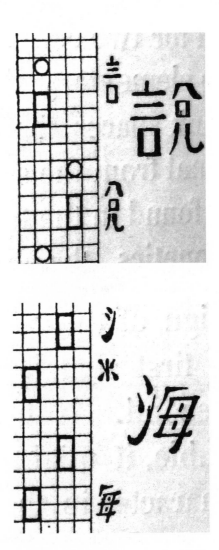

travels over the next decade would take him to Spain, Portugal, Britain, Germany, Switzerland, Italy, Tunisia, Libya, Egypt, and finally China. In 1859, Escayrac de Lauture formed part of the Anglo-French expedition to China dispatched after the outbreak of the Second Opium War. Three years later, evidently inspired by his time in China, he composed his piece on the Chinese language and its place within the emerging telegraphic world.[44]

When Escayrac de Lauture authored this piece, his ambitions extended far beyond Chinese. In a second piece by the Frenchman, entitled *Analytic Universal Telegraphy* and released in 1862 as well, Escayrac de Lauture outlined a grand redesign of telegraphic communication as a whole, based upon what might be understood as an early foray into automated or "machine translation," as it would come to be known one century later. "The telegraph wants a language that is more succinct and intelligible to all people," he argued, as he took stock of the still-novel technology of telegraphy as it extended its reach across the globe. "I will demonstrate that such a language is not a utopia—that it is not only possible, but simple, proper, and necessary."[45] Here was a vital way in which Escayrac de Lauture's puzzle departed from those of Staunton, Gamble, Pauthier, and Legrand, beyond technical questions pertaining to differences between surrogacy, combinatorialism, and common usage. These individuals had, whether in common usage or divisible type, puzzled Chinese in such a way as to bring it within the "compass" of European movable type—to reconcile Chinese to Europe and the West. Escayrac de Lauture's puzzle was different. By focusing upon the question of Chinese telegraphic transmission—this most challenging of telegraphic languages—his goal was not to subordinate Chinese script under one or another Western and alphabetic information technology, but to make it a laboratory case through which to imagine how all languages and scripts—alphabetic and nonalphabetic alike—might have equal footing by being subordinated within a shared, universal telegraphic language. For Escayrac de Lauture, to puzzle Chinese meant puzzling the entire semiotic structure of telegraphy itself, root and branch.

As Escayrac de Lauture dreamt of a telegraphic language "more succinct and intelligible to all people," the international telegraphic community was traveling along a very different trajectory—one that did not please this Frenchman in the slightest. At telegraphy's inception, the entrepreneurial Samuel Morse referred to the new invention as "the American telegraph" and, even more intimately, "my telegraph."[46] Even as Morse eagerly promoted the technology in Russia, western and southern Europe, the Ottoman Empire, Egypt, Japan, and parts of the African continent, the telegraph code upon which it was based remained fundamentally connected to the Latin alphabet and the English language—that is, to the fabric of Morse's linguistic world.[47] With the short "dot," the long "dash," and code sequences ranging primarily from one to four units in length, the code was originally designed to accommodate thirty discrete units: just sufficient to encompass the twenty-six English letters, with four code spaces remaining. Essential symbols—such as Arabic numerals and a select few punctuation marks—could then be relegated to the less efficient realm of five-unit code sequences (later expanded to the even less efficient six-unit sequences in "Continental Morse").[48]

While the code was ideally suited to handle English, the same could not be said for other languages—even alphabetic ones. With its thirty letters, German bumped up against the limits of the code's capacity, while French and its multiplicity of accented letters spilled out beyond it. Nevertheless, such Anglocentrism was further reinforced by the International Telegraphic Union in its original list of signifiers permitted for telegraphic transmission. At the ITU conference in Vienna in 1868, the collection of acceptable symbols was confined to the twenty-six unaccented letters of the English language, the ten Arabic numerals, and a small group of sixteen symbols (being the period, comma, semicolon, colon, question mark, exclamation point, apostrophe, cross, hyphen, e-acute (é), fraction bar, equal sign, left parenthesis, right parenthesis, ampersand, and guillemet).[49]

The expansion of telegraphy's authorized list of transmittable symbols was an extremely conservative and slow affair, moreover. It was not until 1875, for example, that the Saint Petersburg conference of the ITU finally expanded the original list of twenty-six letters to include a twenty-seventh: the accented "e" (é), now no longer sequestered to the specialized list of "signes de ponctuation et autres."[50] The conference further stipulated that, for those using Morse code, it would now be possible to transmit six other special, accented symbols: Ä, Á, Å, Ñ, Ö, and Ü.[51] It was not until the London conference of 1903, almost two decades later, that this supplemental list of accented letters was granted admission into the "standard" semiotic repertoire.[52]

Just as Escayrac de Lauture was authoring his essay, moreover, the 1860s witnessed the accelerated expansion of modern colonialism and with it a rapid development of the telegraphic network beyond the Latin alphabetic world. In 1864, cables were laid in the Persian Gulf that, when connected to the existing landline system, put India into direct tele-graphic connection with Europe.[53] In 1870, a further rapid expansion saw cables laid from the Suez to Aden and Bombay, and from Madras to Penang, Singapore, and Batavia.[54] Such growth brought telegraphy into contact with scripts it had not been originally designed to handle, rais-ing a profound question: Would the inclusion of new languages, scripts, alphabets, and syllabaries prompt a radical reimagination of telegraphy itself, or would they be absorbed and subordinated to the logic and syn-tax of existing approaches?

Escayrac de Lauture, for one, was optimistic about the possibilities of just such a radical redesign—an optimism that inspired his Chinese foray in 1862, nearly a decade before the Qing empire was incorporated into the international telegraphic network.[55] While the physical technology of the telegraph—the machinery of it—was an achievement that granted humans a power bordering on the godlike, the Frenchman argued, the semiotic architecture of telegraphy remained crude and bounded, too closely connected to actually existing human languages (namely Eng-lish—but also alphabetic languages more broadly). Escayrac de Lauture called for the development of a perfect universal symbolic language whose sophistication would measure up to the brilliance of the machine, rather than hold it back. The test case for this universal telegraphic lan-guage would be Chinese, whose eventual integration into global teleg-raphy would form a crossroads: Would Chinese simply be absorbed into the existing, all-too-human and all-too-English Morse code, or would humanity finally develop a new, universal language worthy of this new technology?

It was in his second essay, *Analytic Universal Telegraphy*, that Escayrac de Lauture ventured from the question of Chinese telegraphic transmis-sion to the question of universal telegraphic transmission. In this fifteen-page treatise, Escayrac de Lauture outlined his vision for the creation of an "algebra of language," as he termed it—one that would enable teleg-raphers around the world to traffic directly in meaning, despite the lan-guage barriers that separated them.[56] Specifically, he mapped out a series of tables in which was positioned an abbreviated lexicon of terms and phrases consisting of between 450 and 600 words. "Discourse is like a calculation that uses words," Escayrac de Lauture wrote, arguing that "we must find the algebra of this calculation." "We must find the common measure of thought and human discourse," he continued. A "language of facts and numbers," he called it: "a language without poetry, hovering above the vulgarities of common life."[57]

Escayrac de Lauture set out to create a "catalog of principal ideas" (*idées principales*) that established the core meaning underlying every possible human utterance, which would then be qualified by recourse to "accessory ideas" (*idées accessoires*).[58] To transmit the term "indifference," for example, Escayrac de Lauture's system would transmit the principal idea "affection," qualified by the accessory idea "negated." To convey "hatred," by contrast, involved sending "affection" qualified by "opposite." Scales or levels of intensity could also be communicated, enabling one to use "affection" to express adoration, love, repulsion, horror, and even execration through the appropriate use of modifiers.[59]

Through the development of corresponding tables in all the languages of the world—tables like those in Escayrac de Lauture's Chinese telegraph code—telegraphic transmission would, he felt, finally achieve its full global potential by enabling humans to traffic directly in meaning: to transmit a telegram, one need only identify the address of the principal and accessory terms within one's own vocabulary, and one's interlocutor would be able to locate the corresponding meanings—reconstituting them just as one would the character *shuo* (說) upon receipt of *yan* (言) and *dui* (兌). Reception of foreign transmissions would be straightforward, Escayrac de Lauture felt certain. "Without knowing even one of these words," he assured his readers, "one could, with the aid of a simple vocabulary, establish the meaning of a sentence with certainty."[60] The telegraph language Escayrac de Lauture proposed, he felt sure, "would be more appropriate for international communications than any known language."[61]

DOUBLE MEDIATION: THE CHINESE TELEGRAPH CODE OF 1871

In 1871, the growing network of telegraphic communication reached the shores of the Qing empire, with a single line opened between Shanghai and Hong Kong in April of that year. Carried out by two foreign companies—the Great Northern Telegraph Company of Denmark and the

Eastern Extension A&C Telegraph Company of the United Kingdom—the installation of this line marked the initial step in the construction of an empire- and then nationwide communications web, woven one filament at a time. A line was installed between Saigon and Hong Kong in June 1871, another between Shanghai and Nagasaki in August, and a third between Nagasaki and Vladivostok in November. In the ensuing years, this network expanded to encompass Amoy, Tianjin, Fuzhou, and other cities throughout the empire.[62] Chinese authorities and companies would steadily gain ownership of this web, and expand it to a total length of approximately sixty-two thousand miles by the middle of the Republican period (1911–1949).[63]

With the entrance of China and the Chinese language into international telegraphy, however, what ensued was not a reimagination of the modes or syntax of telegraphic transmission, as contemplated by Escayrac de Lauture. Instead, the Chinese telegraph code of 1871—invented by two foreigners—left the global information infrastructure of Morse code unaltered, while placing Chinese script in a position of structurally embedded inequality. Developed by a Danish professor of astronomy named H.C.F.C. Schjellerup and formalized by the French harbormaster in Shanghai, Septime Auguste Viguier, the code of 1871 drew its inspiration from the common usage mode.[64] A group of approximately 6,800 common usage Chinese characters was organized according to the Kangxi radical-stroke system, and then assigned a series of distinct, four-digit numerical codes running from 0001 to 9999. Approximately 3,000 blank spaces were left at the end of the code book, and a few blank spaces left within each radical class, so that individual operators could include otherwise infrequently used characters essential for their work.[65] To transmit a Chinese telegram using this system, the telegrapher began by looking a character up in the code book, finding its four-digit cipher, and then transmitting this cipher using standard Morse signals (figures 2.9 and 2.10).

The code designed by Schjellerup and Viguier thus placed Chinese script within a fundamentally different relationship with telegraphic transmission protocols than alphabetic and syllabic scripts. If Escayrac de Lauture had imagined a universal telegraphic language in which every script would be governed by the same, shared protocols of encoding and

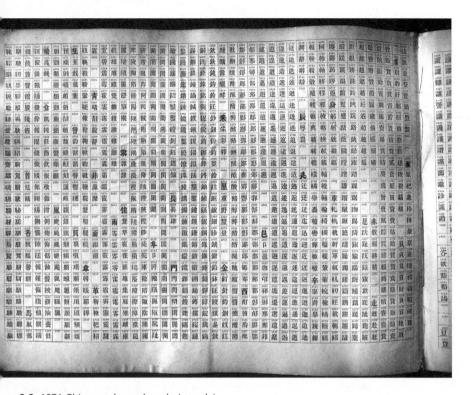

2.9 1871 Chinese telegraph code (sample)

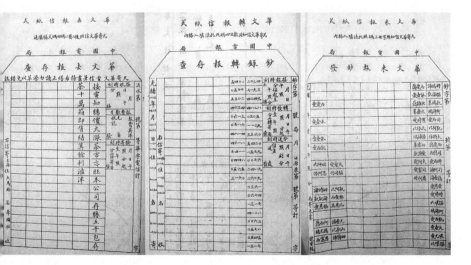

2.10 Sample of Chinese telegram and encryption process

transmission, the 1871 code was premised upon an *additional* or *double mediation* of Chinese: a first layer mediating between Chinese characters and Arabic numerals, and a second layer mediating between Arabic numerals and the long and short pulses of telegraphic transmission. By contrast, the transmission of English, French, German, Russian, and other languages involved only one layer of mediation—from letters or syllables directly to the machinic code of dots and dashes. In order for Chinese to enter the machine code of telegraphy, then, Chinese script would first need to pass through an additional *foreign* semiotic layer—in this case, that of Arabic numerals, but also conceivably the letters of the Latin alphabet. Chinese script would be doubly governed: first by the same dot-dash protocols as all other languages in the domain of telegraphy, but prior to that, by the Latin alphabet as it operated in English and Morse.

With the rapid demise of Escayrac de Lauture's dream, the politics of surrogacy played out in deleterious ways for the Chinese script. First and most basically, the only symbolic units to be employed in Chinese telegraphic transmission were Arabic numerals, which themselves were the longest and most expensive code units within Morse (figure 2.11). The shortest numerical code (the number "5" with its five short pulses) was already five times longer than the shortest letter ("e" with its single short pulse). The numeral "0" was the longest transmission sequence in the entire code, requiring five long pulses. Overall, then, its status as a purely numerical system retarded the Chinese telegraph code from the outset.[66]

Beyond the question of cost, the double mediation of Chinese exerted a second, even costlier penalty upon the script. Owing to its reliance on numeric encipherment, the 1871 Chinese telegraph code inadvertently rendered Chinese vulnerable to an ever-changing and proliferating number of legal and tariff penalties originally designed, not with Chinese in mind, but rather "secret" or "encoded" forms of telegraphic transmission. By the time Chinese script entered the world of telegraphy, the use of enciphered and encoded transmission within the alphabetic world was practically ubiquitous, being the rule rather than the exception in global telegraphy.[67] Decades earlier, telegraphers concerned with secrecy, and more so with saving money, began to develop a wide array of telegraph codes that could be used in coordination with Morse.[68] The dominant

A	.-	**M**	--	**Y**	-.--	**6**	-....
B	-...	**N**	-.	**Z**	--..	**7**	--...
C	-.-.	**O**	---	**Ä**	..--.	**8**	---..
D	-..	**P**	.--.	**Ö**	---.	**9**	----.
E	.	**Q**	--.-	**Ü**	..--..	**.**	.-.-.-
F	..-.	**R**	.-.	**Ch**	----	**,**	--..--
G	--.	**S**	...	**0**	-----	**?**	..--..
H	....	**T**	-	**1**	.----	**!**	..--.
I	..	**U**	..-	**2**	..---	**:**	---...
J	.---	**V**	...-	**3**	...--	**"**	.-..-.
K	-.-	**W**	.--	**4**	....-	**'**	.----.
L	.-..	**X**	-..-	**5**	.....	**=**	-...-

2.11 Morse code

objective of codes was to save money on transmission by creating code words that represented longer sequences, including entire sentences. Using a code book from 1885, for example, receipt of the single word "toothbrush" conveyed the much longer phrase "Telegram has been delayed in transmission." Meanwhile "gasping" conveyed "Send the goods that are ready, and expedite the remainder."[69]

Confronted with the mid-century spread of coded languages, governments and cable companies pushed for amendments to the international telegraphic system, in terms of both permission and pricing. A maze of regulations was designed to police and limit the use of such encrypted transmissions. Codes and "numbered languages" (*langues chiffrées*) were repeatedly barred from use in telegram addresses and in many other types of communications. In terms of pricing, "secret" transmissions were almost universally assessed at significantly higher per-word costs, so that cable companies could recuperate the revenue they were otherwise losing. As a practical consequence of such regulations, telegraphers operating in English, French, German, or any of the other telegraphic languages were thus presented with a constant choice: namely, whether to transmit a message in code or "in the clear"—otherwise known as "plaintext."

None of these regulations had anything to do with Chinese script in the slightest, and yet they would come to have profound implications for

Chinese characters immediately upon the language's entrance into global telegraphy. By virtue of its behavior as a "numbered language" or *langue chiffrée*, the Chinese telegraph code—and with it the Chinese language—was unrecognizable to the international telegraphic world as anything other than a *secret language*. Phrased differently, whereas all other telegraphic languages were understood to exist in one of two conditions—as "plaintext" or as "secret"—Chinese was the one telegraphic language which was legally understood as *inherently* secret—a script that possessed *no plaintext version*. This status was not, it bears emphasizing, because of any inherent properties of Chinese script—but solely because the international telegraphic union had opted, understandably but nonetheless harmfully, to "absorb" Chinese into the existing Morse regime "as is," rather than taking the globalization of telegraphy as an occasion to reimagine and universalize the semiotic protocols of telegraphy in a thoroughgoing, genuine sense—as Escayrac de Lauture had once imagined. In order for Chinese script to participate in the telegraphic revolution, it alone would have to pass through a semiotic terrain of Arabic numerals and Latin letters over which it possessed no authority—a space where the Chinese script would be exposed to prerogatives, punitive tariffs, and usage restrictions that had nothing to do with Chinese itself, but rather with the foreign symbols upon which Chinese script now had no choice but to rely.[70]

LIVING UNDER ENCRYPTION: EXPERIMENTS IN HYPERMEDIATION

Beginning in the twentieth century, Chinese authorities and companies would steadily gain ownership of the physical infrastructure of telegraphy in the empire—and then the Republic.[71] Invited to join the international telegraphic conference in 1883, the Qing eventually dispatched a representative to the Lisbon conference of 1909.[72] In 1908, China's Ministry of Communications would take over control of the Chinese Telegraph Administration, originally formed in 1882. In 1912, China was home to 565 telegraph offices, a figure that expanded to 1,094 by 1932. The length of line nearly doubled over the same period, from 62,523 kilometers in 1912 to just over one hundred thousand kilometers in 1932.[73] At the same time, however, the Chinese telegraph code of 1871 and its derivations continued to embed China and the Chinese script within a place of comparative disadvantage long after "telegraph sovereignty" (*dianxin zhuquan*) had been redeemed in a political and economic sense. Long after Chinese authorities took possession of the materiality of cables and towers, as well as financial and legal authority, the problem of *semiotic* sovereignty remained unresolved.

As representatives of the Qing and the Chinese Republic began to take a direct role in the international telegraphic community, attempts were made to improve the disadvantaged position of Chinese. In particular, Chinese ministers and engineers spearheaded an initiative through which they secured a peculiar legal exemption for the four-digit Chinese telegraph code, removing it from the category of "coded," "secret," and "enciphered" scripts, and securing for it the status of *de jure* "plaintext." In a moment both sublime and brilliant, Chinese representatives successfully established the four-digit code as the *artificial equivalent* of Chinese "plaintext"—despite the fact that it was not "plain" at all by prevailing standards of telegraphic communication. Beginning as early as 1893, any four-digit Chinese transmission recognized as a "bona fide" Chinese character—that is, a four-digit code that corresponded directly and meaningfully to one of the characters found in the 1871 code book—was to be counted as a single "word" when transmitted between stations in China.[74] A Chinese transmission would henceforth only be considered "secret" if the four-digit codes had in some way been further manipulated or permuted by the transmitter for reasons of privacy.

From this moment on, Chinese telegraphy underwent a profound yet subtle transformation. Phrased simply, Chinese telegraphers became "code conscious" in ways not true of their alphabetic brethren elsewhere in the world, adopting a bifocal relationship with the Chinese script itself. Through one eye, Chinese was Chinese: character-based script whose transmission over telegraphic cables constituted the primary objective of the telegrapher. Through the other, however, Chinese was *code*: sequences of digits that, save for rare savants who had managed to memorize every one of the 6000-plus codes by heart, always required the telegraph code book to decipher. For example, hearing the pulse sequence ----- -.... .---- ..... a telegrapher in China would not have leapt over the intermediary step of numeric encipherment and directly written out the "plaintext" character 北 (*bei*, meaning "north"). Instead, his first inscription on the paper would have been a *code*: the symbols to which this Morse code pattern corresponded—0-6-1-5. Only after this would he be in a position to translate this second-order code into "plaintext" by locating its corresponding Chinese character in the 1871 telegraph code book. By contrast, trained telegraphers in the alphabetic world did not need to undertake this additional step, and could instead translate the dots and dashes they received into plaintext alphanumeric messages. In the alphabetic world, the coded nature of telegraphic transmission was digested and disappeared inside the body of the telegrapher himself, contributing to a kind of "myth of immediacy" prevalent in so many discussions of alphabetic telegraphy both then and now. For Chinese telegraphy, the fundamentally coded nature of transmission—one might say of language itself—could never be ignored or denied. By consequence of the 1871 code, the trained telegrapher in the Chinese world was forced to *stay in code*.

While undoubtedly inefficient, and while undoubtedly placing Chinese script at a notable disadvantage when compared to alphabetic transmission, nevertheless this condition of "staying in code" or "living under encryption" exerted a subtle yet unmistakable influence on the kinds of experimentation and innovation undertaken by everyday Chinese telegraphers in the course of their work. Over the course of the late Qing and early Republican periods a vibrant, local-level sphere of innovation within Chinese telegraphy focused on making modest yet profound adjustments to the use of the four-digit code. A loose network of telegraphers, code book publishers, and entrepreneurs dedicated themselves to the development and refinement of a wide array of experimental methods designed to make the Chinese telegraph code faster and more efficient. To an even greater extent than their elite and metropolitan counterparts, such individuals lacked the political power to transform the thick and alienating medium that was built into the technological and legal framework of telegraphy. They knew, that is to say, that there was little chance of their day-to-day operations giving rise to a radical and revolutionary overhaul of the global information infrastructure itself—one that would reduce the layers of mediation through which Chinese had to travel, and bring Chinese into parity with other telegraphic languages.

Counterintuitively, they began to create *additional* layers of mediation, above and beyond the two already embedded within Chinese telegraphy. Their experiments and workarounds focused on *mediating mediation*—that is, inserting additional practices and devices of their own design between themselves and the Chinese code so as to render their relationship with the code more favorable, workable, memorizable, suitable, or otherwise advantageous. These mediations of mediations— *hypermediations* for short—were decidedly personal, corporeal, local, and immediate, encompassing a variety of novel *aide-mémoire*, forms of training and bodily practice, and reorganizations and repaginations of the Chinese telegraph code book, among other strategies. Such hypermediations, while they would seem prima facie to increase the time and effort needed to use the code, often rendered it faster, less expensive, and yet fundamentally unchanged in terms of its root-level semiotic architecture. Unable to develop a Chinese telegraph code from scratch that would be deeply attuned to the affordances and limitations of Chinese script, they left the underlying semiotic architecture of the Chinese telegraph code untouched and instead engaged in radical reimaginations of their *relationship* with, and pathways through, that architecture.

One of the earliest experiments in hypermediation involved retrofitting a relatively new technique of three-letter or trigraph coded transmission. Developed in the United Kingdom and elsewhere, this code sought to exploit the Roman alphabet as a base-26 system, wherein a sequence of three letters could be used to transmit a total of 26^3 or 17,576 units. Such a system would have proven far more efficient for Chinese than the four-digit system, insofar as the trigraph system required one fewer code unit (constituting a 25 percent reduction at the outset), and because it made use of letters rather than numerals (with letters being faster to transmit than numerals within Morse code, as explained above). In a newly published code book from 1881, steps were taken to employ this alternate encoding system by pairing each of the conventional four-digit sequences with a unique three-letter code. The code "0001" was paired with "AAA"; "0002" with "AAB"; and so forth. A third layer of mediation acted in tandem to this second one, moreover, being a set of twenty-six Chinese characters used by the editors to represent the Roman alphabet and thereby simplify and render more approachable graphemes that were still foreign to most in China circa the 1880s (table 2.1 and figures 2.12 and 2.13).[75]

2.12 Chinese character-based mediation of Roman alphabet (letter/character/pinyin)

Table 2.1

Chinese character-based mediation of Roman alphabet (letter/character/pinyin)

愛	ai	e	依	yi	i	藹	ai	m	姆	mu	q	摳	qu	u	尤	you	y	喂	wei
比	bi	f	夫	fu	j	再	zai	n	恩	en	r	阿	a	v	霏	fei	z	特	te
西	xi	g	基	ji	k	凱	kai	o	窩	wo	s	司	si	w	壺	hu	x	時	shi
諦	ti	h	鴟	chi	l	而	er	p	批	pi	t	梯	ti	x	時	shi			

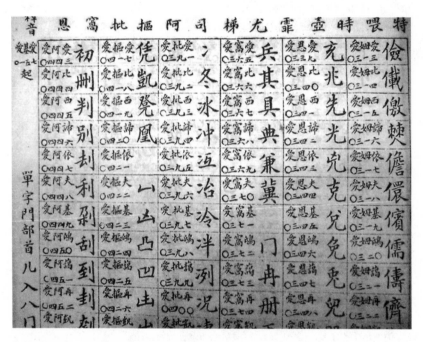

2.13 Sample page from character-mediated version of code book

Using this additional layer of mediation in concert with the second, a Chinese telegrapher could in theory carry out his work *entirely in Chinese* even as he operated with and through the Roman alphabet. Phrased differently, if we imagine being able to wiretap the electrical pulses that coursed through telegraph cables in China at this time, what might have sounded to us as the three-letter code sequence "D-G-A" would, for both the sending and receiving agents, have been understood in terms of the Chinese character sequence "諦基愛" (*di-ji-ai*). Even "alphabetic order" itself could be emulated using such a mediation technique, with telegraph operators being able to memorize the sequence "愛、比、西、諦、依 ..." (*ai, bi, xi, ti, yi ...*), rather than A, B, C, D, E. Within this complex and multilayered interplay of mediations, the very dynamics and valences of signification could thus be reimagined, all while leaving the underlying architecture of the telegraphic code unchanged.[76]

As the ambient level of familiarity with Arabic numerals and the Roman alphabet increased in China, Chinese character mediation schemes such as these began to disappear from code books. Being more comfortable with the likes of "1, 2, 3 ..." and "a, b, c," the average telegrapher presumably no longer needed or found useful the mediations of "一二三" and "愛、比、西." Mediations of other sorts soon replaced them, however, one of the most subtle and successful centering on a clever repagination of the Chinese telegraph code book itself. In contrast to early code books, in which each page started with the code sequence ---1 and concluded with ---0 (e.g., 0101 to 0200 in code books that featured 100 characters per page, or 0101 to 0300 in those that featured 200 characters per page), publishers in the Republican period began to reorganize the code book such that each page ran from code sequence ---0 to --99 (e.g., from 0100 to 0199, 1200 to 1299, etc.).[77] Although seemingly trivial, this alteration produced a secondary effect eminently useful within the everyday work of the Chinese telegrapher, as the page numbers of the code book itself were hereby transformed into *aide-mémoire* that corresponded to the first two digits of the codes featured on any given page. Repagination of the code book made possible a new "memory practice," one in which a telegrapher using a code book circa 1946 knew, for example, that the code "1289" could be found on page 12, code "3928" on page 39, and code "9172" on page 91—with the corresponding page numbers being featured in bold red letters at the top and bottom of each page.[78] To understand how telegraphers engaged with the Chinese code, mediating this relationship in novel and experimental ways, we therefore must also pay close attention to the social history of the code books themselves.

At the microhistorical level, each of these efforts was oriented toward a variety of facilitations: enabling Chinese telegraphers to work within a foreign and unfamiliar alphanumeric context, and to accelerate the process by which telegraphers could track down a given character or code, among other objectives not considered here. On the macrohistorical level, however, these local-level efforts added up to something much broader: a historical process connected to what has been theorized in this chapter in terms of "semiotic sovereignty." Through the creation of these many different "mediations of mediations," telegraphers were not only undertaking pragmatic processes pertaining to time and money, but were also establishing a novel relationship with an information infrastructure that, as we have seen, placed Chinese within a position of structurally enforced inequality. Telegraphers were taking "symbolic possession" of the four-figure code—playing with it, mediating it, suiting it to their own linguistic and physical preferences, capacities, and limitations. With these incremental, highly localized acts, telegraphers began the process of surrounding a system that had once surrounded them.

In the next chapter, we leave behind the worlds of telegraphy and movable type printing to arrive, at long last, at the age of the Chinese typewriter. As we do, we will reencounter each of the three puzzles examined in this chapter—common usage, combinatorialism, and surrogacy—only in a new technolinguistic context. With the advent of typewriting as an exciting new technology of inscription, people of a new generation began to contemplate the puzzle of typewriting for the Chinese language. As they did, they revisited and in some cases rediscovered the three modalities examined here, migrating them—as well as the politics embedded within them—into a new technolinguistic terrain.

3

RADICAL MACHINES

The machine must then produce, not letters, not parts of words, but complete words at each stroke.
—Devello Sheffield, 1897

I realized at the very beginning that the design must be fundamentally and radically different from any of the existing American typewriters. Any such idea as providing one key for each character was little short of absurdity.
—Zhou Houkun, 1915

One decade after its first appearance, Christopher Latham Sholes's invention began to circle the globe, transported and adapted for non-English languages by typewriter companies such as Remington and Underwood. As we saw in chapter 1, these encounters with Arabic, Hebrew, Cyrillic, Mongolian, Burmese, and many other non-Latin scripts led typewriter engineers and entrepreneurs to focus their energy on making the minimum number of necessary modifications to the original, English-language device so as to bring these orthographies within the compass of Western typewriting technology. As the device stretched to absorb these writing systems and their particular qualities—the right-to-left orientation of Hebrew or the orthographic variation of Arabic letters, for example—there was one market that remained frustratingly elusive: China. Because Chinese characters were nonalphabetic, engineers at Remington and elsewhere found themselves conceptually ill-equipped to solve the "problem" of Chinese typewriting by means of selective mechanical adjustments, the way they had with more than one hundred other languages. Try as they might, Chinese was never enveloped into the broader history of these Western multinational corporations.

With the globalization of the Western typewriter, the allure of this apparatus nevertheless became firmly established in China, as powerfully here as elsewhere in the world. China needed a typewriter of its own, many felt, not only because of its utility as a business appliance but as a symbol of modernity. With each passing year, it seemed as if China might soon become the only country left on earth *not* to possess one.[1] "No Chinese Typewriters," one article in 1912 abruptly read, driving home this increasingly matter-of-fact point. This lack of a Chinese typewriter—and more importantly, the growing conviction that the very existence of such a machine was impossible—was being harnessed by critics of the Chinese language who would see Chinese characters abolished altogether. To build a Chinese typewriter, then, would achieve much more than simply bringing Chinese business practices up to speed; in the ongoing trial against Chinese script, such a machine would constitute irrefutable evidence that Chinese was compatible with modernity.

To build this key piece of evidence would not be simple, however. As examined earlier, the conventional, shift-keyboard typewriter was poorly equipped to handle Chinese—a fact that, as we also saw, was blamed on Chinese writing itself, rather than on the machine. To build a "Chinese typewriter," then, would require engineers, designers, linguists, and entrepreneurs to depart from the conventional, Western typewriter form at a time when this form enjoyed unprecedented prestige and popularity—when more and more people worldwide believed it to be truly universal. They would need to subject the very idea of typewriting to kinds of reconceptualization that Remington and other companies were unwilling, and perhaps unable, to do. In a very real sense, they would need to rescue typewriting from the monoculture of single-shift keyboard machines, either by resurrecting earlier, forgotten forms of typewriters, or perhaps by inventing new types of typewriters altogether. In their effort to bring China into the age of technolinguistic modernity, then, such "Chinese language reformers" were by necessity "typewriter reformers" as well.

This kind of thoroughgoing reimagination posed risks, however. On the one hand, if in the pursuit of Chinese typewriting Chinese script were transformed beyond recognition, in what sense would the resulting machine be a *Chinese* typewriter at all? In 1913, a contributor to the *Chinese Students' Monthly* voiced an impassioned rejection of the abolitionist argument, but also a cautionary word against those who would prioritize the pursuit of modern information technologies at the expense of Chinese script.[2] "Many foreigners as well as a few extreme types of the mission educated Chinese," the author explained, "are in favor of a fundamental change in the Chinese language." Such individuals called for the abolition of Chinese writing, and perhaps even its substitution with English, "giving as one of the reasons that a typewriter cannot now be invented to suit the Chinese script." The author continued:

We Chinese wish to say that the privilege of a mere typewriter is not tempting enough to make us throw into waste our 4000 years of superb classics, literature and history. The typewriter was invented to suit the English language, not the English language the typewriter.

While the "materialistic education of the West has taught many a young Chinese to judge everything by its earning capacity," the article inveighed, "we hope that they will not judge the value of their own civilization in the same way. ... They must not forget that the preservation of the national language, which is the life and soul of a nation, is among the first things to which they should dedicate their service and devotion."[3]

There was a Charybdis to this Scylla, moreover. If, in the pursuit of Chinese typewriting, the machine itself were transformed beyond all recognition, in what sense would the resulting device be a Chinese *type-writer*? Inventors knew full well that their machines could not hope to mimic a Remington or Underwood, and yet they were also well aware that the West and *its* typewriter would remain, to some extent, the judge-in-absentia of their efforts. Phrased differently: If the "typewriter" as understood in the West (and by this point in much of the rest of the world) could not be "adopted" in China in any straightforward way, but would instead need to be critically reimagined, might not the resulting machine prove entirely illegible and unrecognizable to the Western eye? And if unrecognizable to the world as a typewriter, would it be a "typewriter" at all? Such were the questions confronting those who sought to bring Chinese into the new age of technolinguistic modernity.

THE AMANUENSIS MACHINE: DEVELLO SHEFFIELD AND THE FIRST CHINESE TYPEWRITER

O.D. Flox could hardly have known what to expect when he boarded a small vessel bound for Tongzhou at the northern terminus of the old Grand Canal. He was a member of the Western Civilization Union, an organization based in the United States whose stated aim was to "improve the social condition of men in heathen lands by introducing every variety of labour-saving machines," and he had learned of an American inventor who perhaps held the key to just such a machine: a typewriter for the Chinese language.[4] "The idea of a Chinese type-writer," he speculated, "the object of which was to save men from going crazy over the attempt to memorize the bewildering mass of hooks and crooks in Chinese characters, seemed to me to be a bold and original one."[5]

O.D. Flox's voyage had been prompted by a pair of intriguing articles appearing in the *Chinese Times*. Entitled "A Chinese Type-writer," the first appeared in January 1888 and reported briefly but glowingly on the American inventor and his device. "It is a marvel in the aid which it affords a foreigner in rapidly writing in beautiful clear characters," the article read. "It is astonishing how rapidly you learn and settle your doubt about characters and tones. ... You are learning like a child with its alphabet blocks, while at the same time you may be communicating with a Chinese friends [*sic*] or composing a book."[6]

The second report struck a different tone. In a March 17 letter to the editor, facetiously titled "That Chinese Type-writer," Henry C. Newcomb of the Islands' Syndicate for the Promotion of Useful Knowledge expressed sharp doubts as to the supposed achievement of the American's contrivance, relaying word of an unnamed "friend" who had paid a personal visit to the inventor's studio. "His type came to hand in huge bundles nearly a cubic foot in size," the letter continued, "and before the types could be used at all they must be sorted. This looks like a simple process; and so it would be indeed, if a man's life were not limited to three-score years and ten."[7] "The fact is," Newcomb concluded, "that this contrivance is not of much use for ordinary people, unless they have a teacher at hand to give the points." "And if one is to have a teacher," he continued, "why not make *him* do the writing in the first place? Why should one 'keep a dog, and do his own barking'?"[8]

In a *Heart of Darkness*–like quest, Flox set out on a "minute specimen of river craft and allowed myself to be dragged by ropes for days." When he reached Tongzhou, he found his man—but not at all as Flox expected him to be. Born on August 13, 1841, in Gainesville, New York, Devello Zelotes Sheffield served briefly as a teacher before being enlisted in the 17th New York Volunteer Infantry in the opening year of the American Civil War.[9] Sheffield served in the Army of the Potomac for two years, being promoted to the rank of commissary sergeant, before being invalided home and "retaining through the rest of his days the traces of the army experiences and illnesses," as his obituary would later recount.[10] In the years to follow, missionary work—and China in particular—beckoned. "China is the field to which my attention has been especially called," he wrote to his brother in March 1868, taking up residence in Tongzhou the following year with his young bride, Eleanor.[11] To Flox, he seemed "the farthest remove from the popular idea of a missionary, who is supposed to be a well-fed easy going man, who is always careful to make frequent reports to his society of 'progress' in his work." To the contrary, the five-foot-eleven Sheffield still bore the scars of a near-death encounter years earlier, when he was attacked and left for dead by a Chinese carpenter in his employ.[12] "He is yet in middle life," Flox reported, and yet "looks as lean as Pharaoh's second herd of kine, and certainly gives no indication that his machine has thus far loaded his table with the abundant fruits of the earth."[13]

The China to which the newlywed Sheffields voyaged was a fast-changing one. Only nine years prior, in October 1860, the Qing court lost the second of two swift and humiliating military conflicts with the British empire, the first having erupted two decades prior in the Opium War of 1839–1842. With China compelled to sign the Treaty of Nanjing in 1842 and the Treaty of Tianjin in 1858, multiple Chinese cities were opened as treaty ports for foreign traders, and Christian missionaries were legally authorized to expand their activities across the Qing imperium.[14]

At the time of Flox's visit, Sheffield's newly invented contrivance was not a mechanical typewriter so much as a system for rapidly inking and stamping Chinese characters. In 1886, he had crafted an experimental set of wooden stamps, drawing upon the carpentry experience he gained from his father, and no doubt aware of William Gamble's work on the common usage Chinese type case we examined in chapter 2. Arranging his Chinese character stamps in alphabetic order, according to the Peking Syllabary—a Romanization system developed in 1859 by Sir Thomas Francis Wade—the missionary was able to locate, ink, and stamp one character at a time in rapid succession. "I found in experience," he wrote of his method, "that I could use this system of tabulated stamps in writing as rapidly as Chinese scholars usually write their characters, and for five years I employed them constantly in composition."[15]

Flox portrayed Sheffield's process in even more enthusiastic terms:

The inventor then turned to his case of type and with the proud air which may always be noticed in one who knows that by his genius he has wrung some secret from nature, he touched the characters as with the hand of magic, and whole sentences of beautiful Chinese flowed forth, the characters arranging themselves with the precision of soldiers in their ranks. On seeing this exhibition of the real merits of the machine, my eyes filled with tears of genuine emotion. I grasped the hands of the inventor and said to him, "My dear sir, you are a benefactor of the human race. The resources of the Western Civilization Union may be depended upon for introducing this beautiful machine throughout China, and we will take care that your reputation as a great inventor and as a true philanthropist is never sullied by the winds of ignorant or malevolent criticism."[16]

In the same year Sheffield developed his new stamping method, he also purchased a broken-down Western-style English typewriter in the city of Tianjin, hiring a Chinese "clock-tinker" to repair it so the missionary could use it in his composition of English-language materials. "I cannot write much faster yet than with a pen," Devel'lo explained in a letter to his parents, "but shall soon learn to do so. The great advantage is that I can now write in the evening without fear for my eyes. This I have never done in China."[17]

Sheffield's acquisition set in motion a new quest for the missionary: to devise a "similar" machine for writing in Chinese. Inspired by the emergence of typewriting technology in the United States, Sheffield began to ponder how he might turn his Chinese stamps into a unified, mechanical apparatus. The question remained, however, as to how such a machine was to be built, particularly given the fact that Chinese was not an alphabetic script. "In printing Western alphabetical languages," Sheffield reasoned, "a full key-board for capitals, small letters, figures, etc., need not contain more than eighty keys to meet all possible wants in writing, and with the use of shift-keys some of the best machines are rapidly manipulated with only thirty direct keys." He decided it was untenable, however, to outfit a Chinese machine in such a way. "This suggests the radical difficulty of adapting the Western type-writer to the Chinese language," Sheffield had reflected, "in which each word is represented by its own ideograph."

Sheffield's motives for inventing a Chinese typewriter were complex. While it is tempting to assume that he was driven by a desire to more rapidly produce and disseminate Christian and Western texts among potential Chinese converts, this desire was in fact already well served by existing print technologies at the missionary's disposal. Indeed, Sheffield had already signed his name to many Chinese-language works by this point, beginning with his six-volume magnum opus, the *Universal History* in 1881. This was followed by a number of other Chinese-language publications, including *Systematic Theology* (1893), *Important Doctrines on Theology* (1894), *Political Economy* (1896), *Principles of Ethics* (1907), *Psychology* (1907), and *Political Science* (1909).[18] He was also a frequent contributor of short Chinese essays to journals such as *The Child's Paper*.[19] As all of this suggests, Sheffield's publishing ambitions were more than amply provided for by existing methods and technologies.[20]

When it came to the more intimate act of writing letters in Chinese, however, here Sheffield felt at a loss. "Paul accomplished much in strengthening the work which he had personally established by letters to the churches," Sheffield remarked, suggesting that both he and other missionaries might employ his new apparatus to correspond by letter with Chinese associates. "There is evidently a wide and important field for missionary activity in this direction, which is now largely neglected by reason of reluctance to master the use of the Chinese character as a means of written communication."[21] He did not conceive of his invention in terms of its potential impact on the Chinese economy, Chinese modernity, or other grand and abstract concepts. For Sheffield, his machine signified liberation from his own longstanding dependence on Chinese clerks and secretaries who composed his letters on his behalf—in a word, Sheffield's goal was to develop a kind of Chinese robot, or amanuensis machine, that could output characters composed and cut by native Chinese hands, but at the same time obviate all need for an actual Chinese clerk. While Sheffield and many of his foreign colleagues considered themselves highly proficient if not fluent in Chinese, only with a novel apparatus like this, he believed, would it be fully possible to compose documents by his own hand that upheld the aesthetic standards befitting a man of his erudition and status.

Aesthetics were not the only concern, moreover. "I am convinced," Sheffield wrote, "that to the present time foreigners doing literary work in the Chinese language are under unnecessary bondage to Chinese assistants."[22] There was something decidedly prophylactic about the clerk, he felt: a culturally alien third party constantly intermediating, subtly altering, and ultimately interfering in the foreigner's work. "They usually talk to their writer," Sheffield commented, referring to foreigners and their Chinese assistants, "and he takes down with a pen what has been said, and later puts their work into Chinese literary style." "The finished product will be found," he continued, "to have lost in this process no slight proportion of what the writer wished to say, and to have taken on quite as large a proportion of what the Chinese assistant contributed to the thought."[23] As with many of his colonial and semicolonial contemporaries, then, Sheffield was motivated by a persistent anxiety, a concern that the unavoidable process of translation and amanuensis upon which foreigners relied necessarily led to the loss (and perhaps deft and malicious erasure) of one's intended meaning, and the interpolation of the native clerk's world views and sensibilities.[24]

To drive home his point, Sheffield referenced an unnamed book on botany by a "distinguished Western scholar in Chinese, in which he informs the student that there is in Southern China a certain plant produced from a worm!" "Of course," Sheffield continued, "such an interesting fact in natural history was contributed by the Chinese writer, and in some way survived the ordeal of proof-reading."[25] "No foreigner should be ambitious to write in Chinese for publication without the revision of a good native scholar," he conceded, "but by early and habitual practice in independent composition, which one would be stimulated to by a typewheel, the foreigner I am confident would become master of his own thoughts in Chinese at a much earlier date than at present, and would do his work in composition to suit his own tie without regard to the presence or absence of the Chinese writer."[26] Sheffield's machine would allow the foreigner in China to wrest back control of *meaning itself.*

THE BODY OF CHRIST: THE COMPETING LOGICS OF
COMMON USAGE IN SHEFFIELD'S CHINESE TYPEWRITER

In his pursuit of a typewriter for Chinese script, Sheffield's experimental process did not unfold in a vacuum. As with Gamble, Pauthier, Legrand, Escayrac de Lauture, and others before him, his process was shaped by his own deep-seated beliefs about Chinese characters. "Each character," he proclaimed, "must be treated as an irresolvable individual." Sheffield expanded on this conviction: "The machine must then produce, not letters, not parts of words, but complete words at each stroke. It must be able to bring with rapidity and precision from four to six thousand characters to the printing point."[27]

Sheffield's observations about Chinese characters represented personal convictions, not neutral, objective statements of fact. As we saw in the preceding chapter, divisible type printing regarded characters not as "irresolvable individuals" but as meta- or epiphenomenal entities built up or "spelled" by using still more fundamental, elemental units. In Chinese telegraphy, meanwhile, Chinese characters were treated as referential touchstones, to be addressed but not trafficked in directly. Had Sheffield shared the commitments and dispositions of Pauthier, Legrand, and Beyerhaus, or perhaps of Escayrac de Lauture or Viguier, he might have set off down a very different path in conceptualizing his new inscription technology. Sheffield was aware, it bears emphasizing, of other approaches to the question of Chinese information technology—including the divisible type method we examined in the preceding chapter. Indeed, Sheffield once met the typewriter magnate Thomas Hall in New York, the developer of the index typewriter we met briefly in chapter 1. "On learning that I was trying my hand on the production of a Chinese type-writer," Sheffield later recounted, "[Hall] showed the incredulous interest of the man of experience, who has tried and failed, towards the man of inexperience, who is about to try, and of course to fail in like manner."[28] "He told me that he had already grappled with the problem of making a typewriter for the Chinese language," Sheffield continued in his account, and "drew from a drawer a crumpled piece of paper printed upon in the Chinese character." "He had started out with the thought that he could separate characters into their component strokes, and by having all possible strokes arranged upon a printing-face of his type-writer he could combine these strokes to produce the desired characters."[29] As Sheffield recounted, however, Hall had become discouraged once he "discovered that while the number of possible strokes was not alarmingly great, the size, proportion, and relation, of the strokes in combination were infinite in their variations."[30] Aesthetic concerns also complicated and compromised this system of printing, insofar as the resulting characters often appeared spatially loose and incoherent. "Such a system of writing Chinese characters by ticking off the strokes that unite to form them would bear less likeness to the living character than the dry bones in the vision to living men!"[31]

While we cannot know precisely why Sheffield arrived at the commitments he did, we do know that his beliefs about Chinese characters shaped every step along the path of developing his typewriter. First, in committing himself to the "irresolvable individuality" of characters, the very first question that surfaced for him was, quite naturally: how was he to fit these tens of thousands of individuals on a machine? Riding through the streets of Tongzhou aboard a rickshaw, as he recounted, Sheffield was struck by a realization: to solve the problem of the script's abundance, he would begin by frequenting local foundries and typesetting offices, speaking with Chinese printers who, through their accumulated experience of cutting, forging, and using Chinese typefaces, possessed an intimate, firsthand knowledge as to which characters were most and least frequently used. Based upon these observations—observations reminiscent of Staunton's and Gamble's in chapter 2—Sheffield's typewriter would include only what he termed a "careful selection of characters for general use."[32] As for the many other characters in the Chinese lexicon, they would simply be excluded.

By 1888, Sheffield had news to share. "Have I written you about my new invention?" he rejoiced to his family members back home. "It is certain to attract considerable attention when it is known to the public. It is a Chinese type-writer, a machine with which to write Chinese." His goal was to make the wheel of the machine from wood, but to "send it to America to have it made by a machinist in metal." "I think it will print faster than a Chinese teacher can write," he continued, "and in that case it will be in considerable demand, especially by foreigners in China, as but very few of them are able to write Chinese. They may be good Chinese scholars, reading the language with ease, but they have not taken the time to learn all the complicated strokes that make up the characters."[33]

The typewriter Sheffield had produced looked nothing like the Western machine he had purchased in Tianjin (figure 3.1).[34] Instead, it looked like a "small round table," as he described it, wherein Chinese characters were arranged into thirty concentric circles. Having determined that "the working vocabulary of Chinese scholars is easily within the limit of six thousand," and furthermore that "this list can be reduced to four thousand, with but rare occasion to strike outside of the list to give expression to thought," Sheffield ultimately settled upon a total of 4,662 characters in all.[35] As for the other tens of thousands of characters in Chinese, they would be jettisoned entirely.[36]

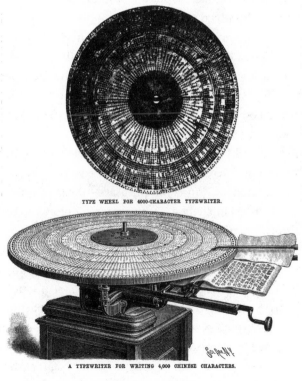

TYPE WHEEL FOR 4000-CHARACTER TYPEWRITER.

A TYPEWRITER FOR WRITING 4,000 CHINESE CHARACTERS.

3.1 Chinese typewriter invented by Devello Sheffield; from "A Chinese Type-writer," *Scientific American* (March 6, 1899), 359

Sheffield's machine departed from Chinese typesetting in another vitally important way. Because his machine inscribed one character at a time, and thus required only one of each, he was able to fit all of his characters within arm's reach, creating a device that achieved William Gamble's implicit yet unreachable ideal of sedentary efficiency. Indeed, having dispensed with any need for ambulatory movement, Sheffield was able to shift his focus entirely to the upper extremities of the body, giving rise to a new ideal altogether: the goal of minimizing even the movement of the operator's hands. To this end, Sheffield further subcategorized his 4,662 characters into four regions, the first comprising 726 "very common characters," the second a set of 1,386 "common characters," the third a set of 2,550 "less common characters," and the fourth with an additional group of 162 special "untabulated characters"—special characters that, owing to their importance to Sheffield and his missionary work, were included and sometimes duplicated within the list of "very common characters."[37] Ideally, if the boundaries of his four regions were set well, most of his time would be spent within the narrow region of "very common characters," allowing his hands to work within ever-decreasing distances. Having made it possible to manipulate the machine from a single, stationary position, Sheffield had invented a new type of Chinese technolinguistic body, sedentary and dexterous. Sheffield had become the first "Chinese typist" in history.

As Sheffield pushed "common usage Chinese" beyond the framework of Jin Jian's printing office, and indeed even the work of William Gamble, the tensions latent within common usage—those of inclusion and exclusion, as examined in chapter 2—became more pronounced. Pulling in one direction was the descriptive imperative of common usage, one which dictated that such a device would need to encompass all of the characters needed to reproduce Chinese discourse as it currently existed. Characters of everyday usage, such as *ta* (他 "he"), *si* (四 "four"), and *shang* (上 "atop"), needed to be present on Sheffield's machine, lest it be incapable of composing even the most basic of Chinese sentences, and also positioned in the easiest-to-reach zone so they could be reached with maximum speed. Sheffield's concerns did not begin and end with the banality of particles, numerals, and common adjectives, however. As with Gamble and other Christian missionaries before him, another imperative pulled in the opposite direction. Sheffield was a harvester of souls, and as such he also wished to intervene and give rise to new terms in the Chinese language—to traffic with Chinese-reading interlocutors in highly *uncommon* concepts. Here we discover in Sheffield's machine a proverbial Noah's ark populated by birds, lions, apes, camels, dogs, and other assorted beasts. His vocabulary was that of the Bible, and as such his typewriter was a land of slaves (*nu* 奴) and hegemons (*ba* 霸), ghosts (*gui* 鬼) and sorcerers (*wu* 巫), deafness (*long* 聾) and blindness (*mang* 盲), funeral (*sang* 喪) and blessing (*guan* 盥), blood (*xue* 血) and excrement (*fen* 糞), father (*ba* 爸) and son (*zi* 子).[38]

No two characters were more important for Sheffield than *ye* (耶) and *su* (穌), for in their union they formed *Yesu* (耶穌), the Chinese translation of *Jesus*. These two characters posed a unique challenge for Sheffield, however, being pulled in diametrically opposed directions by the logics of reproduction versus aspiration. By itself, *ye* was a frequently used particle in literary Chinese, and so could claim its rightful place within the list of 726 "very common characters." *Su*, by contrast, was far less common, typically appearing only as a variant of the more standard form *su* (蘇), as in the place name *Suzhou*.[39] In the tension between the descriptive and prescriptive imperatives, then, the literary Christ-body of *Yesu* was being pulled apart. If Sheffield obeyed the imperative of mere description, *su* would either have to be placed in a separate region from *ye*, or perhaps excluded from the machine altogether—since, after all, his character set already represented but a fraction of the total Chinese lexicon. To obey the second imperative would involve "promoting" *su* far above what lexical evidence recommended. The first imperative would quite literally divide Christ, locating his two halves in separate domains, and thus embedding within the very structure of the machine a need to perpetually exert energy to reconstitute him. The second imperative would place the unity of the linguistic Christ-body above the concerns of the mundane world.

In the end, Sheffield was decisively indecisive. He included two copies of *su* on his machine, placing one where it belonged *empirically*—within the list of 2,550 "less common characters"—and the other where it belonged *theologically*—within the exclusive list of "very common characters." The Christ-body was thus simultaneously intact and divided on

Sheffield's machine, forming a taut, straining distance that in many ways captured the objective of Sheffield's missionary work overall: to begin in a time when *Yesu* was an uncommon word in Chinese and then to use epistolary technologies such as the typewriter to help bring about its commonality and ubiquity. Sheffield wanted to close the gap between the two *su* characters on his machine, so to speak, with the ascent of *su* into the realm of high-frequency characters itself being the barometer of Christ's ascendance in China.

Reports of Sheffield's machine circulated in the American media in 1897, reaching readers in the states of Arkansas, Colorado, Illinois, Kansas, Kentucky, Louisiana, Michigan, New York, and Wisconsin, among other locales.[40] "A Chinese typewriter has been invented by Rev. Mr. Sheffield," the *Daily Picayune-New Orleans* reported. "It is said to be a very remarkable machine, and is exciting a great deal of comment over there."[41] The machine, it was reported, "is said to exceed so far the speed of the swiftest Chinese writer that its value is assured."[42] "It turns out to be a great success," the *Semi-Weekly Tribune* reported, "and will relieve both the foreigners and the native Chinese from the necessity of using a paint brush and a pot of ink in conducting their correspondence."[43]

Perhaps inspired by these reports, a curious change of heart took place in Sheffield wherein he clearly began to think of his machine as one perhaps capable of "liberating" Chinese clerks as well—in this case, liberating them from the requirements of manuscript. "Few as yet see in it anything more than a cunning play-thing," he lamented, however, presumably speaking of Chinese clerks to whom he had demonstrated the apparatus without success. "They do not comprehend how it is that foreigners seem ever to be planning how to save time. Many find time hanging heavily on their hands, and scholars will leisurely copy books containing hundreds of thousands of characters rather than purchase the book. But the world moves, and fortunately for China she is fastened to the world!"[44]

Ultimately, Devello Sheffield never saw his beloved invention become anything more than a curiosity-inducing prototype. He died on July 1, 1913, shortly before his seventy-second birthday, with the whereabouts of his machine remaining a mystery to the present day.[45] Perhaps Sheffield's machine was digested long ago by a colony of white ants, that archenemy of missionary printers and their wooden apparatuses. A more romantic idea is to imagine that it resides unrecognized somewhere in Detroit, Michigan, where a sixty-seven-year-old Sheffield and his wife briefly furloughed in the spring of 1909.[46] A *San Francisco Chronicle* article from the time gives us some cause to hold on to such an idea, mentioning that Sheffield "brings with him a Chinese typewriter of recent invention," and providing a rather detailed description: "The instrument contains 4,000 characters on a large, circular board, within twenty-four circles. The machinery is intricate and is about as large as four machines with the American characters. It is the clergyman's intention to have a large number of the machines manufactured in this country to be shipped back to China."[47] To have carried this to the United States at the age of sixty-seven would have been challenging enough—to bring it back upon his return, even more so. Whether the machine slumbers in a Michigan attic somewhere or has long since been composted back into Chinese soil, it was never transformed into a product of mass manufacture.

The production of the first commercialized Chinese typewriter would take another decade, and would be pioneered, not by an American missionary residing in China, but by a Chinese engineering student residing in America.

MODERNITY IN THREE THOUSAND WORDS OR LESS: ZHOU HOUKUN AND HIS TYPEWRITER FOR THE CHINESE MASSES

Barely one year after Devello Sheffield made his final voyage from the United States to China, a young Chinese student made this passage in the opposite direction. Zhou Houkun arrived in San Francisco on September 11, 1910, completing a one-month journey from Shanghai by way of Hong Kong and Honolulu. He was twenty years old, unmarried, and bound for Boston.[48] With him, the pursuit of common usage Chinese typewriting resurfaced in the opening decades of the twentieth century—albeit under dramatically different circumstances.

Zhou Houkun hailed from Wuxi, in Jiangsu province, and had recently completed schooling at Nanyang College in Shanghai, predecessor to the Jiaotong Engineering University founded in 1921.[49] Among his shipmates on the Pacific Mail steamship were fellow Boxer Indemnity scholars, those chosen in the second annual competition to study abroad, primarily in the United States. Zhou Houkun was among the students announced in 1910, along with upstart notables Hu Shi and Zhao Yuanren (aka Yuen Ren Chao).[50]

After disembarking, these young men headed in different directions. Zhao Yuanren and Hu Shi both proceeded to Cornell University, before themselves parting ways—Hu to Columbia and Zhao to Harvard. Zhou Houkun stopped in the middle of the country, taking up residence at the University of Illinois at Urbana-Champaign, where he studied railway engineering during the 1910–1911 school year.[51] The east coast beckoned him, however, and Zhou transferred the following year to the Massachusetts Institute of Technology. Here he graduated with an MS in aeronautical engineering, the first such degree ever awarded in the United States.[52]

The China that Zhou left behind was in urgent need of modernized railway systems, ships, and aircraft, making Zhou's time abroad precisely the sort of investment motivating the overseas studies program. Connected to the Boxer Indemnity Payment, the program required students to focus upon particular majors considered essential to the modernization of China, including agriculture, English, business, mining, law, political science, natural sciences, and education. Soon, however, he found himself enthralled by a novel venture that would capture his attention for the next five years: Chinese language reform. In this regard, he was of a pair with both Zhao Yuanren and Hu Shi, each of whom had quickly abandoned their more "practical" studies to pursue lifelong careers in Chinese linguistics, literature, and cultural reform. But while Zhao went on to become a towering figure in Chinese linguistics, and Hu Shi a luminary in China's literary and political circles, Zhou Houkun's entryway into Chinese language reform remained inseparable from his enduring passion for engineering. His goal was not to revolutionize the study of Chinese, or its literary politics, but simply to build a machine: a typewriter for the Chinese language.

"We will never throw away that wonderful language of ours," Zhou later wrote, "in order to satisfy the imagination of a few who thought a mechanically inadaptable language ought to go in favor of some other language which would be adaptable to the mechanical device."

The whole idea is so revolting that any further comment would put it on a par with other problems, which privilege it does not deserve. It is the duty of the engineer to design machines to suit the existing conditions, but it should never be the privilege of the engineer to ask for a change of existing conditions to suit his machine.

"Blame the engineer," Zhou argued, "but do not blame the existing language." "An engineer who can not build machines according to reasonable specifications is not an engineer in the best sense of the word."

A turning point for Zhou came in 1912 at an exhibit he attended as a junior undergraduate at MIT. Here in the Mechanics Building, "my attention was naturally directed towards exhibits of mechanical devices," Zhou explained.[53] "Among them, one machine especially attracted my attention. A girl sat in front of a keyboard, touched the keys, punched multitudes of little holes in a long reel of paper, and, when finished, placed the hitter in a machine which produced fresh, clean, clear types of lead all lined up and ready for the printing press. The whole process occupied only a few minutes. The machine worked automatically and incessantly with a rapidity which easily put to shame our method of type-setting in China. I was told that it was a monotype machine."[54]

"My thoughts at once turned back to China," Zhou continued, "and brought before me vividly the scene in a Chinese press room—where typesetters, little trays in hand, travelled to and fro seeking for a particular type in a maze of thousands. The process was slow, tedious, inefficient and has constituted one of the great obstacles to the advancement of general education in China. Something must be done." And so the pursuit of a Chinese typewriter recommenced under very different motivations than in the work of Devello Sheffield. Sheffield had set out to build an amanuensis machine, one that would free him from reliance upon Chinese clerks when composing correspondence in the Chinese language. Zhou Houkun would undertake the same pursuit, but this time with the goal of modernizing his homeland and its language.

"I realized at the very beginning that the design," Zhou wrote, "must be fundamentally and radically different from any of the existing American typewriters. Any such idea as providing one key for each character was little short of absurdity." In certain respects, this passage by Zhou reminds one of the earlier statements by Devello Sheffield, who years earlier also proclaimed the need to depart from the frameworks of Western typewriting in his attempt to build a machine for Chinese. Together with Sheffield's earlier statements, moreover, Zhou's proclamation alerts us to the existence of technolinguistic alternatives that predated the rise of the Remington monoculture but also lived on in small pockets within it. Zhou's statement, however, must also be distinguished from Sheffield's in terms of the degree of commitment—and even resistance—that it represented in light of the time and place in which it was uttered. During the 1910s, to proclaim one's intention to create a typewriter that was "fundamentally and radically different" from American machines was to depart from a paradigm that was not only dominant for the English language, or even for the Latin alphabet exclusively, but practically for all of the world's languages by this point, as examined in chapter 1. In the face of all-encompassing Remingtonian universalism, Zhou's words spoke of radical alternatives.

Like Sheffield before him, Zhou set out to create a machine based on common usage characters—this time reducing the total number to approximately 3,000 in all. In choosing these 3,000 characters, however, Zhou would need to develop a lexicon that differed markedly from his missionary predecessor's. The biblical bestiary of Develio Sheffield's typewriter would quickly be abandoned, his proverbial Noah's ark evacuated, and his lions, camels, dogs, and birds shepherded to the confines of the "secondary usage character box." Likewise, in the era of state-sponsored vernacularization, the literary Chinese particles and pronouns found on Sheffield's machine—characters like the self-deprecatory first-person possessive *bi* (敝, signifying "my humble"), which Sheffield had included as part of "common usage"—would need to yield their position to the vernacular Chinese question particle *ma* (嗎), the second-person singular pronoun *ni* (你), and so forth. Weights and measures were changing as well, and Arabic numerals were appearing in Chinese-language texts in greater numbers—all of which Zhou and any future common usage theorists would need to account for.

Fortunately for Zhou, the 1910s witnessed the rapid growth and indigenization of "common usage" studies in China, providing him with ample empirical evidence upon which to determine the lexical boundaries of his device. No longer dominated by foreign printers or missionaries, moreover, algorithmic "distant readings" of the Chinese-language corpus became a mainstay for Chinese intellectual elites themselves, who set out to subject ever-larger corpora of Chinese texts to the same acid bath of reason we saw in the case of William Gamble and others. Common usage Chinese became a vibrant business, in fact, wherein Chinese educators, language reformers, entrepreneurs, publishers, and statesmen advanced their own hypotheses about the most scientific and practical ways of determining the boundary between utility and uselessness. Although their work was reminiscent of their foreign predecessors', this new chapter in the history of common usage Chinese centered around a new objective: to enable the Chinese "masses" to grasp their own written script.[55]

One of the most formidable participants in this debate was Chen Heqin (1892–1982), who received his education at the Columbia Teachers College in New York, as well as Southeast University in Nanjing.[56] Chen set out to determine a repertoire of what he termed "foundation characters" through the analysis of texts, signposts, contracts, and other Chinese-language materials. In 1928, Chen published what George Kennedy of Yale would later vaunt as the "first work of any large scale" of character frequency analysis. The work, entitled *Commercial Press Characters in the Chinese Spoken Language Listed According to the Frequency of Their Appearance in Recent Books and Magazines,* was based upon the study of a corpus of over a half-million characters, comprising children's books, newspaper periodicals, women's magazines, and what he termed "standard literature"—each of these forming roughly one quarter of the study.[57]

Chen Heqin's statistics were remarkably similar to those produced by William Gamble some seven decades prior. A mere nine characters appeared more than ten thousand times, constituting 14.1 percent of the total. The next most common twenty-three characters occurred 14.7 percent of the time, for a total of between four and ten thousand instances each. The subsequent forty-six characters each occurred from two to four thousand times, together composing 13.1 percent of the whole. And the subsequent ninety-nine accounted for another 15.1 percent. Chen Heqin had arrived—albeit more systematically—at the same observation as the algorithmic readers of the previous century: fewer than 200 characters accounted for well over half of all usage.[58] All of this was good news for Zhou Houkun and his common usage Chinese typewriter. As Chen's work seemed to show, even a radically reduced range of Chinese characters on the typewriter might not necessarily impede the expressive capacities of potential users—precisely because their expressive capacities were already quite limited.

On the other hand, the challenging puzzle was to decide which other characters to include—a deeply political question that set off widespread competition by Chinese educators, politicians, language reformers, and others over how to determine what the "Chinese masses" needed. In 1920, the Ministry of Education ordered primary schools to replace literary Chinese with vernacular, helping spark what Charles Hayford describes as a "commercial war" in the area of educational publishing. A panoply of houses set out to compile and sell new "national language textbooks."[59] In 1922, Changsha became the first site of a major literacy campaign by James Yen, with an estimated twenty thousand copies of his character primer selling within months. The course proceeded in five stages, with students earning a colored stripe for each sequence completed. By the time of graduation, one could proudly show off one's completion by wearing all five stripes, which together formed the five-colored flag of the republic.[60] The educational reformer Tao Xingzhi created his own 1,000-character primer, published by Commercial Press in August 1923. During its first three years, the primer was estimated to have sold over three million copies.[61] Tao outlined an ambitious vision, advocating the sale of the new 1,000-character primers in every rice shop, as a full-scale campaign to replace the traditional *Trimetrical Classic* (*San zi jing*) and *Thousand Character Classic* (*Qian zi jing*).[62] Mao Zedong was also a participant in the business and politics of common usage. In 1923, he oversaw the creation of a new foundational character set, this one designed to address the political commitments and visions of the nascent Chinese Communist party, founded only two years prior.[63]

The flurry of "thousand character movements," as George Kennedy termed them, continued unabated throughout the 1930s.[64] Hung Shen, a director in the Star Motion Picture Company of Shanghai, entered the fray in 1935 with his own book, *How to Teach and Use 1100 Basic Chinese Ideographs*. The North China Language School compiled its own reference, *The Five Thousand Dictionary*. In May 1935, Li Chih published an extension of Chen's research in the *Chinese Journal of Education Research*.

Li's corpus was three times larger than his predecessors', having added elementary school texts, among other documentary sources.[65] In 1938, the Lu-Ho Rural Service Bureau published its own minimum vocabulary, the *2000 Fundamental Everyday Usage Characters* (*Richang yingyong jichu erqian zi*). Studies such as these drove home a consistent argument: "The student cannot afford," as Kennedy phrased it, "to load his mind with useless freight, and, in the early stages of Chinese study at least, an ideograph which recurs no oftener than once in ten thousand must be definitely labelled useless."[66]

In his attempt to create a "popular" Chinese typewriter, Zhou Houkun gravitated in particular toward the work of Dong Jing'an (1875–1944), professor at the Shanghai Baptist College and Seminary and author of the widely popular "600 Character" mass education series.[67] For Zhou, the trend seemed clear. These days, he explained, China had "popular education" (*tongsu jiaoyu*), "popular textbooks" (*tongsu jiaokeshu*), "popular lectures" (*tongsu yanjiang*), and "popular libraries" (*tongsu tushuguan*)—so why not, then, a "popular typewriter tray bed" (*tongsu dazipan*)? In his patent materials, Zhou explained that his "popular" typewriter would be composed of precisely the same characters found in Dong Jing'an's series, along with certain complements. His would be a machine for the masses.

The first prototype having been completed by May 1914, Zhou's machine contained a cylinder measuring roughly sixteen to eighteen inches in length and six inches in diameter, on which a set of approximately three thousand character slugs was arranged in accordance with the Kangxi radical-stroke system—more than the "popular education" list by Dong, but far fewer than the list by Sheffield (of whose machine Zhou was aware).[68] All of these characters were printed in a grid on top of a separate, flat, rectangular finding aid located in the front of the machine. The operator used a metal finding rod to locate the right character on the finding aid: when the tip of the rod was moved into place over the top of the desired character on the printed grid, the other end of the rod brought the corresponding character on the cylinder to the printing position (figure 3.2).[69]

Zhou's machine was received with a measure of international praise and attention. On July 23, 1916, the *New York Times* ran a detailed story about Zhou entitled "Chinaman Invents Chinese Typewriter Using 4,000

藏暉室劄記

藏暉室劄記（續前號）

胡適

四日晨赴習文藝科學學生同業會。(Vocational Conference of the Arts & Sciences Students) 鄭君。萊主席先議明年本部同業會辦法衆舉余爲明年東部總會長力辭不獲又添一重擔子矣胡君宣明謂一文論『國家衛生行政之必要及其辦法之大概』極動人其辦法尤爲井井有條廁省工業大學周厚坤君新發明一中文打字機鄭君請其來會講演圖式如下。

周厚坤君新發明中國打字機字機

其法以最常用之字（約五千）鑄於圓筒上（A）依部首及畫數排好機上有銅版可上下左右推行竟得所需之字則銅版可推至字上版上安紙上有墨帶另有小椎一擊則字印紙上矣其法甚新惟覓字頗費時然西文之字長短不一長者須按十餘次始得一字今惟覓字費時既得字則一按已足矣吾國學生有狂妄者乃至倡廢漢文而用英文或用簡字之議其說曰漢文不適打字機故不便也夫打字機爲文字而造非文字爲打字機而造也以不能作打字機之故而遂欲廢文字其愚眞出鑿趾適屨國造之上千萬倍矣又兄吾國文字未必不適於打字機乎宣明告我有祁君者居紐約官費爲政府所撤貧困中苦思爲漢文造一打字機。其用意在於分析漢字爲不可更析之字母。（如一口子

3.2 Photograph of typewriter by Zhou Houkun in *New Youth*

Characters." "A Chinese typewriter of unique design," the article read, "utilizing no less than 4,000 written characters, has just been invented by Mr. Hou Kun Chow, a graduate of the Massachusetts Institute of Technology, one of the first Chinese students to be educated in this country, and now a mechanical engineer in Shanghai."[70] Following its exhibition at the office of the American Consul General in Shanghai, an exhibition Zhou personally oversaw, Consul General Thomas Sammons reported the machine to be "simple in design and portable."[71] His praise was limited, though. "Obviously, however," Sammons continued, "no great speed will be possible in operating the machine as now constructed."[72]

In April 1917, Zhou's photograph was published in *Popular Science Monthly*, the only known photograph of the inventor together with his apparatus (figure 3.3). The machine was shown atop a small cloth-covered table, the glass-encased character guide extending out from the device. Bespectacled in wire-frame glasses and attired in a suit and leather shoes, Zhou sat beside his device, in an attentive posture. With his right hand, he held the pointer gingerly. To his left, on the table, the chassis of the typewriter remained opened and viewable, the character matrix being roughly the shape of a wax phonograph cylinder.[73] Uncertain as to what to call this machine—this typewriter without keys—the article's author was clearly at a loss for appropriate alternatives. "The 'keyboard,'" the article explained, placing the word *keyboard* in cautionary quotation marks, "consists of a flat table surface upon which are duplicated the type characters."

Having developed his common usage machine, Zhou's critical next step was to secure financial and manufacturing support with which to transform his vision for Chinese typewriting into a commodified reality. Emboldened by his reception and propelled by the vigor of youthful reformism, Zhou Houkun returned to China, bringing his prototype with him.

Zhou was not without competition, however. As he ventured to create a Chinese typewriter by seizing upon the common usage approach, another overseas Chinese student—with equal passion and vigor—was racing to beat him to the finish line. For this young inventor, his pursuit would be based upon a set of questions that differed dramatically from Zhou and the common usage approach, asking: What might Chinese

3.3 Zhou Houkun in *Popular Science Monthly*

typewriting look like if it departed entirely from the core assumption of common usage, namely that the indivisible ontological foundation of Chinese writing was the Chinese character? What might happen if this central tenet were relaxed, or perhaps abandoned altogether? This young student was named Qi Xuan, and he pursued a radically different approach to Chinese typewriting—not that of Gamble, Sheffield, Zhou, and the common usage approach, but that of Pauthier, Legrand, Beyerhaus, and divisible type.

THE RETURN OF DIVISIBLE TYPE: QI XUAN AND THE COMBINATORIAL CHINESE TYPEWRITER

On February 20, 1915, the Palace of Fine Arts in San Francisco unveiled an illustrious exposition to celebrate the newly completed Panama Canal. Designed to showcase "the world's progress in art, music, poetry, religion, philosophy, science, history, education, agriculture, mineralogy, mechanism, commerce and transportation," the exhibition treated visitors from around the world to spectacular displays, from a 435-foot "Tower of Jewels" designed specifically for the event to an unveiling of perhaps the first steam locomotive purchased by the Southern Pacific railroad company. As he walked the grounds of the exposition, a young overseas Chinese student named Qi Xuan would have seen a miniature reproduction of Beijing's Forbidden City, and perhaps walked past the Japanese exhibit and its presentation of Formosan tea and Japanese girls in traditional garb. He would have likely walked past the "Underground China" display, as well, with its seedy portrayal of opium dens, gambling, and prostitution, and perhaps also the ostentatious Underwood typewriter display and its oversized, 28,000-pound typewriter in the Palace of Liberal Arts (heralded by the company as "The Machine You Will Eventually Buy").[74] After circumambulating the grounds, though, the young Qi would no doubt have found himself quickly back at the exhibit he had come to San Francisco to man, and the invention he had come to showcase: a typewriter for the Chinese language.[75]

Very little is known about Qi Xuan, save what we are able to piece together from the scattered traces he left behind during his American journey.[76] Born on August 1, 1890, near the city of Fuzhou in southeastern China, he graduated from the Anglo-Chinese College in 1911, just months before the outbreak of the revolution that brought the Qing dynasty to an end.[77] Qi soon traveled to London, where he undertook nine months of study during the 1913–1914 academic year. The twenty-three-year-old Qi arrived in New York in February 1914, and when interviewed by the customs agent, announced his intention to study at Princeton University. His plans did not materialize, evidently, although he managed to land safely at New York University.

At NYU, Qi began work on his new model of Chinese typewriter during the 1914–1915 academic year, receiving a measure of financial support from the Chinese Consul General Yang Yuying, and likely technical support from a young NYU engineering professor, William Remington Bryans.[78] Optimistic and driven, the young Qi hoped to complete and secure support for his invention before his competitor Zhou Houkun did at MIT.

Like so many of the machines in our story, the prototype of Qi Xuan's does not survive. By closely analyzing Qi's United States patent, however, as well as one of the surviving photographs of the young inventor, we can learn a great deal (figure 3.4). Like Zhou Houkun's common usage typewriter, Qi Xuan's machine also featured a cylinder encased in a copper plate, upon which were etched 4,200 common usage Chinese characters. And like Zhou's device, Qi Xuan's machine was a "typewriter with no keys." The machine had only three mechanisms: a backspace, a space bar, and the lever to initiate the type mechanism. To type one of the common usage characters, the operator would rotate the cylinder by hand, bring the desired character into the striking position, and depress the typing key to imprint the character form on the page.

3.4 Qi Xuan and his typewriter

While this description of Qi's machine suggests an identical design to that of Zhou Houkun, there was one profound difference. In addition to the 4,200 common usage characters on his cylinder, Qi also included a set of 1,327 *pieces* of Chinese characters that the operator could use to assemble or "spell" out less frequent characters piece by piece, in a manner akin to composing an English-language word letter by letter.[79] Central to Qi's design, then, was the same quasi-alphabetic reconceptualization of Chinese characters we saw in the context of nineteenth-century divisible type printing. Like Pauthier, Legrand, and others before him, Qi set out to decompose characters so as to render them amenable to what can be thought of as "shape-spelling"—conceptualizing "radicals" as China's orthographic analog to the alphabetic letters of other world languages. Quite importantly, it was the identification and design of these "pieces" that constituted the focus of Qi's patent application, far more than the mechanism itself. "My invention relates to a system," the inventor explained, "of arranging and separating certain Chinese characters into new and novel radicals, and combining said radicals to form various words, and to a machine capable of carrying out not only the system devised by me, but also the separation and combination of the radicals forming the Chinese characters now in use."[80]

Like his predecessors, Qi Xuan also departed from the conventional set of 214 character components of the *Kangxi Dictionary* that had formed the taxonomic basis of Chinese dictionaries, indexes, catalogs, and retrieval systems for centuries. To transform Chinese radicals from taxonomic rubrics into productive modular forms, he would need to determine the precise number of variants necessary for his machine to produce all possible Chinese characters. Although we have no indication of whether Qi Xuan was familiar with the work of Legrand, Beyerhaus, and Gamble before him, the similarities were unmistakable: 1,327 components on Qi's typewriter, as compared to the 1,399 divisible type in the Beyerhaus font.

For all the continuities that linked Qi Xuan's approach to those of his nineteenth-century predecessors, however, there were marked differences. Notably, the transition of combinatorialism from typography to typewriting exacerbated the spatial politics of divisible type examined in the preceding chapter. In contrast to the divisible type operator, who at the very least had been able to assemble characters in advance and transfer them

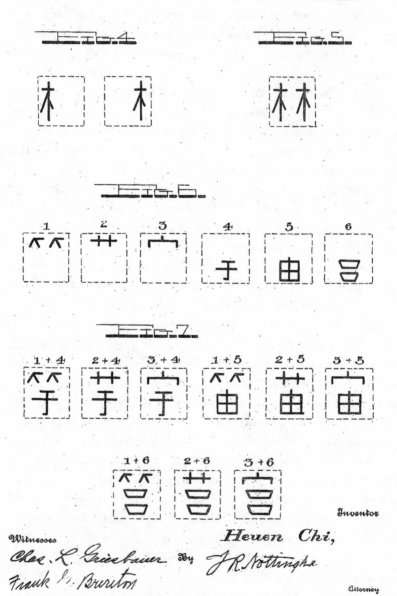

H. CHI.
APPARATUS FOR WRITING CHINESE.
APPLICATION FILED APR. 17, 1915.

1,260,753.

Patented Mar. 26, 1918.
3 SHEETS—SHEET 3.

3.5 Qi Xuan United States patent

as a locked form to the chase bed, combinatorial typewriting necessarily took place in a sequence, the interstices within which becoming places of potential error. As part of the mechanism of the typewriter, the modules on Qi's device were all "moving parts," a subtle yet profound change that could only have increased the margin of locational error.

Qi's patent application also reveals important aesthetic shifts at play within the combinatorial approach. Whereas Pauthier and Legrand had attempted to design their divisible type system such that it would conceal as fully as possible all evidence of the principles upon which it was based—their goal, as we recall, was "to solve the problem of representing the figurative language of China, with the fewest possible elements, without, however, altering the composition of the symbols"[81]—Qi Xuan's writings were far less anxiety-ridden about adhering to the dictates of Chinese calligraphic beauty. In his patent documents, the mechanical quality of his sample characters was unmistakable, each described using mathematical formulas that emphasized their status as epiphenomenal effects rather than "irresolvable individuals"—to invoke Devello Sheffield's parlance. Through the lens of Qi's machine, the character *yu* (宇) was not primarily a semantic-formal-phonetic complex signifying "eave, house, or universe," but the outcome of a simple additive process Qi expressed as "3+4" (where 3 referred to ⁀ and 4 referred to 于). Similarly, what Qi expressed as "2+5" produced the *effect* of *miao* (苗, signifying "sprout") by the combination of 艹 and 田. Perhaps counterintuitively, then, it was not Pauthier, Legrand, or other cultural "outsiders" but rather Qi as a cultural "insider" who exhibited greater willingness to depart from conventional notions of beauty and to embrace the machinic qualities of the divisibly typed Chinese character. During the first half of the twentieth century, moreover, Qi Xuan's experiment would be far from the last—or the most extreme—embrace of these new machinic aesthetics. In Maine, a scholar of Greek philosophy patented a Chinese typewriter of his own design, despite being unable to read or speak the language. Based on a set of geometric shapes—or "values" as the inventor Robert Brumbaugh referred to them—the typewriter would enable an operator to compose Chinese characters by means of successive impressions made upon a static, non-advancing platen (figure 3.6).[82] In Hong Kong, Wang Kuoyee patented yet another combinatorial Chinese typewriter, this one

Oct. 24, 1950   R. S. BRUMBAUGH   **2,526,633**

CHINESE TYPEWRITER

Filed Sept. 25, 1946   2 Sheets–Sheet 2

FIG.2

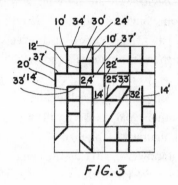

FIG.3

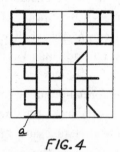

FIG.4

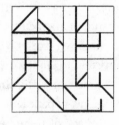

FIG. 5

Robert S. Brumbaugh,
Inventor,

By William D. Hall.
His Attorney.

3.6 Figure from Robert Brumbaugh patent application (filed 1946, patented 1950)

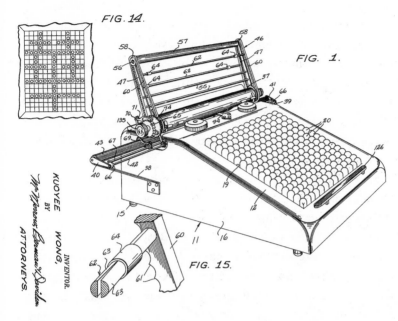

3.7 Figure of the character *hua* (華 "China") from Wang Kuoyee patent application (1948)

replacing strokes and radicals with a dot-matrix-esque system in which the operator used typewriter keys to create graphic representations of Chinese characters by populating a 13 × 17 grid with small circular dots (figure 3.7).[83]

The seismic shift we see in the work of Qi Xuan and others in the first half of the twentieth century must be read in light of the era's sweeping cultural iconoclasm, a trend marked in many circles by an interrogation if not outright rejection of "traditional" aesthetics and orthography. One need only peruse the bold, even startling new Chinese advertising and display fonts that began to appear in the 1920s, 1930s, and 1940s to appreciate the powerful aesthetic transformations then underway. To be sure, amid the innumerable new "-isms" of the day—*modern-ism, real-ism, expression-ism, anarch-ism, Marx-ism, social-ism, commun-ism, femin-ism, fasc-ism,* and more—surely one of the most powerful was that of *mechan-ism*: an embrace of new "visual logics" of mechanistic rationality.[84] With the rapid development of Chinese print capitalism, moreover, as well as the rise of high modernist experimental typography elsewhere in the world, Chinese publishers and advertisers deployed stunning and often radically stylized Chinese fonts to catch the eye of a new, self-styled class of urbanites and cosmopolitans.[85] It would seem that the aesthetic politics of divisible type printing—a politics Pauthier and Legrand attempted to overcome by subordinating their new technology as much as possible to the logic of Chinese orthographic aesthetics—might in Qi Xuan's era be resolving itself, now that this aesthetic logic was at least in part becoming subordinated to technology.

On March 21, 1915, in New York, Qi Xuan delivered the first known presentation of his machine to journalists, as well as to his supporter, Consul General Yang Yuying.[86] The *New York Times* dispatched a journalist to the young student's apartment on the Upper West Side, near the intersection of Amsterdam Avenue and 115th Street.[87] Here the complex challenges of combinatorial typewriting were laid bare. As reported in the following day's article, Consul General Yang performed something of a ribbon-cutting ceremony at the event, using Qi's apparatus to type a letter to the Chinese minister in Washington—to spell Chinese characters using these more than 1,300 components.[88] Although the letter contained only one hundred characters, it took Yang approximately two hours to complete—a fact in which the newspaper report clearly delighted. Qi Xuan attempted to explain to the journalist the reasons for this poor performance, contending that "this slow speed was due to the unfamiliarity of the operator with the characters."[89] "The inventor believes that with practice a speed of 40 words a minute may be obtained," the article continued, "which is quite good for Chinese." The damage had been done, however. When the article posted the following morning, any hope of attracting the interest of manufacturers would arguably have evaporated, effectively bringing an end to the young inventor's foray into the world of typewriting. We can only surmise the mixture of conflicted and competing feelings Qi may have experienced when reading the article's extensive title: "4,200 Characters on New Typewriter; Chinese Machine Has Only Three Keys, but There Are 50,000 Combinations. 100 Words in TWO HOURS. Heuen Chi, New York University Student, Patents Device Called the First of Its Kind."[90]

Like the title itself, the article oscillated between moments of praise (e.g., lauding the young inventor's intellectual pedigree) and denigration (comparing his machine to the "little tin typewriters made for children's toys").[91] Its overall tone was one of tragic comedy—of a talented young Chinese student led down a fruitless path by his own quixotic ambitions. Once the narrative of Qi's machine entered the public realm, moreover, it began to circulate more broadly. The same condescending tone pervaded media accounts in the years following, as in a 1917 article in the *Washington Post* announcing a newly patented model of Chinese typewriter (most likely Qi Xuan's machine). Entitled "The Newest Inventions," the article tellingly placed the new machine alongside such absurdities as a "dancing radiator doll" and "a mouse trap for burglars."[92]

WHAT WILL CHINESE WRITING BE?

By the year 1915, China was no longer a country in search of a typewriter. To the contrary, there now existed two *types* of Chinese typewriter, each putting forth a different approach to the question of Chinese techno-linguistic modernity, and each departing dramatically from the Western typewriter form. Manufacturers in China and abroad would now have to choose between them—to decide which, if either, constituted a promising path forward. The common usage approach to Chinese typewriting tapped into widespread concerns and longstanding research among Chinese elites into the question of the "foundational" vocabulary of a modern Chinese citizenry, and in that sense could perhaps render this machine understandable and comprehensible to customers already familiar with the notion of "common usage" in other contexts. Yet the common usage approach could never hope to encompass the entirety of the Chinese language, which for an inscription device meant something quite different than in the context of Chinese-language pedagogy. To learn and memorize a core set of "foundational characters" in no way prevented a student from eventually expanding beyond those lexical boundaries, whereas in the case of the typewriter, common usage formed a more or less impassable horizon. In this respect, modern Chinese information technologies premised upon the common usage model would necessarily be marked by an irresolvable restlessness, with elites, educators, and entrepreneurs forever battling one another over the authority to define the boundaries separating first- and second-class characters, and the boundaries separating which characters to include and which to banish. The price of technolinguistic modernity so conceived would be that of division, restlessness, and the constant policing of boundaries.

Qi Xuan and his divisible type machine put forth a strikingly different answer to the question of Chinese writing. In contrast to the common usage model, this machine offered up a vision of modern Chinese inscription that would embrace frequent and infrequent characters alike, quieting the mania and incessant lexical self-improvement of common usage systems, and thereby unifying the Chinese script within a new technolinguistic domain. But technolinguistic modernity of this variety would come with its own compromise, as well. To achieve this unity and "leave no character behind," it would first be necessary to shatter Chinese characters into pieces, giving up on the idea that characters—and even the beloved *brushstroke*—constituted the ontological foundation of Chinese writing. Instead, both strokes and characters would have to cede the throne to "the radical," the once taxonomic and etymological entity now reimagined as the productive "root" of Chinese writing itself. This ontological revolution necessarily brought with it a politics over which the common usage model never had to fret: the politics of Chinese aesthetics in the age of mechanical reproduction.

In this competition over the future of Chinese script, Zhou Houkun and Qi Xuan did not escape the attention of language reformers and entrepreneurs in China, moreover.[93] Hu Shi, the towering figure of the nascent New Culture Movement, witnessed Zhou Houkun's invention in person during his travels to Boston, and heard about Qi Xuan's machine through media reports. Seizing upon their inventions, Hu authored an indictment of the abolitionist position. "Our country has some arrogant students," Hu wrote, "ones who go so far as to recommend throwing out Chinese and replacing it with English, or using simplified characters." Hu continued:

They say that Chinese doesn't fit the typewriter, and thus that it's inconvenient. Typewriters, however, are created for the purpose of language. Chinese characters were not made for the purpose of typewriters. To say that we should throw away Chinese characters because they don't fit the typewriter is like "cutting off one's toes to fit the shoe," only infinitely more absurd.

Hu Shi was not the only one to take notice of the work by Zhou Hou-kun and Qi Xuan. "Few people conceive of the possibility of inventing a typewriter for the Chinese written language, which is not alphabetic," C.C. Chang wrote in 1915, immediately upon viewing Qi Xuan's combinatorial device. "The success of Heun Chi [Qi Xuan] of New York University in devising a machine for the Chinese language definitely proves this possibility and opens the way to further invention and improvements along this line."[94]

Perhaps most importantly, both Zhou Houkun and Qi Xuan caught the attention of Zhang Yuanji (1867–1959), president of the most important center of print capitalism in China at the time—Commercial Press in Shanghai. In Zhang's diaries, the first mention of Zhou Houkun came in March of 1916, just after Zhou had returned to China from the United States. On May 16 of the same year, Zhang recorded a brief entry about Qi Xuan as well, commenting on second-hand reports of Qi's apparatus. Apparently, Zhang noted, it produced Chinese characters of reasonable quality and as a typewriter might surpass that of Zhou Houkun.[95]

Zhou Houkun and Qi Xuan were well aware of each other's work, as well. Indeed, before manufacturers in China had decided where to invest their resources, this pair of young entrepreneurs set out in a public debate over the merits of their respective approaches. Zhou was the first to strike. In "The Problem of a Typewriter for the Chinese Language," a piece from 1915 in which Zhou both introduced his machine and disparaged his competitor's, he adopted a discursive strategy identical to those we saw in chapter 1.[96] To assert the impracticality of a combinatorial Chinese typewriter, Zhou began by emphasizing the type of morphological, scalar, and positional variation that took place within Chinese radicals—all in an effort to prove the impossibility of his competitor's idea. As there were four different sizes, shapes, and locations in which a radical might appear—for a total of sixty-four possibilities (four to the third power)—one would in fact need to multiply the total number of radicals—roughly two hundred—by this total number of possible varieties. As Zhou explained, one was left with the absurd proposition of a keyboard requiring no fewer than 12,800 keys.[97] "Many men went about the problem by attempting to resolve the Chinese characters into 'radicals,'" Zhou concluded. He continued:

The idea comes natural when we see that the Chinese characters are composed of "radicals," of which there are over two hundred. It was reasoned that by having two hundred keys to which are attached two hundred types of the "radicals," they can produce a machine on the same principle as the American Typewriters, which will be able to print any character by the combination of these "radicals." The scheme looked well on the face of it; but they forgot, that the same "radical" in different characters differs not only in size but also in shape, and, furthermore, they occupy different positions in a character.[98]

Zhou's criticism of Qi Xuan's machine was predicated upon feigned ignorance if not purposefully misleading language. Qi's machine was combinatorial, indeed, but it did not encompass 12,800 keys—nor even a keyboard, a fact that Zhou would undoubtedly have known. Zhou would also have known, given his extensive work on the Chinese language, that the level of variation among Chinese radicals did not approach anything like his "four-cubed" equation. As evidenced in the work of divisible type printers, and by Qi Xuan's own patent documents, the number of variant radicals required did not exceed 2,200—less than the total number of characters on Zhou's common usage device.

Qi Xuan wasted no time in mounting a response to Zhou's attack. "Not in the least boasting of my own success," Qi protested, "I should say frankly that my device is the only yet known practical and scientific method of making a typewriting machine for our mother-tongue—that is to split the characters into 'radicals' in such a way as to be able to make by combination, words equal in size and proper in form."[99] He took aim at Zhou's arithmetic, offering up the possibility to his readers that Zhou had simply erred in this most basic of multiplication problems. "Well, Mr. Chow certainly has a good memory of some algebraic law," Qi insinuated bitingly, "but I am sorry to say that he has unfortunately misapplied it, so that he is discouraged from proceeding further along the correct path by the startling figure of 12,800!"[100] "Therefore, I venture to conclude that Mr. Chow appears either to lack sufficient mechanical knowledge or intelligent study of the 'radical system,' although he writes as if he were a great authority on the problem of Chinese typewriter invention."[101] At this point, Qi Xuan returned fire against Zhou's machine itself. "I cannot help thinking," Qi wondered aloud about all common usage machines, "that their inventions are merely 'imperfect printing press machine' [sic], which has little mechanical advantage and commercial value."[102]

Ultimately, however, it was Zhou Houkun who prevailed in this struggle. Whether because of his more active self-promotion, or because the design of his machine was premised upon common usage—an approach with a longer pedigree in China than that of combinatorialism—Zhou captured the lion's share of attention by companies back in China, most notably Commercial Press. Meanwhile, information about Qi Xuan's device was far sparser and less reliable. An article from the Chinese press in 1915, for example, even mistook Qi Xuan's name, reversing and mistranscribing it to read "Xuan Qi" [宣奇]—all of which suggests that many of the first introductions to Qi Xuan in China were being filtered through the English-language, American press.[103] Zhang Yuanji himself was not even clear about the inventor's name, miswriting Qi Xuan's name in his diaries as late as 1919.

Commercial Press established an initial working relationship to bring the bright young MIT graduate and his common usage device into the company. Zhou Houkun had been on his way to Nanjing, to head a newly formed industrial institute at Nanjing Normal College, but his term of employment would not commence until July. Taking advantage of this gap, Commercial Press succeeded in bringing Zhou and his typewriter to Shanghai, to begin overseeing the machine's development and manufacture.[104] To this new venture they lent their considerable reputation and expertise, vying to become the company that would succeed where global giants like Remington and Underwood had failed: to produce and sell a mass-manufactured typewriter for the Chinese market.

At long last, China would have its typewriter.

4

WHAT DO YOU CALL A TYPEWRITER WITH NO KEYS?

The Chinese typewriter manufactured by the Commercial Press solves a serious problem in office administration in China. The machine has all the advantages of a foreign typewriter.

—Brochure for the Shu-style Chinese typewriter at the 1926 Philadelphia world's fair

With its more than one hundred acres of botanical gardens, the Huntington in San Marino, California, transports hundreds of thousands of visitors each year into an otherworldly terrain populated by the Mexican Twin-Spined Cactus and the South American Heart of Flame. Inside, the library and museum draw scholars from around the world thanks to a world-renowned collection of rare books and artifacts on the American West, the history of science, and a range of other subjects. The Huntington also enjoys a rare distinction that few people are aware of: it is home to one of the oldest surviving Chinese typewriters in the world, the "Shu-style Chinese typewriter" manufactured by Commercial Press in Shanghai in the 1920s and 1930s (figure 4.1).

The machine belonged to You Chung (Y.C.) Hong (1898–1977), a Chinese-American immigration attorney who served the Los Angeles Chinatown community. Establishing design patterns that would be followed by all mass-manufactured Chinese typewriters in history, it was a

4.1 Shu-style Chinese typewriter, Huntington Library

"common usage" machine featuring a rectangular tray bed containing approximately 2,500 character slugs. As in movable type printing, these character slugs were not fixed to the machine, but rather free-floating. If one were to remove a Chinese typewriter tray bed and turn it over, that is, the characters would spill out onto the floor, "pieing the type."[1] And, as with every Chinese typewriter ever mass-manufactured, it was a typewriter with no keyboard and no keys.

The tray bed of the Shu-style machine was divided into three zones based upon the relative frequency of characters: a zone of highest-usage characters, located centrally on the tray bed in columns sixteen through fifty-one; secondary usage characters, located on the right and left flanks of the tray bed, in columns one through fifteen and fifty-six through sixty-seven; and special usage characters, limited to columns fifty-two through fifty-five. Less commonly used characters, referred to as *beiyongzi*, were placed in a separate wooden box, a kind of lexical storage facility from which the typist could select characters and place them temporarily in his or her tray bed using tweezers. In this wooden case were housed approximately 5,700 additional characters.[2]

By the time of my visit in 2010, the characters on the tray bed of the Huntington Library's specimen had long since rotted, exhibiting the consistency of weathered graphite or limestone. They had become so fragile, in fact, that when a curator demonstrated the typing mechanism to me, the gear accidentally shattered the faces of three of them. As for the characters located in the box of "lower-usage characters," these had simply fused together into a mass of gray.

Even in this aged condition, though, the machine still bore the signature of what had clearly been a deeply personal relationship between the device, the typist who used it, and the man in whose employ this typist worked. As I moved around the machine, shifting my angle of view, fleeting constellations emerged from out of the brittle, charcoal-colored mass—shimmers of characters exhibiting greater reflectivity than their neighbors on account of being replaced more recently, likely because they had been used more frequently when the machine was still in service. Shifting to my right, and leaning toward the machine, the first shimmer of characters came to light: *qiao* (僑 "emigrant"), *yuan* (遠 "far away, distant"), and *ji* (急 "urgent"). With a slight adjustment of my orientation this first constellation melted back into the charcoal bath, and a second constellation came into view: *zao* (遭 "to encounter hardship"), *xi* (希 "to hope for, long for"), and *meng* (夢 "dream, to dream"). Each of these constellations was accompanied by more commonplace terms as well: *yi* (一 "one"), *bu* (不 "no, not"), *shang* (上 "atop"), and *qu* (去 "to go"). Not unexpectedly, one of the crispest characters on the tray bed was *hong* (洪), Y.C. Hong's surname. Traces of the life this machine had lived still lingered, even decades later.

In this chapter, we leave behind the technical blueprints of inventors, linguists, and engineers, and explore instead the lived histories of Chinese typewriting. The Chinese typewriters in this chapter will not be prototypes nor the illusions of foreign cartoonists, but fully formed commodities that companies manufactured and marketed, that schools and training institutes explained and taught, and that typists used in the course of their careers. We will focus on the many worlds of the machine during the early twentieth century, from the advent of its mass manufacture at the Shanghai-based Commercial Press—the preeminent center of print capitalism in Republican China[3]—through the proliferation of Chinese typewriting schools in which young women and men studied this new technology and later employed it in a widening arena of government offices, banks, private companies, colleges, and even elementary schools. It is here that we will strain to see and hear the Chinese typewriter within its own linguistic context, attending to a history all but drowned out by the clichés of China's technolinguistic backwardness we have examined thus far.

Even as we begin to observe the Chinese typewriter in these more local-level and intimate contexts, however, at no point will we find it entirely "at home" or in its "natural habitat." Even with its unique mechanical design, its training regimens, its type and extent of usage, the gender makeup of the clerical workforce who used it, and the iconography and culture of the machine itself, at no point would either the Commercial Press machine or its later competitors ever have constituted the stable, accepted "counterpart" or "equivalent" of the Western machine—although not for lack of trying. Throughout the period, manufacturers, inventors, and language reformers alike remained acutely aware that the Chinese typewriter was at all times being measured against the "true" typewriter: the machines of

Remington, Underwood, Olympia, and Olivetti, which steadily strength-ened their hegemonic grip across the globe. Perhaps if the Chinese type-writer had never been conceptualized as a "typewriter" in the first place—if, instead, it had been described as a "tabletop movable type inscription machine," or as some other niche apparatus disconnected from the larger, global history of modern information—constant comparisons between it and the "real" typewriter might never have been invoked. But this did not happen: the machine was a *daziji*—a *typewriter*—and by consequence was inescapably enmeshed within this broader, global framework.

During this period, the tensions between the Chinese typewriter and the "real" typewriter were pronounced. To function and make headway in the Chinese-language environments of government, business, and education, the Chinese typewriter would need to attend completely to the real-world necessities and practicalities of the Chinese language itself—and yet, as a "typewriter," it would need to be legible to an outside world wherein resided the sole authority to decide which machines did or did not merit this title. The Chinese typewriter, and Chinese linguistic modernity more broadly, were thus caught between two impossibilities: to mimic the technolinguistic modernity that was taking shape in the alphabetic world, or to declare complete independence of this world and set out on a path of technolinguistic autarky. With neither option being feasible, Chinese linguistic modernity was caught irresolvably between mimesis and alterity.

FROM MOVABLE TYPE TO THE MOVABLE TYPEWRITER: THE SHU ZHENDONG MACHINE

Following the debut of his prototype, and particularly after he took up his position within Commercial Press in Shanghai, Zhou Houkun enjoyed a modicum of celebrity. On July 3, 1916, he demonstrated his machine at the Chinese Railway Institute in Shanghai, one of his first appearances as a representative of Commercial Press.[4] Weeks later, Zhou continued his demonstrations at the Jiangsu Province Education Committee Summer Supplementary School, where the young engineer was praised for creat-ing a machine that would print 2,000 characters each hour, rather than the 3,000 per day achievable by hand.[5]

Commercial Press remained hesitant about bringing Zhou's device into production, however. While no sources clearly identify the reasons for this equivocation, one factor was undoubtedly the design of the apparatus itself. As examined in chapter 3, Zhou Houkun's common usage machine featured a cylindrical character matrix upon which were etched Chinese characters in permanent and unchangeable form. In stark contrast to common usage as it functioned within its original context of typesetting, or even the earlier Chinese typewriter designed by Devello Sheffield, the characters on Zhou's machine were completely fixed—impossible to adjust to different terminological needs and contexts. This posed a problem. As examined in the preceding chapter, there had been some hope that the tension within common usage might be resolved—at least for the Chinese masses—by means of promoting greater usage of "foundational character" sets within the contexts of both formal and informal education. If mass literacy could be achieved by statistically determining a limited set of characters that people needed to know, or would tend to employ, then in theory it might be possible to design a "foundational" Chinese typewriter tray bed to form a perfect fit with this mass vocabulary. For specialists and professionals, however, a fixed set of 2,500 characters would never be enough. For those working in banks, police stations, or government ministries, day-to-day language varied widely, which meant that Zhou Houkun's original, unchangeable design would inevitably limit the device's utility. To transform the common usage model into a workable technolinguistic solution for Chinese typewriting demanded that these devices offer up a measure of flexibility and customizability.

As early as winter 1917, relations between Zhou and Commercial Press began to sour, with the engineer and his corporate patrons drifting in separate directions.[6] Zhou proposed a visit to the United States, where he hoped to inspect American typewriter production and develop an improved version of his machine to better suit the needs of potential customers. Zhou offered to cover the cost of his own travel, but requested that Commercial Press commit to providing the cost of manufacturing an improved machine he planned to complete upon his return. Zhang Yuanji declined, calling the financial commitment untenable. Zhou countered, offering to cover all expenses himself, but requesting that

Commercial Press agree to sell and distribute the machine on his behalf. This offer was rebuffed as well. "I sense that it would be best for us to cancel our old contract," Zhang recommended to his associates at the press, "and that [Zhou] handle everything on his own from here on out."[7] Thus came to an end Zhou Houkun's brief relationship with Commercial Press, as well as his long-held dream of being the first to design a commercially successful typewriter for the Chinese language. In the years to come, Zhou returned to his first love of airplane and ship construction, to serve his country by other means.[8] Circa 1923, he went on to join the Hanyeping Steel Company as their director of technology.[9] Meanwhile, by May 1919, Commercial Press was in possession of Zhou Houkun's prototype, but had not pursued plans of production.[10]

Commercial Press may have lost interest in Zhou Houkun, but not in the cause of Chinese typewriting. The company launched a Chinese typewriting class in 1918, further demonstrating its commitment to the new technology.[11] After Zhou's departure, more importantly, the company continued its pursuit under the guidance of another engineer: Shu Changyu, who would come to be known by his pen name, Shu Zhendong. Having studied steam-powered machines, and made forays at both the Maschinenfabrik Augsburg-Nürnberg (MAN) factory in Germany and the Hanyang factory in China, Shu joined the company circa 1919, receiving his first Chinese typewriter patent in the same year.[12]

One of the first steps Shu took was to abandon the Chinese character cylinder in Zhou's original design. Shu replaced it with a flat, rectangular bed within which character slugs would sit loosely and interchangeably. With this change, the common usage machine would still be outfitted with the same number of characters as before, but now typists would be able to customize their machines to fit different terminological demands—as had been possible on Sheffield's first prototype decades earlier. As free-floating metal slugs, characters could be removed and replaced using nothing more complicated than a pair of tweezers.

Following this important change in design, Commercial Press's hesitation evaporated. The company dedicated a substantial financial investment to its new Chinese typewriter division, which occupied a reported forty rooms. Employing more than three hundred workers, and entailing some two hundred pieces of equipment, the process of building the new

。形情機造體全
The main part of the factory.

。作工字排機鑄
Fitting the type-writer.

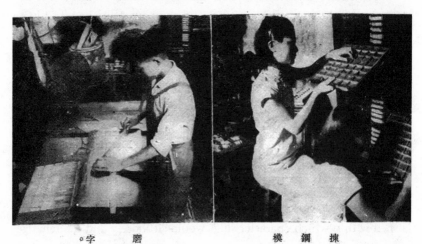

。字　　磨
Adjusting the letters.

揀　鋼　模
Checking the models.

4.2 Roles and tasks in Commercial Press typewriter manufacturing plant

4.3 Commercial Press typewriter manufacturing plant

machine was divided among a variety of roles: melting lead to be used for character slugs, casting the character slugs, error-checking the tray bed character table, fitting the machine chassis, setting character slugs in their designated locations on the tray bed, and more (figure 4.2).[13] If Bi Sheng had invented movable type nearly one millennium earlier, Shu Zhendong and Commercial Press now set out to manufacture a *movable typewriter* (figure 4.3).[14]

MARKETING THE MACHINE AT HOME: COMMERCIAL PRESS
AND THE FORMATION OF A NEW INDUSTRY

China's first-ever animated film was entitled "The Shu Zhendong Chinese Typewriter," and it was an advertisement for the new Commercial Press machine—the first mass-produced Chinese typewriter in history.[15] Released in the 1920s, the film was produced by Wan Guchan and Wan Laiming, foundational figures in the early history of Chinese animation.[16] The film is no longer extant, unfortunately, and yet a plethora of promotional materials from the era help us speculate as to its content. "It is said that the fastest speed is 2,000-plus characters per hour," one article in 1927 read, "or three times faster than writing by hand."[17] Claims such as these focused on three merits of the machine in particular. First, it saved time as compared to composing a manuscript. Second, it produced more legible characters than did writing by hand. Finally, and most importantly, it could be used in concert with carbon-infused paper to produce multiple copies of a single document.

Song Mingde, Commercial Press employee and one-time head of the company's Chinese typewriter division, placed particular emphasis on the machine's function within textual reproduction, situating the device within the *longue durée* narrative of Chinese writing. In the time of Cang Jie, Song explained, paying homage to the apocryphal inventor of Chinese characters, Chinese writing had been limited to strictly ideographic forms carved upon the surface of bamboo. With the invention of the brush and paper, Song continued, this rendered it possible to "copy and record by hand" (*yong shou chaolu*). "Compared to bamboo, this was already ten thousand times more convenient." From here, Song leapt over a large expanse of history to praise the third central invention within his abbreviated history, the printing machine (*yinshuaji*). With the advent of printing technology, he explained, the reproduction of many tens of thousands of copies was now well within reach.[18]

This leap from manuscript to mass reproduction had left behind a vast gap, however. Printing press equipment was expensive, Song emphasized, and required extensive preparation before usage, justifiable only in cases when large quantities of material were required. Still unresolved was the problem of smaller-scale, day-to-day textual reproduction of the sort required by modern businesses, whether in the form of short-run reports, office memos, or legal record-keeping. In cases when only 10 or even 100 copies were called for, the printing press was hardly an option. But neither were handwritten documents an attractive alternative, Song argued, since they "took time" (*feishi*) and were "irregular" (*bu zhengqi*).

Commercial Press thus advanced its typewriter as part competitor, part complement to both the human hand and the printing press.[19] Early signs of commercial interest were encouraging, moreover. In his diary entry from April 16, 1920, Zhang Yuanji made note of a potential order from the Chinese postal division for one hundred machines.[20] By 1925, Chinese consulates as far away as Canada were reported to have purchased a typewriter for their affairs.[21] In 1926, Huadong Machinery Factory listed the Chinese typewriter as one of its best-selling and most widely known products.[22] Commercial Press sold upward of two thousand units between 1917 and 1934, for an average of roughly one hundred per year. Commercial Press, in turn, attempted to raise awareness of the machine, not only by creating its pioneering animated film, but also by providing extensive demonstrations. In November 1921, Tang Chongli of Commercial Press included the Shu Zhendong machine as part of a demonstration of new technology and machinery by the Department of Forestry and Agriculture.[23] On May 3, 1924, Song Mingde was scheduled to depart on a six-month voyage to present-day Southeast Asia. During his travels, he would promote the Shu-style Chinese typewriter to overseas Chinese merchants in Luzon, Singapore, Java, Saigon, Sumatra, Siam, and Malacca.[24]

The national or civilizational significance of the Chinese typewriter was also a common selling point, both for promoters of the machine and for the inventors themselves. In the journal *Tongji*, Shu Zhendong reflected upon his pathway to Commercial Press and his development of the Chinese typewriter, lamenting as well that fewer and fewer Chinese compatriots "place importance on writing" (*zhongshi wenzi*).[25] "A country's writing is a country's pulse," Shu protested. "If the pulse dies out, what once was a country is no longer a country."[26] As for those who urged the abolishment of Chinese characters on account of the technological challenges of typewriting, he likened this to "refusing food for fear of choking."[27] For Shu Zhendong himself, as for his predecessor Zhou, the Commercial Press Chinese typewriter served as a tangible refutation of the idea that Chinese writing was incompatible with the demands of the modern technological age.

THE MISSING "TYPEWRITER BOY": THE AMBIVALENT GENDER OF THE CHINESE SECRETARY

When the first Chinese typewriter rolled off of the Commercial Press factory floor, a fully formed Chinese typewriting industry did not miraculously spring into existence. The formation of a new industry would depend in equal measure upon the development of an entirely new clerical labor force: a battery of trained "Chinese typists" (*daziyuan*) to take up posts and put the machine to use in government, education, finance, and the private sector. In short, the development of Chinese typewriting required *students* of the Chinese typewriter—individuals whose bodies and minds would need to be trained to meet the requirements of this new machine, and to exploit its capacities.

Beginning in the 1910s and 1920s, a constellation of privately owned, one- and two-room typing institutes were established to meet this requirement, and to make money in the process. In Shanghai, Beijing, Tianjin, Chongqing, and other metropolitan areas, students received training in the new technology, typically over the course of terms lasting between one and three months, at the cost of fifteen yuan or less. Eager to secure employment as the first wave of human capital in this new clerical profession, these students would also play an important promotional role for Commercial Press itself—and eventually, as new entrepreneurs entered the market to compete with Commercial Press, for its competitors as well (a subject we will turn to later). By opening schools that specialized in a particular make or model of Chinese typewriter, and then securing employment for graduates, a manufacturer could begin to make inroads for his machine in the private, educational, and government sectors.[28]

When thinking about the history of typewriting from the context of the United States, where the profession was steadily and rapidly gendered as *women's work*, one might be tempted to assume that Chinese typing steadily feminized as well. Perusing Chinese newspapers and magazines from the era, moreover, one would indeed have encountered a new figure in the Chinese periodical press that might reinforce this assumption: the "typewriter girl" (*dazinü* or *nüdaziyuan*). Typically in their teens and early twenties, these young women were featured alongside other "modern girls" of the era, encompassing painters, dancers, athletes, violinists, and scientists. They formed part of the still little-understood world of Republican-era professional women, entering not the blue-collar worlds of textile manufacturing, match factories, flour mills, and carpet weaving, but the white-collar worlds of administration and office work.[29] Whether framed *en plein air*, as in a group photograph of Chinese typists in Beihai Park, or poised before their machines in traditional dress and finely parted hair, such representations in many ways accord with what one might expect at a moment in global history when clerical work was undergoing a process of feminization worldwide—the "typewriter girl" was, as has been argued, an "American export" (figure 4.4).[30]

4.4 Photograph of female Chinese typist, 1928; *Eastern Times* photo supplement (*Tuhua shibao*) [圖畫時報] 517 (December 2, 1928): cover

When we look beyond representations of Chinese typists in the periodical press, however, the archival records of these same Chinese typing institutes from the period reveal a strikingly different history. In data compiled on just over one thousand typing students enrolled in a variety of institutes between 1932 and 1948, more than 300 (or 30 percent) were, in fact, young men—far greater than found elsewhere in the world at this time.[31] In other words, whereas mass media representations of Chinese typewriting suggested a profession that mimicked the prevailing gender norms of typewriting globally, the actual practice of Chinese typewriting did not. Typing in China remained a mixed labor pool throughout the period, with women constituting a larger, but by no means exclusive, share of the labor force. Young men, predominantly teenage and lower- or middle-school-educated like their female counterparts, entered these schools as well and sought training in the new technology. They graduated from typing institutes in sizable numbers. They served as typists. In certain cases, they went on to found typing schools of their own.[32] One such example is Li Zuhui, a young man hailing from Wujin in Jiangsu province who himself was a graduate of the Chinese-American Typing Institute (*Hua Mei dazi zhuanxiao*). Li later went on to direct the Mr. Hui's Chinese-English Typing Institute (*Hui shi Hua-Ying wen dazi zhuanxiao*)— a program founded circa 1930 in the Shanghai International Settlement with the stated aim of supplying "typing talent" (*dazi rencai*) to the Shanghai business community. In between, he served as a typist at the Zhili Province Bureau of Negotiations and the Tianjin Customs House.

To gain a better sense of the typing profession in its early formation, we must begin by looking at the schools where young people received training on this new device and from where they entered this new profession. At one end of the spectrum were small-scale training operations, such as the "Victory and Success Typing Academy" in Shanghai.[33] Founded in May 1915 and originally focused on English typewriting, the two-room tutoring company eventually expanded into Chinese typewriting with the purchase of two machines. Circa 1933, the school enrolled six students in its English-language typing class and eight students in its Chinese typing class. At the other end of the spectrum, larger typing institutions were also in operation, such as the Shanghai-based Huanqiu Typing Academy with its hundreds of students. Founded in the fall of 1923, and originally focused exclusively on English-language typewriting as well, the institute expanded into Chinese typewriting during the fall of 1936, offering training on the Shu-style machine manufactured by Commercial Press.[34] Founder Xia Liang, once an employee of the Standard Oil Company in Shanghai, developed his institute around a team of experienced professionals. The principal of the school, Chen Songling, was a graduate of the Shanghai-based Nanyang University, as well as a former employee of the Shanghai Customs House in legal translation. Working under Chen Songling were four tutors: Chen Jie, a graduate of Shengfangji Middle School in Shanghai, and a former employee of Standard Oil as well; Xia Guochang, a graduate of the Shanghai Business English Academy and a former employee of the Shanghai Telephone Company; Xia Guoxiang, formerly an employee of Robert Dollar and Company (*Dalai yanghang*); and Wang Rongfu, a graduate of the Shengfangji Middle School.

This overlapping network of pedagogical, entrepreneurial, and technological centers and practices combined to expand the network of the Chinese typewriter and Chinese typists into companies, schools, and government offices across the country. Schools placed their graduates in metropolitan and provincial governments throughout China, including the Nanjing Inspectorate, the Fujian Provincial Government, and the Sichuan Provincial Government; and major Chinese corporations such as the Chinese Soap Company, Macao China Bank, and the Zhejiang Xingye Bank.[35] Graduates of the program also went on to teach Chinese typing in elementary schools. The Henan Provincial Government directed every department to dispatch secretaries to enroll in two-month courses in Chinese typewriting, for example, after which they would return to their original positions.[36] Such training would make people better at their jobs, authorities argued, and would "contribute to emerging enterprise in China" (*xinxing gongye*).[37]

If the gender makeup of Chinese typewriting was a complex affair, and one that diverged sharply from its highly feminized counterparts in the United States and Europe, nothing within the contemporary Chinese periodical press would have suggested it. Whether in photographic spreads or news reports, male Chinese typists were conspicuously absent, with the new industry of Chinese typewriting presented as one dominated by young, often attractive female students and clerks.[38] In 1931, the journal *Shibao* featured a photograph of the young female student Ye Shuyi as she practiced diligently at her Chinese machine.[39] In a 1936 spread in *Liangyou*, a young female typist—in this case using an English-language typewriter—was featured alongside photographs of other "new women" (*xin nüxing*): women aviators, radio announcers, telephone operators, and beauty parlor owners.[40] In a 1940 spread in *Zhanwang*, young women were shown at work on Chinese typewriters, accompanied by the caption "Type-writing is a congenial occupation for women."[41] Placing the Chinese typewriter girl in her broader context, the same spread included other "congenial" occupations such as nurse and flower seller. To emphasize the modern condition of Chinese women, though, the professions of lawyer and police officer were included in the mix as well. At all times, though, femininity and motherhood were honored, as stressed by another photograph and caption: "Mother teaching her child how to be a good girl."

Male typing students sometimes did show up in photographic representations of Chinese typing, but in such cases their presence was subtly written out of the story by means of the contexts in which such photographs appeared, and the captions that accompanied such visuals. In 1930, for example, *Shibao* featured a portrait of eight young typing school graduates—six women and two men—with the caption "Graduates of Hwa-yin Type-writing School, Peiping." While appearing gender-neutral at first, with the photograph including both men and women, and making no mention of the "typewriter girl," nevertheless the surrounding photographs on the same page reveal the editor's understanding of the typing profession as one with a distinctly feminized valence: "Modern Drill by Girl Students at Tsinghua University," "Morning Drill by Girl Students at Nan-Kai University," as well as photographs of three young female athletes.[42] Still other images were starker in their erasure of male typing students, as in a photographic spread from *Great Asia Pictorial*, also in 1930. In this photograph, twenty-two Chinese typing students were shown—fifteen women and seven men—and yet the caption read: "Photograph of First Entering Class of Female Students at the Liaoning Chinese Typing Institute."[43]

An important tension emerged, then, between the lived experience of Chinese typewriting and the images that these schools and Chinese media outlets chose to represent it. What accounts for this discrepancy? Here we must once again return to the global context within which Chinese typewriting was taking shape and was at all times nested. At the very moment China began to form its own typing industry, there already existed such a thing as a "typewriter girl," a robust trope that had found expression globally in a wide variety of cultural and socioeconomic contexts. By contrast, nowhere on earth did there exist the trope of the "typewriter boy," in the sense of a comparably powerful discursive or representational formulation that captured this reality in the form of stereotype. In the United States, the displacement of male typists and stenographers was part of the history of industrial mechanization, a history in which routinized forms of work were increasingly delegated to young women starting at the end of the nineteenth century.[44] Similarly, typewriter manufacturers worldwide had by this point long encouraged this trend toward feminization, targeting women both as potential consumers and as vehicles for the popularization of their new machines. The Remington Company went so far as to encourage consumers to purchase machines and donate them to women as a kind of charity. "No invention," the company's advertisement boldly proclaimed in 1875, "has opened for women so broad and easy an avenue to profitable and suitable employment."[45] Companies also employed young female typists during sales calls. In 1875, for example, Mark Twain purchased his first typewriter from a salesman who employed a "type girl" to demonstrate the apparatus.[46]

By contrast, the male Chinese typist was discursively invisible, with typewriter companies in China perpetuating the idea that the sine qua non of a modern office was a young female typist—echoing the Western world even though Chinese realities were more complex.[47] To be sure, parties in China could in theory have *invented* a stereotype of a male typist, were they sufficiently committed to fabricating new tropes that better captured the on-the-ground realities of the Chinese typing school and typing pool, and yet the periodical and archival records suggest strongly that no such discursive enterprise took place.

CHINESE TYPEWRITING AS EMBODIED MEMORY

When young Chinese women and men entered a typing school and encountered a Chinese typewriter for the first time, one question would have quickly surfaced: How were they to memorize the locations of the more than 2,000 characters arrayed in front of them on the tray bed? What kinds of memory practices were young students to employ, and what kinds of typing pedagogies were available to familiarize them with this new technology? While the relationship between technolinguistic forms and embodied practices has received considerable attention in the West, remarkably less is known about the modern Chinese context. As Roger Chartier has demonstrated, the advent of new linguistic technologies and material forms in Europe enabled and constrained the body in unprecedented ways. The codex, as a form, permitted modes of pagination impossible with the scroll, which in turn rendered possible the creation of indexes, concordances, and other referential paratechnologies. Within this new technosomatic ensemble, the reader was now able "to traverse an entire book by paging through."[48] Unlike the scroll and its accompanying requirement of two-handed reading, moreover, the codex "no longer required participation of so much of the body," liberating one of the reader's hands to, among other things, take notes.

Far less is known about the somatic dimensions of modern Chinese linguistic technology—and, in the case of Chinese telegraphy, typewriting, stenography, braille, and typesetting, hardly anything at all.[49] In the case of the Chinese typewriter, a new type of body came into being: a novel coalescence of physical postures, dexterities, forms of coordination and visualization, and types of corporeal and psychological stress, all of which diverged from more familiar contexts of clerical work in the West and offer us a vibrant counterpart with which to think more expansively about forms of physical and mental discipline. Republican-era typing schools offer us a window into this history, particularly in those schools where students had the opportunity to study both Chinese and English typing. When comparing the two curricula, we find that Chinese typewriting encompassed its own distinct physical regimen, in terms of the demands it placed on memory, vision, hands, and wrists.[50]

Students of Latin-alphabet typing in the Chinese schools were required to take courses in "practicing fingerwork" (*lianxi fenzhi*) and "blind typing" (*bimu moxi*) still familiar to Western students even today. By comparison, students of Chinese typewriting took an utterly different battery of classes which included "character retrieval methods" (*jiancha zi fa*) and "adding missing characters" (*jiatian quezi*), among others. While certain properties of alphabetic scripts—in particular, the limited number of modules—factored heavily in the development of "blind" typing as an ideal, there was no such commitment to blind performance in Chinese typewriting. If the QWERTY operator was trained to type without looking at the keyboard, the Chinese typist was encouraged to rely heavily on his or her faculties of vision, both direct and peripheral, and to push them to ever higher and subtler states of refinement.

Chinese typewriters and English typewriters differed as well in terms of the somatic regimens that their operators followed in order to achieve proficiency. From a very early stage, the alphabetic typewriter had been brought into conceptual and practical relationship with the piano and piano playing, not only in terms of the keys that bedecked both instruments, but also in the formulation of training modules that resembled musical *études* (hence the frequent proposition in the West that women with backgrounds in piano playing made ideal candidates for clerical careers). Ideas and practices of posture were likewise borrowed frequently from the realm of piano playing, as were notions of the relative strength and dexterity of individual fingers. Both pianist and typist were meant to maintain a posture of composure and lithe fixity in their torsos and necks, with wrists neither slumping downward nor protruding upward; and both were supposed to concentrate their attention upon the orchestrated movements of the outermost extremities of their fingers. Central to Western typewriting was a hierarchy of strength, dexterity, and utility between the index, middle, ring, and pinky fingers.

The training of Chinese typewriting was, by contrast, oriented toward helping the student bring the tray bed of his or her machine "into play"—to warm up the cold gray surface of the character matrix by walking around in it and learning its cartography—not entirely unlike the training of Chinese typesetters. To this end, typing lessons were centered around a distinct set of repetitive drills, beginning with common Chinese terms and names. These drills helped familiarize one with the absolute locations of individual characters on the tray bed, but more subtly they imprinted in memory the spatial relations between those characters that tended to go together. In the very first lesson of a typing manual from the 1930s, for example, students drilled by repeatedly typing the common two-character words "student" (*xuesheng*), "because" (*yinwei*), and two dozen others. By practicing with such words, the lexical geometries of Chinese became imprinted within the muscle memories of the student. One might not recall precisely where *xue* and *sheng* or *yin* and *wei* were in the absolute sense of *x-y* coordinates, but over time one would recall the general sensation of these altogether common two-character compounds.[51]

Rather than simply seeking out a character on the tray bed, moving toward it, performing the type act, and then repeating this same process for the subsequent character, the trained Chinese typist was urged to refine a sensitivity to, one might say, the *instantly immediate future*—that is, the very next character. Upon setting out toward the first desired character, the typist was to have already begun the process of transferring a portion of her concentration to the next, performing each type act such that, at any given moment or state, she embodied a psycho-corporeal anticipation of the next.[52] Typists were also trained to become sensitized to the materiality of signifiers.[53] Each time the typist depressed the selection lever, the force of each type act had to be finely attuned to the weight of each character, a measurement that corresponded directly to the character's stroke count. Should one type the single-stroke (and thus lighter) character *yi* (一 "one") with the same force as the sixteen-stroke (and thus heavier) character *long* (龍 "dragon"), one would quite likely puncture the typing or carbon paper and have to begin the document anew. To type *long* with the same force as *yi*, however, would result in a faint, illegible registration (also making it ill-suited for carbon-paper copying). Trained operators necessarily varied the force with which they typed different characters so as to maintain a chromatic consistency across the text, and to avoid puncturing the paper.[54] With this technology, then, the long-standing Chinese concept of "stroke count" was translated into the physical and corporeal logics of mass, weight, and inertia.[55]

4.5 Chinese typewriter training regimen (sample), showing typist's movement from character to character within typewriter tray bed

WAITING FOR CADMUS: THE RISE AND FALL OF "CHINESE PHONETIC ALPHABET" TYPEWRITERS

In light of the strides made by Commercial Press in the launch of their new typewriter division, anyone paying attention to developments in China would have witnessed the unmistakable signs of a vibrant new industry, complete with an emerging network of training institutes. By the 1920s and 1930s, indeed, there existed across China a disciplinary practice in which typists used the machine, and further brought their own bodies into accordance with it. It would not have been surprising, then, to have encountered the 1920s report in the *Remington Export Review* that proclaimed, "After a great many years of futile experimentation, a typewriter for the Chinese language has at last been perfected." The Western world, it would seem, was finally taking notice.

Remington Export Review was not referring to Commercial Press, however, nor to Shu Zhendong's machine. The company newsletter was referring to Remington itself. "Robert McK. Jones, head of the typographical department of the Remington Typewriter Company and a Remington man for thirty-seven years," it continued, "is responsible for its production."[56] A photograph of a white-bearded Jones accompanied the piece, beneath it the caption "Who Developed the Remington Chinese Typewriter." The title of the piece said it all: "At Last—A Chinese Typewriter—A Remington."

Robert McKean Jones was Remington's principal developer of foreign-language keyboards and director of the development department.[57] Born in 1855 in the Wirral in northwest England, near the border of Wales, he operated a workshop in New York where he designed an estimated 2,800 keyboards for a diverse collection of scripts. Jones was one of the foremost technicians behind the globalization of the Remington typewriter examined in chapter 1. He personally adapted the single-shift keyboard machine to a reported eighty-four different languages, a feat that earned him the honorific of "master typographer." With his command of sixteen alphabets, a trade journal reported in 1929, "there is hardly a language spoken of which he has not at least a working knowledge."[58]

Having completed an Urdu keyboard in 1918, and then keyboards for both Turkish and Arabic in 1920, Jones was an obvious choice to lead the company's new Chinese initiative.[59] In the winter of 1921, Remington set out to create a "Chu Yin Tzu-mu Keyboard," also known as the "Phonetic Chinese" keyboard, assigning this project to Jones.[60] The Chinese typewriter of Jones's invention was unlike anything we have encountered thus far, however, for one remarkable reason: *it contained almost no Chinese characters*. With the exception of Chinese numerals, the remainder of his keyboard was dedicated exclusively to symbols from the recently developed Chinese Phonetic Alphabet. Jones offered brief details regarding the language reform efforts then afoot in China upon which his invention was premised. "The Chinese Government has officially adopted and is promulgating a new phonetic alphabet known as the 'Chu Yin Tzu-mu,' or in English, national phonetic alphabet. A formal edict commanding its use by Government officials and requiring the teaching of the system in schools has been published."[61] Jones explained that the new alphabet was "devised by a council of learned men from all sections of the country" and that its objective was to "simplify the ancient complicated system of ideographic characters and promote literacy among the people generally."[62] Jones and his patrons at the Remington typewriter company submitted their own application for a National Phonetic Chinese typewriter on March 12, 1924, even before the government of the Republic of China issued its formal promulgation of *zhuyin fuhao* (figure 4.6).

The cognitive dissonance of this machine—a "Chinese typewriter" with no Chinese characters—was not lost on those journalists who first wrote about the invention. "A Chinese typewriter?" began Paul T. Gilbert in his piece for *Nation's Business*. "Well, I suppose you might call it that; but don't look for any of Wang Hsi-cheh's 5000 symbols on the keyboard," referring to Jin dynasty calligrapher Wang Xizhi (303–361). Gilbert concluded by assuring his readers, however: "It would have been impossible to devise a keyboard which would lend itself to the typing of the Chinese language."[63]

Robert McKean Jones was not the first typewriter inventor to seize upon the dream of a "Chinese alphabet" or "phonetic Chinese" as an entry point into the vast and still untapped Chinese-language market—and neither was Remington. One decade earlier, in the winter of 1913, two UK-based engineers had submitted their own joint patent application geared toward "adapting the Chinese language to the production of printed or

4.6 Robert McKean Jones/Remington Chinese typewriter (1924/1927)

typewritten matter."[64] John Cameron Grant and Lucien Alphonse Legros explained in the course of their patent application that their machine would not feature Chinese characters per se, but rather a phonetic Chinese "alphabet" that was currently in circulation in China. "Within the last few years," Grant and Legros explained, "a new Chinese alphabet, or more strictly speaking, syllabary, has come into certain vogue and semiofficial use." This development would come as a relief both to China and to the wider world, the inventors explained, insofar as a character-based typewriter was itself an impossibility. Regarding China's character-based script, "its adaption for modern machine composition is entirely out of the question, as it would be quite beyond the range of practical possibility to cut and apply such a number of matrices to any known form of machine composition, or indeed, to bring the whole language easily into the compass of any known power of manual composition at case." Meanwhile, however, any attempt by foreigners to develop "Romanized" versions of Chinese was equally bound to fail. Foreign Romanization schemes "have grave disadvantages, firstly, from the fact that the alphabet itself is foreign, and therefore objectionable," and secondly that they require the usage of additional symbols to convey Chinese tones.[65]

But the script that would form the basis of their typewriter was something apart—it had been invented *in* China and *by* Chinese. In the wake of the 1911 revolution that overthrew the Qing dynasty (1644–1911), the newly established Ministry of Education of the Republic of China under Cai Yuanpei announced plans to convene a Conference on Unification of Pronunciation. Under the direction of Wu Zhihui, the meetings commenced in February 1913 and were attended by many of the country's most prominent language reformers. Over the span of approximately three months, and a contentious and complex set of discussions, the committee announced an agreement that centered around a phonetic notation system for Chinese known as *zhuyin zimu*—or "phonetic alphabet."[66] Here, then, was the opportunity that Western typewriter companies had been waiting for: a Sino-foreign script, invented "by and for" the Chinese, but employing a Western system of script.

Legros and Grant were not alone in imbuing this new "Chinese alphabet" with a sense of possibility and promise. In the very same month they submitted their patent application, so too did J. Frank Allard, an engineer working in collaboration with the Underwood Typewriter Company to retrofit a standard Underwood machine with the new phonetic symbols.[67] In April 1920, the *Telegraph-Herald* announced that the "Chinese Alphabet has been reduced."[68] "Chinese of future generations," the article began, "will write in phonetic script and use a typewriter with only 39 characters instead of plying a brush to draw 10,000 or more hieroglyphics if mission workers succeed in an effort they are making to revolutionize handwriting in use in China for more than 4000 years." The system had been developed in 1903 in England by Wang Chao, the article explained, and had already been the subject of interest for designers of a Chinese typewriter. Reverend E.G. Tewksbury of Shanghai, it was reported, "has put the new script into use on American typewriters with complete success," with the font being provided by Chinese engravers. In August 1920, *Popular Science* featured an advertisement for yet another phonetic Chinese typewriter: the Hammond Multiplex.[69] By the time Robert McKean

Jones signed his name to the Remington Phonetic Chinese typewriter, then, his was already one in a long genealogy.

Even as the dream of a "Chinese alphabet" seemed to be coming true, a stark reality confounded the aspirations of these Western typewriter manufacturers: *zhuyin* was never meant to replace Chinese characters. It was never meant to become a "Chinese alphabet" in the sense understood by Remington, Hammond, Underwood, and others. Rather, *zhuyin* was meant to serve as a pedagogical, paratextual system through which to inculcate in the Chinese people the standard, nondialectal pronunciations of Chinese characters. This fact was lost, however, on many foreign observers, most notably foreign inventors eager to develop phonetic Chinese typewriters and hot-metal composing machines. Much like Vladimir and Estragon waiting endlessly for Godot, observers of China persisted in a kind of imperturbable anticipation, poised to wait endlessly for a *Cadmus*—a figure preordained to come to China at some point (always in the near future) and bring with him the wonder and salvation that is the alphabet. As the mythological founder of Thebes and brother of Europa, Cadmus was credited by Herodotus as being responsible for introducing Phoenician script to the Greeks—the forerunner to the Greek alphabet, that invention to which Walter Ong, Eric Havelock, and many others have attributed causal power as a contributor to the Greek Miracle, as discussed earlier. If only Cadmus would hasten to arrive here in the Celestial Kingdom, all of the many woes that had beset the world's last great noncenemic script would be resolved.

Returning to Remington and Jones, we see how the many "false Cadmus sightings" in the twentieth century may have dented, but in no way destroyed, confidence in his eventual arrival. ("I can't go on like this," Estragon protests. "That's what you think," Vladimir responds.) Not five months after Remington submitted its application to the Patent Office, a full-page advertisement appeared on the back cover of China's premier language reform journal, the *National Language Monthly* (*Guoyu yuekan*), which regularly featured the writings of Zhao Yuanren, Li Jinxi, Qian Xuantong, Zhou Zuoren, Fu Sinian, and other influential figures in the country's language reform movement. The advertisement featured the Underwood Chinese National Phonetic Typewriter (*Entehua Zhonghua guoyin daziji*), offering a photographic preview of the keyboard, and the assurance that "its construction is the same as the English language Underwood Typewriter" (*qi gouzao shi yu entehua yingyu daziji xiangtong*). Along the bottom edge, readers saw the output of the machine, in a single line of *zhuyin* script: ㄅㄊㄜㄏㄨㄚ ㄍㄨㄛ|ㄅ ㄅㄚㄗㄐ| (*entehua guoyin daziji*) or "Underwood National Phonetic Typewriter."[70]

Despite such giddy optimism within the industry, all attempts to market and mass-produce phonetic Chinese typewriters failed. Western companies never understood—or perhaps even bothered to understand— the limited orientation of these phoneticization movements among the "Celestials." Instead, in their never-ending wait for the annunciation of the "Chinese alphabet," Remington, Underwood, and others stood poised and alert, ears perked to hear the footfalls of an approaching Cadmus before their competitors did, so as to be the first to meet him at the gates and capitalize on his arrival.

So entrenched was this idea of a phonetic Chinese typewriter that, indeed, even the utter failure of these phonetic Chinese machines escaped the attention of Western media. When Jones died in 1933 in Stony Point, New York, his obituary made no reference to his Arabic machine, nor to the keyboards he designed for Urdu or other scripts. Instead, the notice of his death spotlighted perhaps the only keyboard in the inventor's career that ended in utter failure: "Robert McKean Jones: Inventor of Chinese Typewriter Was Able Linguist." "The development of the Chinese typewriter twenty years ago by Mr. Jones," it continued, "was considered an outstanding achievement. The many characters in the language were believed to constitute an insuperable handicap. Mr. Jones, an accomplished linguist, worked for years in consolidating the various Chinese characters on the keyboard of a machine in a manner that is legible and intelligible to citizens of that country where there are many dialects and many alphabets."[71]

AT THE WORLD'S FAIR: CHINESE TYPEWRITING BETWEEN MIMESIS AND ALTERITY

Whatever strides the Chinese typewriter might have been making at home, the domestic industry alone was clearly not enough to cement China's place within the global family of modern technolinguistic countries. As long as the "real" typewriter existed out in the wider world, and as long as the world outside China remained unprepared to conceptualize the Chinese typewriter as anything but an absurdity, the status of this machine would remain in question—particularly its eligibility to bear the moniker of "typewriter." In 1919, the journal *Asia* featured a photograph of Fong Sec, then editor in chief of Shanghai Press, standing before a Commercial Press machine. "Dr. Fong Sec," the caption read, "stands for the best type of the efficient, modern Chinese," in what would begin as a glowing report. "As an editor of the Shanghai Press, the largest and best equipped of Chinese printing firms, responsible for the publication of almost all Chinese text-books and the dissemination of literature throughout Central China, he holds a position more significant from the point of view of the enlightenment of the people than that of any official of state." What praise was afforded Fong Sec himself was promptly withdrawn, however, once focus shifted to the contrivance before which he stood. "In this picture he is shown with the elephant of typewriters," the article continued, "conceived by a son of the fathers of invention."[72]

During the 1920s, Commercial Press set out on its first concerted effort to win over the Western world and change its views of the Chinese typewriter. The machine's global debut took place in Philadelphia at the Sesquicentennial International Exposition of 1926, mounted in celebration of the 150th anniversary of the signing of the Declaration of Independence. On display would be the cultural and industrial heritage of the global community, with pavilions dedicated to Japan, Persia, India, and more. The Chinese pavilion was overseen in part by Zhang Xiangling, the Chinese consul general in New York, a man for whom the presentation of his country on the global stage was of the utmost concern.[73] Three years prior to the exhibit, circa 1923, Zhang had paid a visit to the Philadelphia Commercial Museum and its display dedicated to Chinese products. "They were disappointed," a report from the period read, referring to Zhang and a group of Chinese businessmen who had accompanied him, "to find that the exhibit, large and handsome as it is, represented the China of the past rather than the new China with its diversified production." "For instance, there were relatively few specimens illustrating the modern industries of the country and the manufactures which China has developed within recent years." To remedy this situation, Consul General

Zhang promised to the museum curators that he would reach out to parties back home and secure a "representative series of samples of Chinese products" for the museum. A few months later, Zhang's promise was fulfilled, with seven cases of materials arriving in Philadelphia, sent by the Shanghai Chamber of Commerce. Included in the collection were samples of silk fabrics, wicker furniture, tobacco and cigarettes, tooth powder, and hosiery, among many other items.[74]

The Chinese typewriter by Commercial Press was selected as one such exemplary object—a device to display the progress of the nation in the industrial arts, while at the same time situating this machine in a longer framework of Chinese civilizational history. Upon entering the pavilion, one immediately encountered the American and ROC flags, draped side by side, as well as fine silks, ornate umbrellas, tea pots, porcelain vases, five-colored glass, calligraphic scrolls, and landscape paintings. Along the rear wall, just right of center under an image of George Washington and a painting of a wildcat poised upon a tree limb, was a display case with an engraved plaque: "Chinese Typewriter" (figure 4.7).[75]

4.7 Shu Zhendong Chinese typewriter (lower left) at the Philadelphia world's fair

In preparation for the machine's global debut, Commercial Press crafted an English-language brochure for the Shu-style machine. The tone was carefully phrased for a foreign audience, taking the machine out of its indigenous framework of manuscript and printing, and in a rare instance suggested that it was in fact on par with the real typewriter of the Western world. "The Chinese typewriter manufactured by the Commercial Press solves a serious problem in office administration in China," the brochure read. "The machine has all the advantages of a foreign typewriter" (figure 4.8).[76] As commissioner general representing China at the Philadelphia world's fair, Zhang must have been pleased by the typewriter's performance and reception.[77] Commercial Press received a "Medal of Honor" for "Ingenuity and Adaptability of the Chinese Typewriter."[78] "It is a miraculous achievement of the inventor that, although there are 3,000 characters in the font," a Philadelphia guidebook read in a rare celebration of a Chinese machine, "yet a typist, after the practice of one or two weeks, will be able to locate any character instantly. 2,000 characters can be written after two months' training, and greater speed can be obtained by longer practice."[79] News from Philadelphia began to reach China starting in the winter of 1927, moreover. On January 12, *Shenbao* relayed communications from Zhang, containing information on those Chinese companies that received honors and awards. Among those awarded was Commercial Press, specifically for the Shu-style Chinese machine.[80]

A medal of honor in Philadelphia was insufficient to quiet criticism and denigration, however, not only from foreigners but also closer to home. In 1926, the same year as the world's fair, linguist and vociferous character abolitionist Qian Xuantong once again railed against the inefficiency of character-based systems of categorization, reproduction, and transmission, leveling a critique against a wide range of objects. Beginning with a critique of character-based dictionaries, catalogs, and indexes, Qian argued that "Chinese characters offer no effective solution, whether it be stroke-count, rhyme schemes, or by relying on the most dog-fart of them all, that radical system from the *Kangxi Dictionary*."[81] Qian reserved his most strident criticism for the Chinese typewriter:

And then there are typewriters on which one cannot have less than two to three thousand characters. The surface area of two to three thousand characters

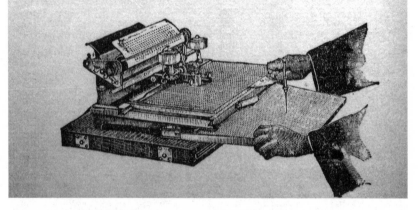

CHINESE TYPEWRITER

The Chinese typewriter manufactured by the Commercial Press solves a serious problem in office administration in China. The machine has all the advantages of a foreign typewriter.

4.8 Commercial Press brochure for the world's fair

is not small, mind you. When typing, no matter how familiar one is with these two to three thousand characters, one has little choice but to search for each character one by one. The first character is all the way in the northeastern corner. The second character is over in the southwest corner, eighth from the bottom. The third character is, once again, up in the northeast, in the third column just off center, eleventh from the top. The fourth character is up in the northwest corner, down a bit. The fifth character is once again in the middle, just a bit southeast of central. And so on. It's really enough to leave one bewildered (*mumi wuse*). And when you come to a character that's not on the machine, and when you can't find it in the "rarely used character tray bed" (since the rarely used character tray bed doesn't have all the characters either), and when you have to write in the character by hand, then you'll see for yourself how much of a hassle it is. Phonetic script has only a few dozen characters and a handful of symbols, so it goes without saying how convenient it is for typing.[82]

Qian's cartographic imagery was playful and devastating. Setting the stage for his critique with the deft use of the term "surface area" (*mianji*)— a term typically reserved for territorial expanses—he transformed the Chinese typist into a lost soul wandering across an expansive landscape of the twice-repeated "two to three thousand characters." To elongate this sense of distance, Qian expressed the location of characters on the tray bed using cardinal directions, the way one might express the location of provinces or cities in China itself. Qian's first hypothetical character was located somewhere in the province of Fengtian, his second in Yunnan or southern Sichuan, his third perhaps in Rehe, his fourth in eastern Xinjiang, and his fifth in Shaanxi.

Meanwhile, back in the United States, the Chinese pavilion in Philadelphia seems to have inspired yet another instantiation of the Chinese

Mr. Underhill: THIS IS OUR LATEST ACHIEVEMENT—A TYPEWRITER FOR THE CHINESE TRADE.

4.9 "Chinese Language Typewriter" in *Life* magazine, 1927

monster we first examined in chapter 1. In the February 17, 1927, issue of *Life* magazine, cartoonist Gilbert Levering presented his vision of a "Chinese Language Typewriter": a colossal contraption featuring a keyboard roughly thirty keys wide by thirty-five keys deep—for a rough total of 1,050 keys in all (figure 4.9).[83]

Twenty-seven years since making his first appearance in the popular press, it would seem, "Tap-Key" was alive and well.

5

CONTROLLING THE KANJISPHERE

Typists more than anyone must follow the times.
—Chinese typewriting manual, Manchukuo, 1932

It should ... not be overlooked that these Japanese typewriters could also be used for correspondence in Chinese.
—Memo from treasurer to secretary general, Occupied Shanghai, 1943

"Pray for Free China." So reads a sticker affixed to the Chinese typewriter I acquired in the summer of 2009. The machine came from a Christian church in San Francisco, where it had been used for years to compose the weekly church bulletin. "They were going to throw it away," explained the man who had learned of my work and reached out by email. "I saved it in hopes that someone might find some way to appreciate such past technology."[1] The machine was a Double Pigeon Chinese typewriter produced on the mainland by the Shanghai Calculator and Typewriter Factory. With its signature pale green color, this was *the* typewriter of the Maoist period (1949–1976), and manufactured well into the 1980s and early 1990s.

The Californians who owned this machine had subverted the device's national and party affiliations, however. On the sticker that "prayed for Free China," the flag of the Nationalist, or Guomindang, party was featured prominently, pledging political allegiance to the Taiwanese government-in-exile (or "renegade province," in Beijing's parlance). The machine's allegiance was further pledged to the United States, with a second sticker on the machine featuring the American bald eagle, grasping arrows in its talons. Poised menacingly next to the pair of unarmed pigeons, the device's official logo, the eagle stood as testament to the late Cold War period in which the machine had been in service.

In what ways can objects be said to have politics, however, not in the technolinguistic sense explored thus far, but in conventional terms of party politics, national identity, and the like? Does it matter who builds an inscription device, in the sense of the engineer's or company's place of origin or ideological leanings—or are technological devices inherently neutral, nationless, cosmopolitan citizens of the world? Did it matter in any way, in other words, that a machine manufactured in the People's Republic of China had been employed by a pro-Taiwan, Christian Chinese church in the United States? Or are technological objects by their very nature agnostic with respect to these all-too-human concerns? These are some of the questions we will take up in this chapter.

The following year, another episode brought these same questions surging back to my consciousness in unexpected ways. Summer 2010 found me in London on a congested ride from Hounslow Central Station to Covent Garden, with an unremarkable matte brown suitcase to my left. Inside its stiff acrylic walls, sheets of crumpled wallpaper samples held in place the central component of another Chinese typewriter, this one built in the 1960s. Neither this suitcase nor its contents had belonged to me just hours earlier, but instead to a Malaysian Chinese-British family living in the suburbs of London along the outermost stretch of the Piccadilly Line. This typewriter had been purchased in Singapore, employed in Malaysia, shipped to London, and was now en route to my office at Stanford University. This would become the second machine in my growing collection, and one whose national identity was even more complicated than that of the first, I had discovered.[2]

When I arrived at the home of the Tai family, I was greeted warmly and guided to a room in the rear of the house. Evelyn sat down at the desk and began to demonstrate the machine I had been invited to see. "I haven't touched it in twenty years," she began. "I'll try my best." Evelyn began to type, and I asked her how she had come to learn the locations of more than two thousand characters. Her response was modest, as if to dismiss the feat of memory. "I just remember it." She expanded somewhat on this answer, explaining how Chinese radicals, or *bushou*, were the key, just as we saw in chapter 4 when we examined typewriting schools and curricula in early twentieth-century China. When learning the layout of the machine, she explained, she had started by learning that all characters of such-and-such radical were located in such-and-such zone. Over time, though, this conscious act of study and memorization had clearly given way to muscle memory. "Say I want *xin* [新 'new']," Evelyn remarked casually, beginning to move the character selector across the typewriter tray bed. "Straightaway you know *xin* is here."

Everything about Evelyn's machine seemed deeply personal. In a small, four-by-three-inch plastic box, she kept a special set of characters she could reach and, using tweezers, place on her machine in one of the empty slots in the matrix. Like the characters on the Huntington machine discussed in chapter 4, whose character tray bed still testifies to the life and times of its owner, Evelyn's characters testified to her Christian faith and the clerical work she performed as a member of her church: *immortal/saint* (仙 *xian*), *to save* (拯 *zheng*), the exceedingly rare third-person pronoun reserved for God (祂 *ta*). Others were not so easily deciphered, as in *gao* (糕) and *qiu* (丘)—most likely the surnames of church members who appeared regularly in the weekly bulletins that had been typed on this machine.

The machine was deeply personal for Evelyn's daughter, as well. During her mother's demonstration, Maria stood just behind us, listening to her mother's explanations. Unlike the rest of her immediate family, all of whom were born in Malaysia, Maria had been born in Singapore after the first of the family's emigrations. There Maria's father had worked as an assistant lecturer at Trinity College, while her mother worked as a kindergarten teacher at a school associated with a local Methodist church (they themselves being Presbyterian). Evelyn purchased her Chinese machine in Singapore and used it to type up lecture materials at Trinity, as well as church programs and orders of worship. While Maria was still an infant, her mother had taken a three-month intensive course in Chinese typewriting administered by the Adult Education Board of Singapore, passing the certification exam in October 1972. The family moved again, this time to London, where her father was to serve as a chaplain at the United Reform Church. They disassembled and shipped most of the machine, all except for the platen. This they had packed inside a matte brown, acrylic suitcase—the very piece of luggage that they would soon turn over to me.

Throughout these many travels, the sound of her mother's typewriter was a constant for Maria, providing some measure of continuity during an itinerant childhood. At home, with her daughter sometimes helping her, Evelyn typed up Chinese-language programs for her church, making copies using stencil paper. Other organizations and outfits requested her help as well, such as local community centers and nearby schools and churches. When she was not helping her mother, Maria explained, she could still hear the muffled yet distinct sound of the machine through the walls of the house, late at night. The sound would accompany her as she dozed off in her bed. Here in 2010, as she watched her mother demonstrate the machine to me, she interjected almost inaudibly: "It makes me fall asleep, that sound."

When first I set out on my research, little could I have known that a conversation in the suburbs of London would present the opportunity to delve into an intimate history of the Chinese typewriter. Over the span of those few hours, I was able to put aside the many questions of politics we have examined thus far—the politics of how Chinese typewriting was understood in the alphabetic world, the politics of the common usage approach to Chinese technolinguistic modernity, the politics of divisible type printing, and more—and instead catch strobing glimpses of the place this one particular apparatus had held in one woman's life for more than two decades. It was one of the "evocative objects" of her world, to cite Sherry Turkle's *a propos* formulation.[3] The machine even shared her name, with an embossed label reading "Evelyn" affixed to the front of the chassis. "This is my one," she remarked to me, "from beginning to end."

As quickly as politics had receded from view, however, they burst forth once again, this time in a remarkably different light. As I looked over the apparatus, a small plaque affixed to the rear of the chassis caught my attention, and instantly complicated the romantic portrait I had been painting in my mind. Upon this plaque was embossed:

"Chinese Typewriter Company Limited Stock Corporation. Japanese Business Machines Limited."

This Chinese typewriter was Japanese.

An entirely new conversation ensued, returning us to the day, many decades earlier, when Evelyn and her husband had first purchased the machine at a shop in Singapore. The store had on offer a selection of Chinese typewriters, she recalled, including the same Double Pigeon model that was waiting for me back at home. Double Pigeon was by far the most widely used Chinese typewriter at that time, both in China and internationally, in large part because it was one of the only models of Chinese typewriter being produced on the mainland.

The other model on offer was the Superwriter, built by Japanese Business Machines, Ltd. Structurally and linguistically, the principles of the machines were identical, both featuring common usage tray beds upon which were arrayed a collection of approximately 2,500 Chinese characters. The typing mechanisms operated identically, as well. Using his or her right hand, the operator guided the character selector device over the top of the desired character, depressing the type bar at the appropriate time. The main question for Evelyn was the reputation of Japanese and Chinese manufacturing more broadly. Should she purchase a Chinese typewriter built in China, or one built in Japan?

As every historian knows, brief and passing moments like this have the capacity to reroute us, hurtling us down unexpected paths that can require years to traverse, but leave us richer for the journey. Working through my archives again, and expanding into new collections in Tokyo, questions came into focus I had not contemplated before: Over the course of modern history, who controlled the means of Chinese textual production and transmission—when, how, and to what ends?

By this point in my research, I knew full well that global giants like Remington, Underwood, Olivetti, and Mergenthaler Linotype had all tried and failed to enter the Chinese-language market, let alone capture it. Standing in Evelyn's living room, the question reemerged: What about Japan? If Western manufacturers had proven unable to bring the Chinese language into their ever-expanding repertoire of world scripts—and increasingly unable to imagine what a typewriter might look like for a nonalphabetic script—what about companies and engineers in Japan, where nonalphabetic Chinese characters—or kanji—formed a core part of the Japanese language? And how might this history relate to the contemporary concept of "CJK," a catch-all term within contemporary computing that refers to "Chinese, Japanese, and Korean" information processing, font production, and more?[4] As I looked upon this Japanese-made Chinese typewriter, was I in fact looking at the "prehistory" of CJK?

In this chapter, we examine the braided histories of *Japanese* typewriting and Japan's takeover of the Chinese typewriting industry. Although the apotheosis of the Western-style keyboard typewriter had placed China and Chinese irretrievably beyond the pale of technolinguistic modernity as understood by multinational companies like Remington, the history of Japanese multinationals was an altogether different one. Japan was home to two distinct approaches to typewriting, one oriented exclusively toward the typing of Japanese kana—the twin syllabaries, hiragana or katakana—and the other oriented toward character-based writing (kanji). In look and feel, the first family of machines was indistinguishable from those built by Remington, Underwood, or Olivetti. The second family, meanwhile, was indistinguishable from those built in China. Occupying this liminal position, Japanese engineers were in many ways less imaginatively confined than their Western counterparts, never restricting themselves to a keyboard-based system incapable of handling characters. Instead, Japan succeeded where Remington failed by developing the very same tray bed–based common usage machines with which Chinese typists had become familiar.

Japanese companies made inroads into the Chinese market as early as the 1920s, a process that accelerated quickly with the expansion of Japanese empire-building in northeast China in 1931, and then with the outbreak of full-scale war in 1937. By the late 1930s and 1940s, Japan dominated the Chinese typewriter market, and would continue to exert influence well into the early postwar period (as exemplified in Evelyn Tai's choice of Superwriter). CJK, it turns out, has a violent past, inseparable from the rise and fall of the Japanese empire and the horrors of the Second World War.

BETWEEN TECHNOLINGUISTIC WORLDS: KANA, KANJI, AND THE AMBIVALENT HISTORY OF JAPANESE TYPEWRITING

Beginning in the second half of the nineteenth century, China's neighbor to the east began to wrestle with its own questions of technolinguistic

modernity, as exemplified in the history of Japanese telegraphy, tele-
phony, industrialized printing, post, stenotype, and more.[5] As Ryōshin
Minami has shown, Japan's printing industry was one of the country's
earliest and most thoroughly mechanized.[6] Beginning in 1876, the application of steam power to printing revolutionized the domestic newspaper industry, enabling Japanese publishers to keep pace with a voracious
and growing appetite for daily newspapers.[7] During the second half of
the nineteenth century, telephone and telegraph technologies were
introduced in a timeline that runs roughly parallel to that of the Qing.[8]
In 1871, the same year Great Northern and Cable and Wireless promulgated the Chinese telegraph code, a telegraph code for Japanese kana was
authored as well. Known alternately as the *Japansk Telegrafnøgle* (Japanese
telegraph key) or *Denshin jigō* (telegraph signals), the code assigned short
and long pulses to Japanese katakana syllables, arranged according to the
predominant dictionary sequence of the era, the *iroha* taxonomic system
(named after the eponymous Heian-era poem).[9] With the Japanese state's
establishment of a monopoly over telecommunications and postal services, and with the expansion of Japan's overseas informal and formal
empire in the 1880s and 1890s, this network began to expand rapidly.[10]

The history of typewriting in Japan is inseparable from the broader
history of nineteenth- and twentieth-century language reform and modernization efforts in East Asia, as well as the era's widespread critique of
character-based writing. Indeed, calls for the abolition of characters began
sounding in Korea and Japan before they did in China. As part of "decen-
tering the Middle Kingdom," in the apt terminology of one historian,
a branch of Korean reformers began to particularize and de-universalize
symbols and ideologies inherited from the Chinese cultural sphere. Believ-
ing that a "break with the transnational culturalism of the East Asian past
was necessary," these reformers reserved particular enmity for hanja—
Chinese character-based script that had been imported and applied to
the Korean language many centuries prior—regarding it as a fundamental
hindrance to the project of scientific (read Western) development.[11] Once
regarded as the *chinmun* or "truth script," the character-based component
of the Korean writing system steadily became particularized as foreign—
as *Chinese* characters judged "merely in terms of their merits and a communicative tool."[12] Hanja came to be understood as irredeemably wedded

to the very doctrines of Confucianism and Daoism now under attack as inherently antimodern. Bringing to mind Ernest Gellner and later Benedict Anderson's examination of "truth language"—hierolects such as church Latin, Old Church Slavonic, or "examination Chinese" that were once understood to offer exclusive access to the canon of truth—the privileged position of such language necessarily eroded once this underlying truth was increasingly regarded as false.[13]

Japanese reformers mounted strikingly similar arguments during this same period, attempting to decouple their country's fate from the sick man of Asia and thereby partake in the global project of modernity.[14] In 1866, Maejima Hisoka, translator at the Bureau for Development of Foreign Studies, presented a petition to the Shogun Yoshinobu entitled "Proposal for the Abolition of Chinese Characters" in which he advocated the replacement of kanji with kana.[15] With the objective of increasing the efficiency of writing and accelerating the pace of language education, Maejima established the company Keimōsha in 1873 to produce the all-kana newspaper *Daily Hiragana News* (*Mainichi hiragana shinbun*). Although the paper failed within its first year, advocates of kana such as Shimizu Usaburō, in his 1874 essay "On Hiragana," elaborated further upon Maejima's proposal.[16]

The first Japanese typewriter was a kana-only machine, containing no kanji whatsoever. Patterned after the single-shift keyboard machines growing in prominence worldwide, and designed to type the hiragana syllabary, this machine was completed in 1894 by Kurosawa Teijirō (1875–1953). Kurosawa soon turned his attention to the development of a katakana machine, which he completed in 1901. Basing his production on the Elliott model Smith-Corona machine, Kurosawa went on to name his machine the Japanese Smith Typewriter, or *sumisu taipuraitā*.[17]

Kana typewriting opened the door to Western manufacturers, an opportunity they wasted no time in seizing. As early as 1905, Remington brought Japanese into its immense and expanding repertoire of world scripts, seizing upon kana-only design as a means of entering the East Asian market while at the same time circumventing the intransigent problem of kanji. Salesmen were offered guidance in how to field questions that might come up from customers, particularly those wondering about the conspicuous absence of kanji on the Remington device. "The

Kana, and especially the Katakana," the salesperson was instructed to respond, "represent the ancient Japanese tongue, but Japan received from China many centuries ago most of her classical literature and advanced learning and adopted the Chinese character for her writing." "In a word, it is impossible to make the machine write the multitude of Chinese symbols commonly used in even ordinary daily routine writing by the Japanese."[18] By means of the kana typewriter, Japan gained entry into the broader cultural repertoire of typewriting: the technosomatic discipline, pedagogical regime, aurality, and more. As in Paris, Beirut, Cairo, and New York, manuals on kana typewriting introduced trainees to the "correct typewriting posture" (*tadashii taipuraichingu no shisei*), the "correct form for the fingers" (*tadashii shushi no keitai*), and the "allocation [of keys] to each finger" (*kaku yubi no bundan*).[19] Kana typewriting also made it possible for Japan to take part in the global romance of the Western-style typewriter—the poetry of it, its iconic style, and even artwork produced on it. In a textbook from 1923, for example, we are struck by three works of typewriter art attributed to one "T. Koga": the figure of Rodin's *Thinker*, a visage of Jesus Christ, and a map of North America, each built up from successive keystrokes.[20]

From its inception, the all-kana typewriter was a machine with a politics, wherein modernization was premised on cutting the Japanese language's ties with its Chinese orthographic heritage. The trope of the impossible, monstrous Chinese typewriter examined in chapter 1 was here mobilized in a Japanese context as a kind of cautionary tale (lest Japan too find itself excommunicated from the universe of modernity). "All educated natives of Japan that we have consulted seem to agree," Remington reported, "that the current method of writing Japanese is cumbersome and antiquated, and utterly unsuited to the present needs of their people." What was called for in Japan was "replacing the present badly mixed system with a purely phonetic one." "It is quite within the bounds of possibility that the advent of the Remington Typewriter for writing Katakana may point a way toward bringing about this change."[21]

Remington soon faced competition from other global firms keen on entering the East Asian arena. In February 1915, the Underwood corporation sponsored the patent of its own katakana typewriter, developed by Yanagiwara Sukeshige. Eight years later, Underwood sponsored another katakana typewriter patent, this one by Burnham Stickney, who had served as the patent attorney for Yanagiwara.[22] Remington quickly fired back at its competitor at the Panama-Pacific International Exposition in San Francisco in 1915 (the same expo where a young Qi Xuan demonstrated his experimental Chinese typewriter, to much less fanfare). Remington's pavilion spotlighted an all-kana Japanese machine, operated by an eye-catching young Japanese-American woman named Tsugi Kitahara. "Greetings from the Panama-Pacific to our friends throughout the world," read the company's promotional postcard sporting her photo, "writing the 156 languages for which Remingtons are made.—Tsugi Kitahara."[23]

Kanji typewriters were not developed until fifteen years after their all-kana predecessors. For their part, these machines were deeply connected to a vibrant counterpart to the all-kana movement: the common usage kanji movement. In 1873, journalist, political theorist, and translator Fukuzawa Yukichi (1834–1901) authored a three-volume educational text for children using fewer than one thousand characters. In the explanation of his work, entitled *The Teaching of Words*, Fukuzawa opined that somewhere in the range of two to three thousand kanji characters would prove more than adequate—with the balance of Japanese writing to be expressed in kana.[24] Between the time of Fukuzawa's *Moji* in 1873 and the May 1923 "List of Characters for General Use" presented by the Interim Committee on the National Language, numerous scholars, statesmen, and educators weighed in on the question of "common usage kanji."[25] In his 1887 reference work *The Three Thousand Character Dictionary*, Yana Fumio proposed that around three thousand would be sufficient.[26] Three decades later, a consortium of Tokyo- and Osaka-based newspaper publishers issued a joint statement on March 21, 1921, entitled "Advocating the Restriction of the Number of Kanji."[27] Following publication of the May 1923 issue, many of the same newspapers offered vocal support, pledging to adhere to this repertoire of kanji in their publications, to begin on September 1, 1923. This plan was laid to waste, however, in the destruction of the Kanto earthquake, delaying the question for another two years.[28]

One of the earliest technological manifestations of this branch of Japanese language reform was the kanji typewriter invented by Sugimoto Kyōta (1882–1972). As early as 1914, he reported nearing completion on a working prototype, and in October of the following year he was heralded by the Tokyo Chamber of Commerce for the device that would come to be known quite simply as the "Japanese Typewriter," or *Hōbun taipuraitā*—maintaining the phoneticized loan word employed decades earlier by kana typewriter developers.[29] In November 1916, he filed his invention with the United States Patent Office, receiving his patent one year later.[30] It was not long before Sugimoto's typewriter met with competition, moreover. The Oriental Typewriter invented by Shimada Minokichi soon appeared on the market. This was followed by the Otani Japanese Typewriter, invented by Kataoka Kotarō and manufactured by the Otani Typewriter Company. Toshiba released its own Japanese typewriter around 1935. Premised on common usage like the others, this machine featured a character cylinder rather than a flat tray bed (figure 5.1).[31]

For all their diversity, these machines and their manufacturers shared common design principles and entrepreneurial goals. In terms of structure, each machine included only a limited and carefully chosen collection of common usage Japanese kanji, organized phonetically according to the *iroha* system.[32] Meanwhile, manufacturers centered their marketing campaigns around a set of core principles: accuracy, beauty, legibility, and the saving of labor, time, and paper.[33]

As in China, a pedagogical network built up around the kanji typewriter that was responsible for training a new labor force. Unlike China, however, Japan's typing schools were attended almost exclusively by young women, the industry characterized by a feminization of this secretarial labor force on par with the more familiar contexts of Europe and the United States, as well as with other communications industries in Japan itself (figure 5.2).[34] In Japan, a survey of professional women conducted in Tokyo and Osaka in the late 1920s offers us a glimpse of the country's clerical workforce. Of the nearly thousand women surveyed, more than half were under the age of twenty and over 90 percent were unmarried. Educational backgrounds varied, with roughly 40 percent having no more than lower school education, and slightly more having backgrounds in girls' schools. The plurality of women worked in government or public office, followed by private companies and banks.[35]

During the 1920s, Japanese typewriter manufacturers helped fortify the gendered parameters of the typing profession through the publication of a new periodical, *Taipisuto*. Founded circa 1925, and published

邦文タイプライターの種類

東洋タイプライター會社製
丸型　東洋タイプライター

日本タイプライター會社製
標準型甲號機（H 式）

大谷式和文タイプライター

和文スミスタイプライター

5.1 Japanese typewriters; from Watabe Hisako [渡部久子], *Japanese Typewriter Textbook* (*Hōbun taipuraitā tokuhon*) [邦文タイプライター讀本] (Tokyo: Sūbundō [崇文堂], 1929), front plate

by the Japanese Typist Association (*Hōbun taipisuto kyōkai*), the monthly was part professional journal, part women's magazine, each issue featuring crisp, bold art deco graphics and cover art dedicated to the portrayal of the modern Japanese woman in all her many forms (whether outfitted in smart business attire, athletic outfits, or traditional kimonos).[36] Inside could be found a spectrum of content, ranging from Tanka poetry and beauty tips to explorations of the life and profession of the Japanese typist and long-form essays on issues confronting Japanese women in general (figure 5.3). Advertisements abounded, whether for the Nippon Typewriter Company or for women's consumer products. The journal featured extensive photographic content as well, a common subject being group photos of graduating classes looking ahead optimistically to the work that awaited them.

5.2 Photograph of Japanese typist Kay Tsuchiya, 1937 (author's personal collection)

5.3 Japanese typist magazine *Taipisuto* 12, no. 12 (December 1942—Showa 16): cover

JAPANESE-MADE CHINESE TYPEWRITERS, OR THE ADVENT OF THE MODERN KANJISPHERE

By the early 1930s, Japan was home to a variety of typewriter models, divided into two broad worlds: the world of kana typewriting, which partook in the globally recognizable culture of Remington, Underwood, and Olivetti; and that of kanji typewriting, which, owing to its intimate affiliation with typewriting in China, was excluded from this global technolinguistic culture. The ambitions of Japanese typewriter designers extended well beyond their own country and language, however. In his 1916 patent, Sugimoto Kyōta was careful to point out that his typewriter was "designed for the Japanese and Chinese languages."[37] When inventor Shinozawa Yūsaku of Tokyo filed a June 1918 patent claim, he likewise characterized the new typewriter as one adapted for "a language in which a large number of characters are used, such as the Japanese or the Chinese language."[38] Whether as an afterthought or a catalytic part of their invention process, inventors of machines for the Japanese language made explicit the larger aspirations and stakes of their projects: a market that encompassed East Asia as a whole.[39]

That Japanese inventors should have conceptualized their projects in terms of a broader Chinese-Japanese (and soon Chinese-Japanese-Korean) market is hardly surprising. The Chinese market would have exerted an irresistible pull on the minds of inventors, offering up the prospect that a Japanese kanji machine could, in theory, readily be made to "handle" Chinese as well. Beyond the question of markets, moreover, there existed a longstanding history of shared cultural heritage between China, Japan, and Korea—the very "transnational culturalism" that many Korean and Japanese reformers had attempted to dismantle beginning in the late nineteenth century, but which typewriter inventors and engineers in the 1910s and 1920s were now setting out to rediscover and render profitable. As Douglas Howland and Daniel Trambaiolo explain, Chinese, Japanese, and Korean diplomats and embassies in the eighteenth century frequently employed "brushtalk" (Chinese *bitan*, Japanese *hitsudan*, Korean *p'ildam*) as a medium of written conversation in contexts when oral communication was unavailable. If an official spoke little or none of his counterpart's language, the "solution to this difficulty—indeed the original design for these primarily entertaining encounters—was to converse by writing in hanzi, or kanji, the Chinese characters used by educated men for literary and official communications."[40]

Whereas the very possibility of a "Chinese-Japanese" typewriter derived its conditions of possibility from the long history of this Sinic transnational sphere, the underlying assumptions and preoccupations that motivated this twentieth-century enterprise were starkly different from those found in the translingual exchanges of centuries past. For inventors and engineers, the elision of Chinese, Japanese, and Korean—or "CJK," as it would come to be abbreviated during the second half of the twentieth century—was not motivated by affirmative notions of shared civilization (*tongwen*) or of a "language community defined by their collective competence with the Chinese writing of the brushtalks."[41] It was motivated by the stark, turn-of-the-century logic of the collective technolinguistic *crisis* that Japan and Korea were now understood to face alongside China precisely because of this shared cultural heritage. The nineteenth- and twentieth-century emergence of a radically new and powerful global information order had not only expelled China from the universe of technolinguistic modernity, then, but had also transformed Japan and Korea into unwitting co-participants in China's technolinguistic woes. By virtue of their orthographic heritage—kanji, hanja, and hanzi—Japan, Korea, and China were conjoined in a newly conceived spatial-informational crisis zone. Here, I will refer to this as the *kanjisphere*.

Western perceptions of the Japanese kanji machine further reinforced this notion of a shared zone of technolinguistic crisis. In sharp contrast to the all-kana typewriters described above, which were celebrated globally as Japan's passport to the land of technological modernity, scorn and mockery were heaped upon kanji machines in ways not unlike those we have seen with the Chinese typewriter. In an extensive article for the *New York Times*, for example, Mary Badger Wilson reported that "there are two great languages ... used by many millions of persons, to which our machines cannot be completely adapted. These are the Japanese and Chinese tongues."[42] Writing about a Japanese typewriter she witnessed in operation at the Japanese embassy in Washington, DC, Wilson reveled in what was by then the common idiom of the machine's onerous size, as well as the impressive—if not impossible—demands it placed upon the memory of the operator. "There are some three thousand characters used in the machine," Wilson continued, "and the Japanese typist must memorize the position of each one in order to attain the high speed demanded!"[43] In another article, from 1937, we see more denigrating representations of Chinese and Japanese machines:

Stenographers who think they are overworked should call on Kathleen Tsuchiya at the Japanese chamber of commerce. She pounds out letters in English on an American typewriter then "hunts and pecks" over the 3,500 separate ideographs of a Japanese typewriter to produce a string of hieroglyphics.[44]

As always, however, the most stinging barbs were those of greatest brevity: "Overheard at a literary cocktail party: 'A Russian novel always contains more characters than a Japanese typewriter.'"[45]

If the kanjisphere was defined by the idea of shared crisis, so too was it marked by a powerful and subversive optimism—bringing us to a second vital difference that separates newer and older conceptualizations of the East Asian cultural sphere. Embedded within the practice of eighteenth-century brushtalk was a potent if often unspoken cultural hierarchy, one in which the medium of exchange was itself derived from, and thus privileged, literary Chinese. In the age of the modern kanjisphere, by contrast, China and the written Chinese language enjoyed absolutely no paradigmatic authority. To the contrary, once China and Chinese characters had been reconceptualized as a communicative problem—a puzzle in need of a solution rather than a medium of communicative possibility—this opened up a new, exciting, and lucrative possibility for Japanese and Korean inventors, one in which Japan and Korea could be transformed from the beneficiaries of Chinese cultural inheritance to sites where the puzzle of East Asian technolinguistic modernity might itself be solved. The "Chinese solution" was equally available to, and the purview of, Japanese and Korean engineers—as well as to foreigners working on Japanese and Korean questions from elsewhere on the globe. For inventors like Sugimoto and others, then, the modern kanjisphere was home to an exhilarating prospect: that the solution of one's "own" problem—the creation of a kanji typewriter, or a hanja typewriter—carried with it the lucrative "positive externality" of solving the Chinese puzzle as well, with everything this implied financially, geopolitically, and culturally.

Over the course of the 1920s, Japanese manufacturers competed head-on with Commercial Press, China's foremost center of print capitalism during the era and, as examined in the preceding chapter, manufacturer of China's first commercially successful Chinese typewriter—the Shu-style machine. Tellingly, it was only in this context that Japanese inventors and manufacturers jettisoned the commonly used katakana-inflected loan word "taipuraitā" [タイプライタ] and replaced it with the Chinese-character term *daziji* [打字機], attempting to establish Japanese-built machines as an identical yet superior solution to the puzzle of Chinese typewriting. The potential payoff of such competition should not be understated, moreover. If Japanese companies could succeed where Remington, Underwood, Mergenthaler Linotype, Olivetti, and other Western companies had tried and failed, the market that remained closed to the United States, Italy, Germany, and others might very well open up to Japan. Whereas Japan formed but a small part of the global culture of keyboard typewriting, then, opportunity existed to become the techno-linguistic hegemon of the modern kanjisphere.

Chinese-language typewriters built by Japanese manufacturers were based upon design principles identical to those of their Chinese-built counterparts, Chinese observers would quickly learn. The machines were outfitted with a tray bed matrix of approximately 2,500 characters, arranged into zones of greater and lesser frequency.[46] Abandoning *iroha* organization, moreover, and dispensing with kana symbols, these machines adopted conventional radical-stroke organization, the standard means by which Chinese typewriter tray beds—as well as dictionaries and other reference materials—were organized at the time. Such minor changes notwithstanding, these machines would have struck any observer as effectively identical to the apparatus developed by Commercial Press.

It was not market competition, however, but military might and war that in the end secured Japan's preeminent place in the Chinese type-writer market. On January 28, 1932, Japanese pilots from the Imperial Army flew sorties over the densely populated Zhabei commercial district of Shanghai, dropping six bombs on the offices of Commercial Press, and visiting destruction on virtually the entire facility. The company's machine shop—home to the company's Chinese typewriter division along with other enterprises—was largely spared, but the conflagration assuredly arrested their marketing efforts for a time.[47] Meanwhile, to the north, Japanese military forces consolidated their recent invasion of northeast China through formation of the Japanese-controlled client state of Manchukuo. By force of arms and engineers, Japan set out to become the dominant technolinguistic force in the East Asian kanjisphere.

With the establishment of Manchukuo in particular, the recruitment of secretaries and bureaucrats from mainland Japan quickly followed, as did the establishment of typing institutes. Here, cohorts of Chinese clerks-in-training would be instructed in the use of Japanese typewriters and Japanese-built Chinese typewriters.[48] One such outfit was the Fengtian Typing Professional School, founded circa 1932, where training manuals and curricula resembled those we witnessed in chapter 4, but with certain noteworthy differences.[49] Like their counterparts in Shanghai, Beijing, and other Chinese metropoles to the south, trainees moved through lessons and lexical geometries to help them develop an embodied familiarity with the tray bed and its layout. At the same time, the content of these lessons now reflected a decidedly different political vision—the Japanese vision for Manchukuo. In a 1932 textbook for clerks and secretaries, editor and Fengtian institute affiliate Li Xianyan guided readers through an assortment of the new intra- and interdepartmental forms they could expect to encounter, many of which must have seemed familiar to anyone with prior training. The fourth chapter of Li's textbook was something entirely different, however: a section dedicated to stationery and government forms "for use by the emperor" (*huangdi yong zhi gongwen*). Here in Manchukuo, Chinese typewriters would be used for the first (and only) time in history to write imperial edicts (*zhaoshu*) and imperial rescripts (*chokugo*), in this case on behalf of the Kangde Emperor—better known as Puyi, the child emperor of the Qing who had been deposed following the revolution of 1911, but restored some twenty years later by his Japanese patrons.[50]

Both ethnopolitically and linguistically, typing schools in Manchukuo were complex diagrams of power and conflicting allegiance. Here at the Fengtian institute, and soon in sites across Manchukuo, Chinese typewriters built by Japanese engineers were to be used by Chinese typists, themselves trained in Japanese-sponsored institutes. Their communiqués and memos, in turn, were all in service of the Japanese client-state of

Manchukuo, authored on behalf of, among others, the rehabilitated Manchu emperor of the former Qing dynasty. No doubt aware of the politically charged nature of this complex arrangement, Li Xianyan included a preface in his manual carefully tailored for his Chinese readers. "Every country has its own particular official correspondence, as does each time period," it began. "They evolve in accordance with the condition of the country and with customs."

To that end, in writing a book explaining stationery, one must also follow the times and reform. This is incontrovertible. Typists are personnel whose responsibility it is to record and copy public documents. As such, typists more than anyone must follow the times and learn new forms of stationery. Only then can they fulfill their duties and match the day and age.[51]

Li's was hardly a poetic or impassioned apology for collaboration. Rather, the dryness of his words matched the rather bloodless content of his textbook. Muted though they were, however, Li's comments on stationery and correspondence carried with them an unmistakable message: here in Manchukuo, in this territory carved off of the Chinese nation-state by military force, politics are not what they once were. You are studying to become a secretary in the state of Manchukuo, not China. Typists must follow the times.

PIRACY AND PATRIOTISM: YU BINQI AND HIS CHINESE-JAPANESE-CHINESE MACHINE

Chinese inventors and manufacturers watched as the Chinese typewriter—this hard-won icon of modernity whose tortuous path we have traced out over five decades thus far—steadily became the purview of Japanese multinationals. In 1919, an unnamed contributor to *Shenbao* worried aloud about how this could have happened, laying blame on China's weak patent regime, which left a gateway ajar to Japanese businesses to bring their own "Chinese" machines to market.[52] Come the 1930s, Japan was seizing the market, not only to the north in Manchukuo, but also in major Chinese metropoles. The solution to the puzzle of Chinese typewriting, and the means of modern Chinese textual production more broadly, was falling into the hands of East Asia's emerging power, and the single greatest threat to Chinese sovereignty. What was to be done?

One possible answer came from an unlikely source: a swimmer and Ping-Pong champion by the name of Yu Binqi. Born in Sushan, Zhejiang, in 1901, Yu Binqi went on to graduate from Southeast University of Commerce before pursuing postgraduate training in Japan at the National University of Commerce and the engineering program at Waseda University.[53] Following a brief stint in the military, his professional career took him further into the world of sports and physical education, first as the managing director of the Shanghai Central Stadium, later as a member of the swimming division of the Chinese National Physical Education Federation, and still later as head of the National Ping-Pong Association. Owing perhaps to this long career in sports, and his notoriety as a talented swimmer, Yu also enjoyed something of a heartthrob status, his debonair visage brightening the cover of *Boyfriend* magazine in 1932 (figure 5.4).[54]

5.4 Yu Binqi

Yu Binqi was an amateur inventor as well, patenting creations that included a new model of travel pillow and an economical water heater. His most famous invention, however, was undoubtedly the common usage Chinese typewriter he developed and began manufacturing in the 1930s—a patent that relied on the slightest of adjustments of an existing machine, and which would soon land Yu in political dire straits.[55]

Yu Binqi's son, Yu Shuolin (b. 1925), though only a toddler at the time, recalled memories of the workshop his father established, tucked away in the back room on the second story of their home on Zhoujiazui Road, in the Hongkou district of Shanghai. On the first floor, a meeting room was available to host guests and clients, with a private office in the back. On the second floor of the "foreign-style" or *yangfang* building was a bed-room, and behind the entire structure was a manufacturing workshop outfitted with a kitchen and some dormitory rooms for workers.[56]

In the entrepreneurial style befitting this Shanghai urbanite, Yu Binqi founded his own institute of typing: the Yu Binqi Advanced Chinese Typ-ing and Shorthand Professional Supplementary School—or, in its shorter version, the Yu Binqi Chinese Typing Professional School (*Yu Binqi Zhong-wen dazi zhiye xuexiao*).[57] Classes were held on the first floor of his office, and within a few years the school boasted a small but well-educated staff of five people.[58] Serving as director of Chinese typing was the school's only female member, Jin Shuqing, a graduate of the Zhejiang University School of Agronomy, a former typing instructor at the Nanjing Municipal Professional School, and soon Yu Binqi's mistress.[59] Wang Yi served as director of shorthand, having joined the school in 1935 after graduating from the National Language Shorthand Institute. Wang had also served as a member of the Ministry of Education's National Language Unifica-tion Preparatory Committee.[60]

A typical cohort for Yu's school was roughly ten students. The school encouraged both male and female students to apply, provided that appli-cants possessed a high school–level education or the commensurate level of professional experience. Classes were to be completed over the course of five months, and addressed such subjects as the use of Chinese char-acter indexes, mimeograph, machine repair, and a typing practicum. Tuition varied depending upon one's course of study, with typing classes and shorthand classes each costing thirty yuan. Following graduation, moreover, Yu Binqi and his associates were active in helping students find employment—a cornerstone of his marketing strategy, as with the typing schools we examined in chapter 4. By helping to place his graduates in government posts and private companies, he not only raised the prestige of the institute, but also opened up avenues through which to insinuate the Yu-style machine into the Chinese market. Yu's school boasted an impressive track record in this regard.[61]

At first glance, Yu Binqi would seem to have been China's answer to the Japanese manufacturing threat. Here was a dashing, cosmopolitan urbanite who practically overnight had established himself as a type-writer magnate, competing not only with Japanese typewriter companies but also with his far better funded counterparts at Commercial Press. He developed a multipronged organization—complete with manufacturing, commercial, and pedagogical branches. He was also a charismatic and unrelenting entrepreneur with a flair for flamboyant gestures. When we look closely at the tray bed of the Yu Binqi Chinese typewriter, for exam-ple, we discover that Yu even smuggled his own name into the matrix of common usage characters, embedding his surname *yu* in the matrix at column 69, row 33, and the characters of his name *bin* and *qi* at coor-dinates 61:10 and 56:10. While we can perhaps excuse his inclusion of the character *yu*—a common character on its own—the inclusion of the highly infrequent *bin* and *qi* was nothing short of bravado. That no other typewriter manufacturers would dream of sacrificing precious lexical space for such characters was, indeed, precisely the point—the entrepre-neur's silent expletive to the world, in a move that would anticipate later gestures by coders and their embedded messages in the age of computer programming.

A closer look at Yu Binqi's career reveals a more complicated trajectory than this patriotic story would otherwise suggest, however, one whose Chinese national bona fides becomes less clear the deeper we dig. Yu Binqi called his invention a "Chinese typewriter," and yet a more faith-ful appellation might have been a "slightly modified Japanese machine." Specifically, Yu began studying the H-style Japanese typewriter sometime around 1930, transforming one small component thereof and rechristen-ing it the Yu Binqi Chinese Typewriter. The only part of the original Japa-nese machine to be modified was the "character positioning device," a component of the machine that helped ensure accurate positioning of the character's impression on the printed page. Changes to this component constituted the sole basis of Yu Binqi's successful patent application.[62]

What began as a shrewd business strategy—to pirate a Japanese-made Japanese typewriter in order to compete with Japanese-made *Chinese* typewriters—took an unforeseen and precarious turn for Yu Binqi beginning in the early 1930s. Following the Japanese invasion of northeast China and the bombardment of Shanghai, Yu Binqi became increasingly reluctant to speak to people outside the family about his typewriting enterprise. As his son recalled, guests to the Yu family home were steered clear of the workshop, his father being aware of acute and growing anti-Japanese sentiment in the city. With widespread calls to boycott Japanese goods, any revelation of Yu's secret—that his work was but a retrofitted H-style Japanese machine—could have easily brought undesired attention.

And indeed it did. In 1931, an unnamed informant alerted editors at the widely distributed newspaper *Shenbao* to the questionable origins of Yu Binqi's ostensibly Chinese machine. Flagged as well were the political affiliations of the entrepreneur himself. Although designated a "domestic" product, the informant suggested, Yu Binqi's machine may even have involved secret "collusion" with Japanese merchants.[63] Such an accusation was certain to raise alarms. With the launch of the "Resist Japan, Save the Nation" movement, nationalist consumers had already begun boycotting Japanese products, ranging from fish to coal. The day following the accusation in *Shenbao*, Yu Binqi responded in his own defense. Presenting receipts to the Resist Japan Association, Yu swore that he would willingly die if anyone could prove that he had purchased raw materials from Japan, or had sought out Japanese workers.[64] On November 11, the Standing Committee of the Resist Japan Association met at the offices of the Shanghai Chamber of Commerce to discuss the claims leveled against Yu Binqi, as well as his rebuttal.[65]

January 1932 brought a measure of relief for Yu. *Shenbao* reported that the claims made against him were false, that his typewriter had been patented by the Chinese government in 1930, and that it was a domestic product of high quality that had since been adopted by banks, post offices, and other government institutions.[66] While it remains uncertain how Yu

Binqi managed to secure this positive and definitive response from the paper, Yu squandered no time in cementing his patriotic credentials. In the fall of 1932, he announced his latest technical improvement for the machine, as well as his latest contribution to the causes of domestic Chinese production and resisting Japanese imports: steel character slugs that would replace the lead type on his current typewriter model. Invented in the United States, a September report in *Shenbao* relayed, this new technology was first applied to Chinese typewriters by Japanese, rather than Chinese, inventors. In the process, Japanese merchants had secured large profits, with these more durable and lighter steel slugs producing crisper text. Yu Binqi brought this new technology home, nationalizing it and thus offering Chinese consumers a means of further "resisting" Japanese imports.[67] In the same year, and in the midst of the refugee crisis in China's northeast, the Yu-style Chinese typewriter would be offered at a 10 percent discount, with thirty yuan donated to the northeastern provinces for each unit sold.[68] The company later accelerated the donation process, promising to donate the full thirty yuan on the customer's behalf with only a thirty yuan down payment on a new machine.[69] In the years to follow, Yu continued to donate to national causes, particularly those involving humanitarian crises and natural disasters. In 1935, his company pledged twenty-five yuan to Chinese flood victims for every machine sold between December and the following February.[70]

The strategy paid off, and at a time when Yu needed it most. In the summer of 1933, Yu Binqi appeared to be running short on capital, prompting him to seek collaborations with other Chinese entrepreneurs and factory directors. He made known his desire either to acquire domestic capital support somewhere in the range of ten to twenty thousand yuan, or perhaps to sell his typewriter patent to another domestic inventor. Yu's call was answered the following year, with five Shanghai-based factories joining in the manufacture of Yu-style machines.[71] By the fall of 1934, *Shenbao* ran articles on Yu referring to his impact as a "revolution" in Chinese typewriting.[72] By the close of 1934, the Hongye Company— the national sales agency Yu Binqi worked with—reported selling an average of forty Yu-style machines each month, a figure no doubt helped by Hongye's offer of free training to consumers.[73] *Shenbao* later reported on Yu's successful development of wax duplicating paper, which could be used instead of carbon paper to create upward of one thousand clear copies, when paired with proto-mimeograph machines.[74] In 1936, *Shenbao* went so far as to call Yu's machine "Five Times More Convenient Than Writing and Copying by Hand."[75]

PAPERWORK OF EMPIRE: JAPANESE TYPISTS ON THE CHINESE MAINLAND

By the winter and spring of 1937, it must have seemed to Yu Binqi that he had finally sanitized the technological history of his machine and fortified his own patriotic credentials beyond dispute. In February, Yu was elected president at the Preparatory Meeting of the Chinese Inventors Association—the very association he had a hand in founding (again in true Shanghai entrepreneurial style).[76] He had even found a way to enlist his first passion—athletics—in the service of shoring up his reputation, organizing a Ping-Pong match with fellow table tennis players to raise funds in support of crisis-ridden Suiyuan province.[77]

With the passage of only a few short months, however, everything changed. In July 1937, Japan launched a full-scale invasion of eastern China, unleashing a war that would claim between twenty and twenty-five million lives before its conclusion eight years later. With the fall of Shanghai in November and Nanjing in December, the Chinese Nationalist government retreated to the tri-city complex of Wuhan. Following a brutal and costly campaign for the protection of Wuhan, the city fell in October 1938, forcing the Nationalist government to retreat once again, this time to the city of Chongqing deep in the interior of China.

In the wake of the invasion, Japan took control of ever-larger swaths of China's information infrastructure. This was true not only for the manufacture and sale of Chinese typewriters but indeed for all typewriter models, Western, Chinese, or otherwise. Import statistics from the time period paint a stark picture, charting a trajectory of increasing and soon total domination by Japanese businesses. From 1932 until the close of 1937, the United States had dwarfed all other countries as the leading exporter of typewriters and typewriter parts to China, fulfilling demands for English-language machines by the foreign merchant community, and by English-speaking staffs of foreign concessions. During this same period, Germany ranked a distant second, primarily on the strength of the country's precision engineering. In 1937, this long-established economic pattern began to transform. In the span of just one year, Japan's share of the typewriter import market rocketed, siphoning away market share from the United States each year from the close of 1937 until the beginning of 1941. Following its declaration of war against the United States in 1941, and its simultaneous military occupation of Southeast Asia, Japan achieved near total control over the flow of foreign machines into the Chinese market. American typewriter exports to China plummeted to practically nothing (figure 5.5).[78]

Domination of the Western typewriter import market was but one piece of Japan's emerging hegemony over China's information infrastructure. In the fields of telegraphy and telecommunications, as Daqing Yang has shown, Japan constructed a robust telecommunications network enabling "an unprecedented degree of administrative centralization and

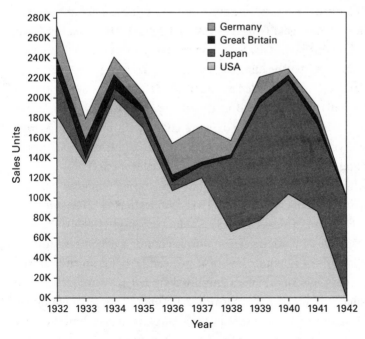

5.5 Typewriter and typewriter part imports to China, 1932–1942

imperial market integration."[79] By 1940, within a few short years of the invasion, Japan was exchanging on the order of twelve million telegrams with parties in occupied China, Manchukuo, and its colonies—a figure ten times larger than the country's combined telegraphic communication with the rest of the world.[80]

China became a vast market for Japanese-language kanji machines, with sales propelled by the bureaucratic demands of empire. Particularly from the fall of 1938 through 1942, during which time Japan shifted from primarily extractive policies toward the attempted creation of a stable colonial regime, tales of patriotic Japanese typists—or *aikoku taipisuto*—began to appear in the Japanese press.[81] On January 4, 1938, readers of the Tokyo *Asahi shinbun* learned of the "Six Patriotic Women" and their arrival in Tianjin, with regular reports coming in as the year unfolded. In 1939, the newspaper followed up with related news of a young typist headed to the South Seas, also in service of her country.[82] Where the Japanese army went, so went these brave young Japanese women, reports explained, risking personal safety to oversee the paperwork of conquest. In the fall issue of *Taipisuto*, one contributor reminisced about her classmate from the Yamagata Girls' Professional School who departed Japan for the Asian continent. Oba Sachiko was her name, and on September 25, 1941, she would be heading off, but not before being toasted at a bittersweet farewell party attended by the head of their division. "I will work very hard!" Oba had exclaimed in terse but potent words. It was a pity to see her go, the author lamented, being such a well-trained and mature typist—and yet it left her with a sense of pride that this classmate would be sacrificing so much in support of the imperial war effort.[83]

The self-sacrificing allure of the patriotic typist achieved perhaps unprecedented notoriety with the 1941 publication of *The Army Typist* (*Jūgun taipisuto*), a novella by Sakurada Tsunehisa (1897–1990) that narrated the tale of a young typist who accompanied Japanese soldiers to Zhangjiakou at the tender age of nineteen.[84] Giving up the security of the Japanese metropole, she committed herself to bearing the perils of Mongolia.[85] In another series featured in *Taipisuto*, a Japanese officer charted his pathway through south China, reveling in his admiration of a Japanese typewriting agency he encountered in an undisclosed location. Coming upon the outfit, he reported, his ears echoed with the nostalgic sound of Japanese typewriting, symbolizing the burgeoning of a new mainland government and the construction of Greater East Asia.[86]

Japanese typing schools were founded in urban centers throughout the mainland occupation zone, Manchukuo, and Taiwan. As one report in *Taipisuto* relayed, by 1941 there was hardly a single company of any size in Taiwan that operated without a Japanese typist, a "typist fever" (*taipisuto netsu*) fed by an annual influx of some five hundred new typing personnel, trained in secondary schools, girls' schools, or institutes associated with the Japanese typewriter company itself. Indeed, the head of the Taibei branch of the Nippon Typewriter Company made known his ambition to have their typewriter "enter every household ... just like in Europe and America."[87]

SAME SCRIPT, SAME RACE, SAME TYPEWRITER: THE WARTIME ORIGINS OF CJK

Japanese imperial expansion was undoubtedly a boon to Japanese typewriter sales. The Japanese-language market paled in comparison, however, to the one that the Nippon Typewriter Company and others stood to gain by investing in the Chinese information technology market—a program they undertook aggressively during the war. By 1942, the company could boast of extensive coverage in China. The company had divisions in Dalian, Xinjing, Fengtian, Anshan, Harbin, Jilin, Jinzhou, Qiqihaer, Shanghai, Beijing, Tianjin, Jinan, Nanjing, Zhangjiakou, Houhe, Taiyuan, Hankou, Jingcheng, and Taibei.[88] Further supported by its divisions in Japan, in cities such as Osaka, Nagoya, Sapporo, Sendai, Niigata, Kanazawa, Shizuoka, Hakodate, Ogura, and Fukui, the Nippon Typewriter Company had taken its place as one of the most important typewriter manufacturers in the world. It had achieved what Remington, Underwood, Olympia, Olivetti, and Mergenthaler Linotype had all failed to achieve, moreover: entrance into, and what is more, domination of, the Chinese-language market. For Commercial Press and Yu Binqi, the decline in market share would have been precipitous.

The flagship machine in Japan's domination of the Chinese market was the Bannō, Wanneng, or "All-Purpose" typewriter, built by the Nippon Typewriter Company. Displayed in a 1940 advertisement from the

Far East Trade Monthly, its verbose yet revealing moniker was the "Japanese, Manchu, Chinese, Mongolian All-Script Typewriter," marking this device as the very materialization of Japan's "Greater East Asian Co-Prosperity Sphere," and its colonial refrains of ethnic harmony and "same script, same race" (*dōbun dōshū*) (figure 5.6).[89] It also marked the first time that a Japanese manufacturer had oriented its machines not only toward the character-based scripts of East Asia, but also the alphabetic scripts of Manchu and Mongolian. The machine quickly became enmeshed within galas spotlighting the ethnic harmony of Manchukuo, including a 1941 "Manchuria-Wide Typing Competition" that brought together typists from the cities of Xinjing, Dalian, Fengtian, Anshan, Benxihu, Mudanjiang, and Harbin.[90]

All of this had a deleterious effect on Chinese manufacturers. Wanneng became the machine of choice and compulsion in China. It quickly displaced both the Shu-style Chinese typewriter manufactured by Commercial Press and the Yu Binqi model. Commercial Press attempted to mount a response with the release of its improved Shu-style machine, featuring a larger platen, the use of an ink ball instead of an ink ribbon, and adjustable print-spacing better suited to alternating between Chinese and Western scripts.[91] Despite such efforts, however, Commercial Press could not compete. As for Yu Binqi, his once celebrated outfit became but a shell of its prior self. While still called a "manufacturing" plant, Yu's

5.6 Advertisement for Japanese-made "All-Purpose" typewriter (with Japanese, Manchu, Chinese, and Mongolian)

factory had neither the capital nor the market to continue building or selling typewriters. Instead, the plant eked out an existence by carrying out small-scale repairs, offering typing services, and melting down metal type.[92] As for Yu himself, he seems to have taken refuge in the world of athletics, appearing in sporadic newspaper reports over the course of the war in connection with various sporting competitions and newly formed athletic organizations. Whether he was granting medals at the Shanghai Ping-Pong competition in 1943, or serving as chief timekeeper at the 39th Japanese Navy track and field competition in May 1944, Yu's life had changed remarkably, as did those of countless others during the war.[93]

ON COMPLICITY AND OPPORTUNITY: CHINESE TYPISTS UNDER JAPANESE OCCUPATION

Without manufacturing alternatives of their own, Chinese typewriter companies had little choice other than to collaborate. As Parks Coble and Timothy Brook have shown, Chinese capitalists operating under Japanese control were less riveted by concerns of patriotism and resistance, and more by the survival of their family operations. Rare were the cases that matched the "heroic nationalist narrative" celebrated in postwar Chinese histories of the era.[94] Only a minority of Chinese capitalists fled Japanese-occupied territories and relocated westward in the path of the Nationalist government-in-exile. The majority who remained, collaborators and non-collaborators alike, worried deeply about rebuilding China's ruined economy, reconstructing the industrial and agricultural sectors, and repairing the country's tax system.[95]

For those in the business of Chinese typewriters in particular, this translated into a variety of economic activities through which to subsist and survive. For example, Chinese typewriters continued to require cleaning during the occupation, services provided by companies such as the C.Y. Chao Typewriting Maintenance Department.[96] Captured within the complicities of the era—Timothy Brook's apt choice of words encapsulating the entanglements between Japanese occupiers, Chinese collaborators, Chinese non-collaborators, and outright resisters—Chinese companies and businessmen serviced the Sino-Japanese clerical world throughout the war, whether the Huanqiu Chinese Typewriter Manufacturing

Company, the Chang Yah Kee Typewriter Company, or the Ming Kee Typewriter Company, among many others.[97]

Facing a shifting political and linguistic context, meanwhile, Chinese typewriter manufacturers stole a page from the Nippon Typewriter Company playbook and began to emphasize the linguistic ambidexterity of their own machines. China Standard Typewriter Manufacturing Company offered a selection of bilingual machines, such as its "Standard Horizontal-Vertical-style Chinese-Japanese typewriter." The capacity to handle both Chinese and Japanese became a selling point of the utmost importance, not only because of the growing Japanese-language bureaucracy, but also to stem the tide of a new threat faced by Chinese companies: Japanese kanji typewriters being retrofitted to handle Chinese, simply by emptying their tray bed matrices and re-outfitting them with Chinese-character slugs.[98] In August 1943, the director of public works in Shanghai received requests for funds to purchase two Chinese machines. While permitting the allocation, the office of the treasurer added: "It should also not be overlooked that these Japanese typewriters could also be used for correspondence in Chinese."[99] In a separate request for additional Chinese typewriters, the response likewise emphasized the potential of converting Japanese machines: "Two Japanese typewriters in the General Affairs Department can also be used to type Chinese after some changes are made in the types. These three typewriters may serve the purpose for the time being."[100]

The complex wartime-era complicities and opportunities were particularly pronounced in the classroom. Chinese typing institutes proliferated during the occupation, sites in which groups of Chinese instructors trained increasing numbers of Chinese students to form a clerical workforce versed in the new technology. At one level, these institutes were spaces of opportunity, possibility, and social mobility—spaces in which young women and men gathered and, in a relatively short span of time and at costs more attainable than extensive formal education, attempted to position themselves for white-collar employment.

The cultures of these small-scale occupation-era institutes are challenging to re-create from available source materials, and yet evidentiary glimpses offer us certain interpretive possibilities. What the archival record bears out is that each of the many typing institutes operating

during the war needs to be understood as an intimate and even exciting place in which young Chinese women and men mixed, and where
many committed themselves to exercising some small measure of control
over their own futures and livelihoods within the confines of this chaotic, destructive period. In Beijing, for example, the Guangde Chinese
Typewriting Supplementary School was founded circa 1938, the year following the outbreak of war. Overseen by twenty-seven-year-old Anfeng
county native and Art Institute of National Beiping University graduate
Wei Geng, one cohort of students in 1938 comprised seventeen young
women and thirteen young men. Female members of the class ranged
in age from sixteen to twenty-seven, with a median age of nineteen.
Whereas the majority of female students boasted a middle-school education, the spectrum varied widely: from those with no more than a lower-
school education, to one lone graduate of the French Catholic Yu Chen
Women's Normal School, founded in 1817 in the present-day Xicheng
district of the city.[101] As for the thirteen men who joined this same 1938
cohort, they were spread across a similar age spectrum, and with similar educational backgrounds. Ranging from seventeen to twenty, most of
the male students in this class arrived with a middle-school education,
flanked by a handful of participants with lesser or greater educational
background.[102]

It was not uncommon for students to continue their studies for more
than one year—well beyond the typical three-month timeframe—sugges-
tive of a practice and a strategy geared toward weathering the economic
uncertainties of the era, or perhaps developing a sense of continuity in an
otherwise chaotic time. In the Jiyang Chinese Typewriting Supplementary School, also in Beijing, one cohort in 1937 comprised eight female
and four male students. These two groups ranged widely in terms of age,
with female members of the class ranging from seventeen to twenty-
seven and male students ranging from eighteen to twenty-eight. Both
groups of students exhibited identical educational backgrounds, with all
but one student boasting a middle-school education. Upon concluding
their training in 1937, all but one member of this cohort reenrolled in
the class in 1938, joined by seventeen new students. While their motivations fall well beyond our grasp, nevertheless the timing of this collective
continuation prompts us to consider whether professional schools such

as Jiyang offered refuge in this tumultuous period in the immediate wake of the Japanese invasion, as well as perhaps a way of maintaining a sense of continued professional identity.[103] Whether they were motivated by these or other factors, nevertheless it is striking to contemplate what bonds must have formed between these eleven continuing students when they reconvened one year later.

At the same time, each of these Chinese institutes also constituted a politically compromised site. Students trained on Japanese-built machines. They worked under the guidance of instructors with greater and lesser ties to Japan. And in the most optimistic of scenarios, they aspired to find employment in the collaborationist government, or in a private sector itself permeated by Japanese interests. At the Guangde school in Beijing, students trained on one of two Japanese-built Chinese typewriters: the Wanneng-style Chinese typewriter, or the Standard-style Japanese typewriter.[104] At the East Asia Japanese-Chinese Typewriting Professional Supplementary School, founded by thirty-eight-year-old Sheng Yaozhang no later than December 1938, students likewise trained on one of four Japanese-built Chinese machines: the Suganuma-style Chinese typewriter, the Wanneng-style Chinese typewriter, the Horizontal-style Chinese typewriter, or the Standard-style Chinese typewriter.[105] Meanwhile, students at the Jiyang Chinese Typewriting Supplementary School in Beijing also worked exclusively with Japanese equipment, studying the Chinese Typewriting Textbook published by the Nippon Typewriter Company for use with the company's Japanese-built Chinese machines.[106]

Like the textbooks and apparatuses used in such schools, teaching staff also had direct and indirect connections to Japan. At the Yucai Chinese Typing School, for example, twenty-seven-year-old principal and Shaoxing native Zhou Yaru was herself a graduate of the East Asia Japanese-Chinese Typing School, and a former typist at Nippon-China Trade, Ltd.[107] In the Xizhimen district of Beijing, twenty-three-year-old Li Youtang oversaw the Baoshan Chinese Typewriting Supplementary School, formed sometime around October 1938. As a graduate of the Tenrikyō School in Japan, Li returned to Beijing to teach at the Beijing Japanese Tenrikyō Association. Shortly thereafter, he teamed up with an associate, surnamed Li, who like him had also come up through Japanese educational circles, graduating from a Japanese language school.[108] Together, the two formed Baoshan one year after the outbreak of full-scale war.

Bringing this history full circle, as it were, was the East Asia Japanese-Chinese Typewriting Supplementary School and its founder Sheng Yaozhang. Hailing from Lianyang county in Fengtian province, Sheng was himself a graduate of the Fengtian Japanese-Chinese Typing Institute, part of the same network encountered above, where in 1932 the instructor Li Xianyan first set down on paper his apology for clerical collaborationism.[109] Scarcely could Li have known the full meaning his statement would ultimately take on: *typists more than anyone must follow the times.*

By 1940, Chinese typewriting had entered a period marked by deep contradictions. As an object and a commodity, the Chinese typewriter itself was thriving, backed by a more formidable manufacturing and marketing network than ever before. As a symbol of modernity, the status of the machine had never before reached such heights, its identity as a technolinguistic advance becoming stabilized, at least in China, for arguably the first time in its history. The ends to which this symbol of technolinguistic modernity were now put, however, diverged sharply from those first imagined by Zhou Houkun, Qi Xuan, Shu Zhendong, and the executives at Commercial Press. This thriving network was created and managed by Japanese multinationals, calling into question the ostensibly Chinese identity of the machine. This symbol had now become enrolled into—perhaps even aligned with—the violence-laden ambitions of Japan's multinational, multilingual empire.

COPYING JAPAN TO SAVE CHINA: THE DOUBLE PIGEON MACHINE

The summer of 1945 witnessed the horrific and precipitous conclusion of the Second World War. With Japanese urban areas now within range of Allied bombers, the winter and spring months witnessed large-scale saturation bombings of metropolitan areas, including the devastating March firebombing of Tokyo. Over the course of the two-day attack, the Allied firebombing resulted in the deaths of an estimated one hundred thousand people. In May, the fall of Berlin and the Nazi surrender precipitated the denouement of the European conflict, freeing the Soviet Union and

Allied forces to concentrate attention more fully on the Pacific theater. On August 6, the United States released the first of two atomic bombs, obliterating the city of Hiroshima and killing somewhere in the order of 90,000 to 160,000 inhabitants. Two days later, the Soviet Union declared war on Japan, opening up a new and dangerous front for the beleaguered Japanese Imperial Army. This was followed on August 9 by the second atomic attack—this time against the city of Nagasaki. On August 15, Japan announced unconditional surrender.

The surrender of Japanese forces precipitated a massive repatriation process that, as Lori Watt has examined, would see nearly seven million Japanese nationals leave China, Manchukuo, and the former colonies.[110] On the Chinese mainland, the communities and economies they had once occupied, and now left behind, were devastated. The Eight-Year War of Resistance Against Japan, as it would come to be known, left behind a Chinese economy in shambles.

Only in the immediate postwar period were Chinese typewriter manufacturers able to regain control of the market. Even this "recovery," however, was far from a straightforward story. In the wake of the Second World War, what had once been the strategy of the lone, debonair athlete-turned-inventor Yu Binqi soon became the collective strategy of the entire Chinese typewriting industry. One by one, Chinese businessmen who had once struggled against the Wanneng-style machine simply began to copy it or sell it directly—all while quietly omitting its Japanese past. Many of these copycat efforts were undertaken by Chinese businessmen who had come of age under Yu Binqi, perhaps in fact inspired by his example. In the late 1940s, Yu's former employee Chen Changgeng resurfaced to open his own typewriter manufacturing plant. Based in Shanghai, this plant would sell the "People's Welfare Typewriter" (*Minsheng daziji*)—the company name lifted directly from the pages of Sun Yat-sen's "Three People's Principles." On the cover of Chen's typing manual, however, the apparatus shown was none other than the Japanese-built Wanneng machine. Perhaps having removed the faceplate reading "Nippon Typewriter Company," supplanting it with one reading "Minsheng," Chen's enterprise was nevertheless premised on the postwar seizure of Japanese typewriter manufacturing and its repackaging as a domestic Chinese product. Yu Binqi had taught Chen well (figure 5.7).[111]

5.7 The "People's Welfare Typewriter"—a Wanneng duplicate

Chen Changgeng was not the only entrepreneur to stake his postwar fortunes on the seizure and Sinicization of the Wanneng machine. In 1949, yet another associate in Yu Binqi's network began to traffic in what he called the "Mr. Fan Wanneng Chinese Typewriter." Fan Jiling, himself a graduate of the Yu Binqi training institute many years earlier, did not attempt to change the name of the machine, and yet he wrote about Wanneng in words that avoided mention of its wartime, Japanese origins.[112] "Ever since the promotion of the Wanneng-style Chinese typewriter," Fan explained in his textbook, "its superior design and manufacture led it to become common everywhere in a matter of a few years, with users singing its praises. Other styles of machine, with their clumsy shapes, gradually became obsolete."

The most effective copycat of all, however, was the newly formed Chinese Communist regime itself, which began to seize Japanese typewriting interests within a few short years of the 1949 revolution and convert them into Chinese-owned enterprises. In 1951, the Tianjin Public Industry Authority took control of the Japanese Typewriter Company, reorganizing it as the Red Star Typewriter Company—a name that, like "People's Welfare" before it, was ideologically appropriate and properly patriotic.

Even with the widespread seizure and nationalization of Japanese outfits, and steep limitations placed on imported machines, the Chinese state and business community could not fully stem the tide of Japanese influence in the domestic typewriter market. In Tianjin, the primary focus of the newly nationalized Red Star Typewriter Company remained the import of Japanese-made typewriters and calculators from Japan. In 1951, by one estimate, more than 4,000 typewriters and calculators were imported, mainly from Japan. "If import statistics from around the country are tallied," the report read, "the loss to the national economy is indeed staggering."[113] "For those of us in the typewriter manufacturing industry, here upon the people's stage of the motherland, this has caused us immeasurable anguish and humiliation."[114]

Beginning in the 1950s, the domestic Chinese typewriter industry partnered with the new regime to mount a coordinated response to Japanese market dominance, consolidating the highly diverse and fragmented network of Chinese companies into larger conglomerates.[115] Ten separate Chinese typewriter companies set out to form what would be known as the Shanghai Chinese Typewriter Manufacturers Association. Han Zong-hai of the Yu-Style Chinese Typewriter Company, Tao Minzhi of the Wenhua Chinese Typewriter Company, Tong Lisheng of the Jingyi Type-writer Company, Hu Zhixiang of the China Typewriter Company, Chen Changgeng of the Minsheng Chinese Typewriter Company, and other associates convened to determine how the merger would take place.[116] Headquartered at 7 Tianjin Road, the consortium would be directed by Han Zonghai, Li Zhaofeng, and Hu Zhixiang.[117] It was this consortium that would go on to build what would become the emblematic typewriter of the People's Republic: the Double Pigeon Chinese typewriter.

In developing the Double Pigeon, the Shanghai Calculator and Type-writer Company would apply the same policy of imitation, or *fangzhi*, as Yu Binqi, Chen Changgeng, Fan Jiling, and others—only this time, they would do so on a national scale and with the backing of the Chinese state. The designers of Double Pigeon explicitly based this machine upon the Japanese-made Wanneng machine.[118] Three phases marked the devel-opment of the Double Pigeon machine. From July to November 1962, the team fashioned and tested four prototypes. From July to November 1963, it repeated this process for an additional forty machines. From January to March 1964, the team made further revisions to the machine, then sub-jected the resulting modified prototypes to further testing.[119] On March 25, 1964, the resulting machine was presented at a conference to a group of company representatives, including those from the Shanghai Machin-ery Import-Export Company and the Shanghai Typewriter market (figure 5.8).[120] "The Double Pigeon DHY-model Chinese typewriter," internal reports explained in blunt terms, "is an alteration of the Wanneng-style Chinese Typewriter."[121]

Like others before them, the Shanghai Calculator and Typewriter Fac-tory and its state patrons soon quietly forgot the Japanese origins of their Chinese machine, and the broader history of Japan's wartime domina-tion of Chinese information technology. Instead, the Japanese-built Wan-neng Chinese typewriter was reconfigured and quietly resurrected as the Chinese-built Double Pigeon machine.

Thinking back to Evelyn Tai and the day she purchased her typewriter at the shop in Singapore, it turns out that the choice she made—between the Japanese-built Superwriter and the Chinese-built Double Pigeon—was

5.8 The Double Pigeon Chinese typewriter

far less stark than I had originally imagined. The history of East Asian information technology during the first half of the twentieth century—and particularly from the 1920s through the 1960s—blurred the lines that otherwise might have demarcated our story into easily discernible national categories. The Superwriter was a Japanese-built apparatus, to be sure, and yet the linguistic and mechanical principles underlying its design, as well as the motivations propelling its development, were inseparable from the deeper history of Chinese typewriting we have examined thus far.

As for the Double Pigeon machine, the lines were hazy as well. Although it had been built by a consortium of Chinese manufacturers, businesspeople, and state authorities, and although it was to become the iconic typewriter of the Maoist period, the history of this machine was itself inseparable from the history of Japanese occupation, the Japanese seizure of China's domestic typewriter market, and Wanneng. For my own part, I quickly realized that the pale green machine that sat atop my dresser at home—ostensibly the most Chinese of all Chinese typewriters—would never again look quite the same.

The Double Pigeon will play a central role in the story to come in mainland China, when we witness how Mao-era typists reimagined it and other typewriters in ways that engineers had not anticipated nor even believed possible. But first we cross the ocean to the United States one last time, to investigate experiments being undertaken in the Manhattan studio of bestselling author, linguist, cultural ambassador, and typewriter inventor Lin Yutang. These experiments, as we will see, would transform the history of modern Chinese information technology forever, giving rise to an entirely new relationship between humans, machines, and language.

6

QWERTY IS DEAD! LONG LIVE QWERTY!

Lin Yutang Invents Chinese Typewriter: Will Do in an Hour What Now Takes a Day
—*New York Herald Tribune*, August 22, 1947

Even though it cost 120,000 US dollars, even though it has saddled us with a lifetime of debt, this creation of my father that he had worked his entire life on, this newborn baby birthed with such difficulty, it was worth it.
—Lin Taiyi, daughter of Lin Yutang

I think this is it.
—Zhao Yuanren on the MingKwai Chinese typewriter, 1948

The Computer History Museum in Mountain View, California, is a temple to technophilia. Twenty exhibit halls and a world-class collection of artifacts chart the history of calculators, punched cards, programming, memory, graphics, and the web, among many other domains of computing. Prominently featured is the UNIVAC I, the Cray-2 supercomputer, and, perhaps most exquisite of all, Difference Engine No. 2—a faithful construction of the unrealized masterpiece by Charles Babbage.

After moving through the first five centuries of the collection, one reaches the Input/Output exhibit hall. It is a cabinet of curiosities in the form of a wearable keyboard "glove," early and now forgotten prototypes of the mouse, and dozens of other fantastical objects. One revealing bit of signage in the Input/Output hall is so modest that it almost escapes notice. It reads:

Keyboards: With so many new elements to create, computer designers were happy not to re-invent text-based input and output. They used existing tele-types and automatic typewriters—including the established QWERTY keyboard.

One can hear a sigh of relief in this explanation, a rare moment of repose in a narrative of computing history that otherwise prides itself on "disruption" and unbroken unrest—an industry that has profited precisely because it does *not leave well enough alone*. Nevertheless, within this wider, agitated history, at least the QWERTY keyboard did not need to be revisited—thank heaven.

For the historian of China, this sign captures a fundamental divergence between the history of IT as it is presented in Silicon Valley and the history of IT examined in this book. In the Valley, the question of text input and output is treated as relatively straightforward—perhaps even uninteresting—particularly when juxtaposed against "hard" questions like stored-program computing, magnetic core memory, network protocols, and more. Imagine for a moment, however, what a museum of Chinese information technology might look like. Much of it would be identical, of course, paying necessary homage to the shared genealogy and common heritage of early calculation, magnetic tape, Ethernet cables, and the like. And, as in Mountain View, there would certainly be an abacus. The Input/Output exhibit, however, would need to change drastically. Unlike the designers of computer systems in the West, who may have been "happy *not* to re-invent text-based input and output," those working in the allied fields of Chinese information technology have had no choice but to regard input and output as among their most sophisticated challenges. No history of Chinese information technology would be possible without dedicating ample space and attention to a question for which no sigh of relief or well-earned complacency is available. In China, the question of text input and output would need to take center stage.

For many readers it will be surprising to learn that computers in China look exactly the same as those in the United States, right down to the QWERTY keyboard. Were you to seat yourself at a café in Beijing,

Shanghai, or Chengdu, the entrepreneurial millennials you would undoubtedly encounter would be hard at work on QWERTY devices. The QWERTY keyboard in China is not what it seems, however.

In the alphabetic world, QWERTY keyboards are used within a *what-you-type-is-what-you-get* framework. When striking the keys marked T-Y-P-E-W-R-I-T-E-R, one expects these same symbols to appear on the screen. In most other language contexts, the same holds true: to depress a key marked "щ" or "ж," and to have this same symbol then appear on the screen, is the unspoken assumption of computing in most of the world. By virtue of China's position within the global history of modern information technology, however—the position we have been charting out in this study—QWERTY is not and cannot be used in this *what-you-type-is-what-you-get* way. Instead, the QWERTY keyboard is used in China in the context of "input," or *shuru*, a form of human-machine interaction that has been fundamental to both Chinese computing and word processing from the 1950s onward. In contrast to "typing" in the rest of the world, where users assume this one-to-one correspondence between the symbols-upon-the-keys and the symbols-upon-the-screen, Chinese "input" assumes no such identity. If this sounds surprising, we would do well to remember that the keys upon which the symbols "Q," "W," "E," "R," "T," and "Y" appear are simply *actuators*—they are *switches* not unlike those we use to ring doorbells or turn on light fixtures. That the closing of these particular switches should, in most of the world, result in the near-instantaneous appearance of these same symbols on one's screen is the result of a complex cascade of mediations that none but a very few could readily explain. Indeed, batteries of highly trained and well-paid engineers and designers have done their best to hide this process, to cloak this mediation so that it takes place "behind the blip," to invoke Matthew Fuller's concept.[1]

In Chinese computing, to close the switch marked "Q" might invoke the corresponding Latin letter under certain conditions, but more likely it will be used to provide an *instruction* or *criterion* to a piece of software known as an "input method editor" (IME). IMEs operate in the background on every computer in China, intercepting the user's QWERTY keystrokes and, on the basis of these instructions, presenting a menu of Chinese character candidates on the computer screen from which the user can choose. No matter if one is composing a document in Microsoft Word, surfing the web, or otherwise, he or she is constantly engaged in this iterative process of criteria, candidacy, and confirmation. The fundamentally mediated nature of text input is thus never concealed in Chinese computing. At all times, mediation remains explicit, visible, and available to the conscious mind—*before* the blip. The "code consciousness" we first encountered in the context of Chinese telegraphy is alive and well.

To help pin down how input works, let us consider how one would use the QWERTY keyboard to produce a simple three-character sequence using one of China's leading IMEs, the Sougou pinyin input method: 打字机 (*daziji*/"typewriter") (figure 6.1).

Since the Sougou system relies upon phonetic pinyin input, the first key we would strike is "D": the first letter in the phonetic value of the first character, *da*. As soon as the input system intercepts this very first letter, a pop-up window appears on the screen, and the system sets into motion. Searching through its database of characters, it offers up Chinese characters whose phonetic value begins with "D," ranked in order of frequency. First in line is the possessive particle *de* (的), one of the most common characters in the entire Chinese language. The second option in the pop-up menu is *dou* (都), meaning "all" or "every"—another highly common term. In the third position is the character we want: *da* (打), meaning "to strike" or "to hit." At this point, if we wish, we do not need to enter the second letter "A," and instead can select our first desired character by depressing the number key "3" (indicating that we want the third candidate in the pop-up menu of suggestions). With the depression of just two keys, then, we have the first of our three characters.

This is not always the case, of course. Were we searching for a less common term than *da* (打), perhaps the homophonic character *da* (沓), meaning "repeatedly," it might be necessary to complete the phonetic spelling of our desired term, in this case by entering the second letter "A." As soon as the IME intercepts this second letter, the menu of candidates begins to fluctuate again, the IME reshuffling its suggestions and limiting them now to those characters whose pronunciation begins with "da," again ranked in terms of frequency, and from which we would then locate and select our intended graph. This is the nature of input: a recursive and dynamically changing process in which the IME offers up a fluctuating and ever more refined list of potential Chinese characters until the user finds the ones he or she wants.

Upon entering the third letter in our sequence—the letter "Z," which forms the initial phonetic value of our second character *zi* (字), meaning "character"—the input process becomes even more sophisticated. No longer seeking out standalone characters in isolation, now the IME begins to search for multiple-character sequences and compound words. Upon depressing "Z," the input system reloads its pop-up menu with a set of the highest frequency two-character compounds *whose first character has the phonetic value of* da, *and whose second character begins with the phonetic value of* "Z." Atop this list are the first two characters we want: *dazi* (打字), meaning "to type." (Other options in the menu include *dazhong* (大众), meaning "the masses," and *dazhe* (打折), meaning "discount.") To complete the input process, one need only continue these steps with our third and final character *ji* (机), meaning "machine."

As indicated above, the complete phonetic spelling of one's desired Chinese characters is only one of the many possible ways to input—and the longest and slowest way at that. Alongside *d-a-z-i-j-i*-1, there are at least a half dozen other ways to enter this three-character sequence within Sougou input, each one employing different input techniques, and yet all resulting in the same screen output.

d a z i # j #
d a z i j #
d a z # j #
d a z j #
d z j i #
d z i j #
d z j #

For the shortest of these—d z j #—this means that a four-stroke QWERTY keyboard sequence is capable of producing a Chinese term that, when translated into English, requires a total of ten: *t-y-p-e-w-r-i-t-e-r*. Something has clearly changed.

Sougou pinyin is only one of many Chinese IMEs on the market, moreover. Google offers its own IME, as do Apple and QQ. There are also IMEs that are not based on pinyin or phonetic input at all, but instead let operators use the letters of the QWERTY keyboard to represent *structural*

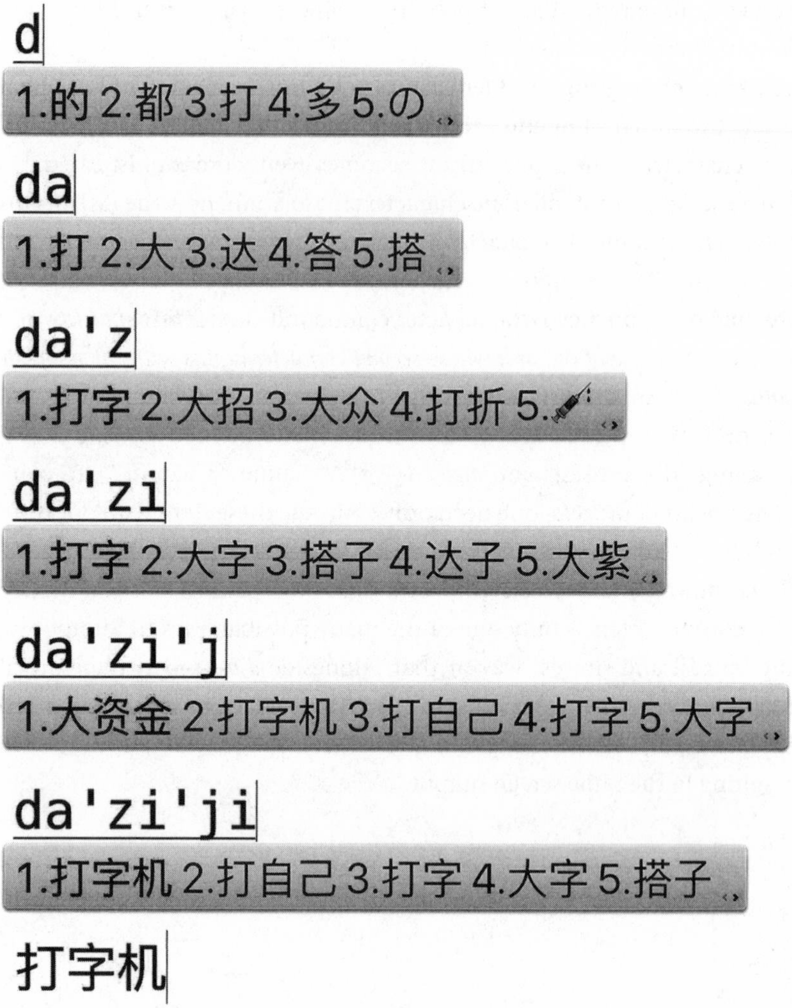

d

1.的 2.都 3.打 4.多 5.の

da

1.打 2.大 3.达 4.答 5.搭

da'z

1.打字 2.大招 3.大众 4.打折 5.💉

da'zi

1.打字 2.大字 3.搭子 4.达子 5.大紫

da'zi'j

1.大资金 2.打字机 3.打自己 4.打字 5.大字

da'zi'ji

1.打字机 2.打自己 3.打字 4.大字 5.搭子

打字机

6.1 Entering "typewriter" (*daziji* 打字机) using Sougou Chinese input

properties of the desired character, such as the radicals or stroke forms out of which a given character is composed.[2] Thus, while the "H" key on one's QWERTY keyboard signifies the eponymous consonantal value, it can also be used to refer to the "tree-radical" (木) when used with structure-based IMEs. Instead of producing a menu of characters that begin with the value "H," then, such a technique would instead result in a menu of characters containing this particular radical. Such input systems are particularly popular among older Chinese dialect speakers who, while fluent in Cantonese or Fujianese, might not be proficient in the standard Chinese pronunciations upon which pinyin input is premised. To enter this same three-character passage for "typewriter" using Cangjie input, for example, the input keylog would read *qmnjnddhn*, each of these letters describing a specific *graphic* feature of the three desired characters—not their pronunciations.

With five commonly used Chinese input method editors, and literally hundreds if not thousands of experimental IMEs developed by independent designers, there exist dozens if not hundreds of different ways for Chinese computer users to input just this one simple three-character sequence. Were we to imagine longer Chinese-language passages, extending into the hundreds and thousands of characters, the number of potential input sequences would quickly become staggeringly large.

By this point, readers will no doubt have noticed a conspicuous absence in the foregoing narrative: Where has the Chinese typewriter gone? Is this perhaps the moment in our history when we at last abandon this hopeless device and turn our attention to the true mechanical savior of the Chinese script in the modern period—the *deus ex machina* that is the personal computer, which, as many have assumed, rescued Chinese characters from the "abyss" separating alphabetic and nonalphabetic script? In fact, the story of this chapter is precisely the opposite: The birth of input—this revolutionary new mode of human-computer interaction upon which modern-day China has laid the foundations of the world's largest IT market and a vibrant social media environment—had nothing to do with computing at all. The first input system in history was, in fact, an experimental Chinese typewriter debuted to the world in the 1940s— the first Chinese typewriter in history to boast a keyboard.

THE UNCANNY KEYBOARD

"We learn with mingled emotions—transcending dismay and yet appreciably milder than despair—that Dr. Lin Yutang, our favorite oriental author ... has invented a Chinese typewriter."[3] So began a 1945 article in the *Chicago Daily Tribune* that revealed to an American reading public the quixotic new pursuit of a celebrated cultural commentator, and beloved author of the bestselling titles *My Country and My People* (1935) and *The Importance of Living* (1937). So allergic were they to this startling news, the authors explained, that at first they simply did not believe it. The news would have been "incredible," they stressed, if it had not come directly from Lin's publisher. "Seeking additional enlightenment," the reporters continued, "we consulted our laundryman, Ho Sin Liu."

"Tell us, Ho, about how big would a Chinese typewriter need to be to cover the whole range of your delightful tongue?"

"Ho, ho!" Ho replied, wittily punning his name into an English exclamation point. ... How indeed shall I answer such an interrogation unless with another? Have you seen the Boulder dam?"[4]

Lin Yutang was born in 1895 in Fujian province, the same year that Taiwan was lost to Japan following the humiliation of the first Sino-Japanese War. Raised in a Christian household, Lin entered St. John's University in Shanghai in 1911, the year a republican revolution delivered the death blow to an already weakened Qing dynasty. His educational career was marked by distinction, continuing at Tsinghua University from 1916 to 1919, and then Harvard in 1919 and 1920. By forty years of age, Lin was a celebrated author in the United States and beyond, becoming one of the most influential cultural commentators on China of his generation.

Years before his breakout English-language debut, Lin Yutang began to contemplate a question that, as we have seen now on more than one occasion, exerted a magnetic pull on the minds of many: the question of how to develop a typewriter for the Chinese language that could achieve the scope and reputation of its Western counterpart. With these inspirations, Lin set off down a path that many years later would lead to perhaps the most well-known, but also most poorly understood, Chinese typewriter in history: the "MingKwai" or "Clear and Fast" Chinese typewriter, announced to the world starting in the mid-1940s.

When MingKwai made its first appearance, the writer at the *Chicago Daily Tribune* and his "laundryman" would be proven wrong: MingKwai was *considerably* smaller than the Boulder Dam. In fact, it looked uncannily like a "real typewriter." Measuring fourteen inches wide, eighteen inches deep, and nine inches high, the machine was only slightly larger than common Western typewriter models of the day.[5] More notably, MingKwai was the first Chinese typewriter to possess the sine qua non of typewriting, a *keyboard*. Finally, it would seem, the Chinese language had joined the rest of the world by creating a typewriter *just like ours*.

MingKwai may have looked like a conventional typewriter, yet its behavior would have quickly confounded anyone who sat down to try it out. Upon depressing one of the machine's seventy-two keys, the machine's internal gears would move, and yet nothing would appear on the paper's surface—not right away, at least. Depressing a second key, the gears would move again, yet still without any output on the page. With this second keystroke, however, something curious would happen: eight Chinese characters would appear, not on the printed paper, but in a special viewfinder built into the machine's chassis. Only with the depression of a third key—specifically one of the machine's eight number keys—would a Chinese graph finally be imprinted on the page.

Three keystrokes, one impression. What in the world was going on?

What is more, the Chinese graph that appeared on the page would bear no direct, one-to-one relationship with any of the symbols on the keys depressed during the three-part sequence. What kind of typewriter was this, that looked so uncannily like the real thing, and yet behaved so strangely?

If the fundamental and unspoken assumption of Western-style typewriting was the assumption of *correspondence*—that the depression of a key would result in the impression of the corresponding symbol upon the typewritten page—MingKwai was something altogether different. Uncanny in its resemblance to a standard Remington- or Olivetti-style device, MingKwai was not a typewriter in this conventional sense, but a device designed primarily for the *retrieval* of Chinese characters. The *inscription* of these characters, while of course necessary, was nonetheless secondary. The depression of keys did not result in the inscription of corresponding symbols, according to the classic *what-you-type-is-what-you-get* convention, but instead served as steps in the process of finding one's desired Chinese characters from within the machine's mechanical hard drive, and *then* inscribing them on the page.

The machine worked as follows. Seated before the device, an operator would see seventy-two keys divided into three banks: upper keys, lower keys, and eight number keys (figure 6.2). First, the depression of one of the thirty-six upper keys triggered movement and rotation of the machine's internal gears and type complex—a mechanical array of Chinese character graphs contained inside the machine's chassis, out of view of the typist. The depression of a second key—one of the twenty-eight lower keys—initiated a second round of shifts and repositionings within the machine, now bringing a cluster of eight Chinese characters into view within a small window on the machine—a viewfinder Lin Yutang called his "Magic Eye."[6] Depending upon which of these characters one wanted—one through eight—the operator then depressed one of the number keys to complete the selection process and imprint the desired character on the page.

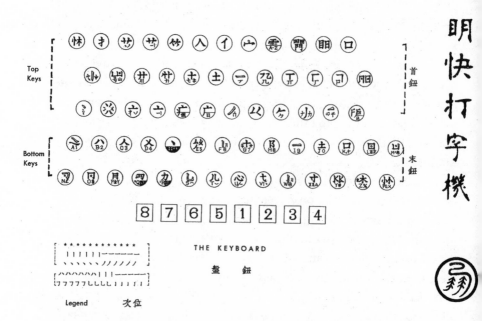

6.2 Keyboard of the MingKwai Chinese typewriter

In creating the MingKwai typewriter, then, Lin Yutang had not only invented a machine that departed from the likes of Remington and Underwood, but so too from the approaches to Chinese typewriting put forth by Zhou Houkun, Shu Zhendong, Qi Xuan, and Robert McKean Jones. Lin invented a machine, indeed, that altered the very act of mechanical inscription itself by transforming inscription into a process of *searching*. The MingKwai Chinese typewriter combined "search" and "writing" for arguably the first time in history, anticipating a human-computer interaction now referred to as *input*, or *shuru* in Chinese.

KANGXI MUST DIE: THE "CHARACTER RETRIEVAL PROBLEM" IN REPUBLICAN CHINA AND THE ORIGINS OF INPUT

Deferring for a moment a more detailed examination of the MingKwai's internal mechanics—of how a three-key sequence resulted in the printing of a Chinese character—most important for us at the outset is to understand the unique historical genealogy of MingKwai as compared to other typewriters in our story—for it is in this history that we uncover the origins of input or *shuru*. When Lin Yutang set out to design his typewriter, he did not draw his inspiration from movable type, telegraphy, or even Western typewriting, but from a language reform debate known as the "character retrieval problem" (*jianzifa wenti*) taking place in China from the 1910s through the 1930s. In this debate, Lin Yutang alongside a host of Chinese library scientists, educators, and linguists fought over and experimented with novel systems of organizing Chinese dictionaries, library card catalogs, indexes, name lists, and telephone books—systems that would help Chinese speakers *navigate* the Chinese-language information environment more efficiently. Of all those who participated in the "character retrieval problem" debate, however, Lin would be the only one to transpose this rigorous discussion of *retrieval* to the domain of *inscription*. To understand MingKwai in particular, and input more broadly, then, we must first delve into the "character retrieval" debates of the early Republican period, and the broader history of China's early twentieth-century information crisis.

In *Too Much to Know*, Ann Blair uncovers many of the ways in which present-day notions of "information society" and "information overload" have a much deeper history than we often assume. Well before the advent of computing and the Internet, purveyors of knowledge and truth in early modern Europe spoke of their age in terms of overwhelming information overload. An array of bibliographers, publishers, and others developed novel approaches and technologies designed to help gain control over the burgeoning information environment that constantly threatened to burst its containers and leave humanity in the ironic condition of over-informed ignorance. In the East Asian context, the work of Mary Elizabeth Berry has revealed parallel trends in early modern Japan. Tokugawan cartographers, bibliographers, compilers, editors, and others deployed a repertoire of repurposed and novel technologies—including maps, digests, encyclopedias, and travel guides—to help them stay afloat in an ever-rising sea of information.[7]

Late imperial and early Republican China was yet another site of "information crisis." With the population boom of the late imperial period, the rise of new forms of state surveillance, and the introduction of novel information technologies such as telegraphy, Chinese elites were growing increasingly anxious about their capacity to keep pace with this new data environment. The history of "information overload" in early twentieth-century China stands apart from the European and Japanese contexts explored by Blair and Berry, however. In China, what vexed elites most were not primarily questions of information quantity or velocity, nor even the country's engagement with new informational genres, but debates over the capacity of the Chinese language itself (or "incapacity," as many argued) to handle "modern information."

Circa the 1920s, banal essentials of the modern information environment became a persistent source of anxiety: how to develop a modern Chinese telephone directory, magazine index, archival index, name list, and indeed all forms of reference materials where one might need to locate some form of information encoded in Chinese characters. In one study from the era, it was found that research subjects using one of the many new experimental character retrieval systems were able to locate Chinese characters between one-tenth of a second and one full second faster than when using China's leading dictionary.[8] Still others determined that Chinese character look-up was consistently slower than its alphabetic counterparts.

While the difference of mere seconds hardly seems to justify the sounding of alarms, it appeared possible to some language reformers that tiny delays such as these might be significant factors within China's broader challenges during the modern age. If information encoded in Chinese took slightly longer to find than in alphabetic look-up, what might this mean when aggregated across the wider textual landscape: indexes, phone books, name lists, concordances, passenger manifests, encyclopedias, commodity inventories, and library card catalogs? It stood to reason that the Chinese-language corpus as a whole was in a very real sense *further removed* from the average Chinese user by thousands of additional minutes, hundreds of additional hours, even added days, as compared to the English-language corpus and its respective user base.[9] Aggregated across the Chinese populace as a whole, it would seem that China's backwardness might indeed be the macrohistorical extrapolation of these countless microhistorical lag times. *China was functioning in slow motion.*[10] As one language reformer claimed, a more advanced system of organizing and retrieving Chinese characters would save every literate Chinese person a full two years in his or her lifetime (assuming a working career of forty years) simply by helping them find information faster.[11]

Propelled by the growing sense of information crisis, the early Republican period witnessed an explosion of experimental organizational methods for Chinese characters. The crisis would come to be known as the "character retrieval problem" (*jianzifa wenti*), and would attract participation by such luminaries as long-time president of Peking University Cai Yuanpei, one-time head of Shanghai Commercial Press Gao Mengdan, pioneering Chinese library scientist Du Dingyou, and dozens of others.[12] By its peak in the early 1930s this "crisis" would give rise to no fewer than seventy-two experimental systems by which to reorganize the Chinese-language information environment—seventy-two new "alphabetic orders" for Chinese script. Sprouting like "bamboo shoots after the rain," as one observer put it, these systems were varied and diverse—and yet they cohered around a growing consensus. The *Kangxi Dictionary* and its attendant radical-stroke taxonomic system were unworkable for the purposes of modern information systems.[13] If Chinese-language information were to be every bit as modern as alphabetic information, *Kangxi* would have to be overthrown.

IN SEARCH OF THE *DAO*

Contenders for the throne of *Kangxi* entered the arena in droves. In 1912, Gao Mengdan—successor to Zhang Yuanji as the chief editor at Shanghai Commercial Press—proposed his *Merged Radicals Method (Guibing bushou fa)* premised upon a sharp reduction in the number of Chinese radicals.[14] In 1922, Huang Xisheng proposed his *Chinese Character Retrieval and Stacking Method (Hanzi jianzi he paidie fa)*. Soon after, library scientist Du Dingyou proposed his *Chinese Character Retrieval Method (Hanzi jianfa)* and *Chinese Character Arrangement Method (Hanzi paizi fa)*. More well known was the *Numerical Retrieval System (Haoma jianzifa)*, later reformulated as *Sijiao haoma* (or "Four-Corner Code"), by Wang Yunwu. It was within the context of this early twentieth-century "character retrieval problem" that Lin Yutang began to develop and propose his own systems. In 1918, he published his *Index System for Chinese Characters (Hanzi suoyin zhi)*,[15] and in 1926 reentered the debate with two new retrieval systems: the *New Rhyme-Based Indexing System (Xin yun suoyin fa)* and the *Last-Stroke Character Retrieval Method (Mobi jianzifa)*.[16]

The sheer number and diversity of experimental retrieval systems, all created within years of one another, raised a startling question: *Even at this late date in history, had the fundamental essence and order of Chinese script yet to reveal itself?* Unlike English and French, in which the sequence of alphabetic letters had long ago been established and stabilized, could it be that the Chinese script remained a wide-open terrain of uncertainty, possibility, and even revelation as late as the third decade of the twentieth century? How was it possible for so many of the leading minds of China to disagree so fundamentally with one another over the Chinese script? Moreover, which of these seventy-plus experimental systems was the true order, if any? Which was most efficient? Which would achieve dominance? Which offered the solution to China's information crisis?

One observer, Jiang Yiqian, took this proliferation of experimental systems as unmistakable evidence that the "fundamental method" (*genben fangfa*) of Chinese had yet to be discovered—and what is more, that Chinese script was clearly more complex than its alphabetic counterparts. "Chinese writing," Jiang wrote, "is not alphabetic, and does not have a set order. As a result, within the history of Chinese culture there

Table 6.1 Retrieval systems invented between 1912 and 1927 (partial list)

Year	Inventor	Retrieval System
1912	Gao Mengdan	Merged Radicals Method 歸併部首法
1916	Otto Rosenberg	Five-Step Arrangement Kanji Table 五段排列法
1918 (March)	Lin Yutang	Chinese Character Index System 漢字索引制
1920 (Dec)	Ministry of Education	Chinese Phonetic Alphabet Retrieval Method 注音字母國音檢字法
1922	黃希聲 Huang Xisheng	Chinese Character Retrieval and Stacking Method 漢字檢字和排疊法
1922	杜定友 Du Dingyou	Chinese Character Retrieval Method 漢字檢法
1925 (June)	王雲五 Wang Yunwu	Code Retrieval Method 號碼檢字法
1925 (Dec)	杜定友 Du Dingyou	Chinese Character Arrangement Method 漢字排字法
1925	桂質柏 Gui Zhibo	26-Stroke Retrieval Method 二十六種筆畫檢字法
1926 (Jan)	Lin Yutang	New Rhyme-Based Indexing System 新韻索引法
1926 (Jan)	萬國鼎 Wan Guoding	Chinese Character "Mother Stroke" Arrangement Method 漢字母筆畫排列法
1926 (Jan)	萬國鼎 Wan Guoding	Revised Chinese Character "Mother Stroke" Arrangement 修正漢字母筆畫排列法
1926 (Feb)	Wang Yunwu	Four-Corner Retrieval Method 四角號碼檢字法
1926 (Oct)	Lin Yutang	Final-Stroke Retrieval Method 末筆檢字法
1927 (Feb)	張鳳 Zhang Feng	Shape-Number Retrieval Method 形數檢字法
1927	張鳳 Zhang Feng	Shape-Number Retrieval Method— Revised Third Edition 三版訂正形數檢字法
1928 (May)	Wang Yunwu	Four-Corner Retrieval Method—Revised Second Version 第二次改訂四角號碼檢字法
1928 (Oct)	Wang Yunwu	Four-Corner Retrieval Method—Revised Third Edition 第三版 第二次改訂四角號碼檢字法

have been all sorts of different character retrieval systems."[17] Drawing upon Darwinian metaphors, Jiang portrayed all of the many competing and divergent character retrieval systems as akin to species that, in their interaction, competition, survival, and extinction, together constituted a churning, collective, evolutionary process that would eventually reveal the essential truth of Chinese script.[18] Among the dozens of systems that had been proposed, he argued,

the question of which is best awaits comparison and corroboration, as well as a long period of use. For the time being, it's impossible to say—and, in fact, there's no need to do so. Over the course of a substantial period of time, and by process of elimination, the best one will naturally rise to the top. ... The best method will also necessarily be the fundamental method of Chinese characters. Since Chinese characters possess a structure, so too then must they possess a way (dao) by which to organize them. Phrased differently, since there must be a system, so too must there be a fundamental method of arrangement.[19]

When discovered, the "fundamental method" would be to the Chinese language what the alphabet was to English: the unambiguous, rational, transparent system by which the Chinese language would finally realize its true order, and in which the swirl of novel experimental systems would finally achieve quietude—its own end of history.

Jiang Yiqian may have been satisfied to await the victor of this evolutionary struggle, but others were less patient. For Lin Yutang, Wang Yunwu, Chen Duxiu, Du Dingyou, and others, the character retrieval crisis was both a professional and deeply personal contest. To the victor went the spoils, not only in economic terms, but perhaps primarily in terms of symbolic capital. Each strove to promote his system within and among government departments, such as the ministries of education and communication. Each strove to develop relationships with China's centers of print capitalism, such as Commercial Press. Above all, however, each participant in the "character retrieval problem" competition harbored a dream that his system would be the one to depose *Kangxi* and to accede to the throne.

All of this raises an obvious question: As Lin Yutang and his contemporaries set out to uncover the "fundamental method" of Chinese script, where did they believe it would be found? When Lin and other participants in the "character retrieval problem" designed their symbolic systems, what governed this process? How, moreover, would they know if this fundamental order had been discovered?

As we examine early Republican debates over character retrieval, we find two primary forces or motifs that shaped this contest: a pursuit to discover the fundamental, orthographic essence of Chinese *script*, and a parallel pursuit to discover the fundamental cognitive or psychological essence of Chinese *people*. In the first of these pursuits the question was, in a sense: How do Chinese characters fundamentally *want to be found*? What was the essential nature of the Chinese character, and its fundamental building blocks? If Russian, Hebrew, Greek, English, and Arabic all enjoyed fundamental, agreed-upon, unambiguous orders, what was the Chinese equivalent? In the second of these pursuits, the question focused on people rather than script: How do Chinese people fundamentally *want to search*? In their efforts to develop a "transparent" system—one that could be used by "anyone and everyone"—Republican-era language reformers ventured into a site of political contestation: namely, the battlefield in which political elites of different stripes vied to define the Chinese "masses" themselves. As part of trying to create systems that "everyone" could use, Republican-era language reforms were thus equally invested in defining who the "Chinese everyone" was—in terms of capacities, limitations, tendencies, and instincts. In setting forth their respective retrieval systems, each system constituted a competing theorization of both Chinese script and the "average Chinese user."

HOW ANCIENT CHINA MISSED THE POINT: CHARACTER RETRIEVAL AND THE ESSENCE OF CHINESE SCRIPT

To begin our discussion with the first of these poles—character retrieval as the search for the *dao* of Chinese script—an illustrative case is that of Chen Lifu (1900–2001). Throughout the Republican period (1911–1949), and into the postwar era, Chen Lifu was deeply involved in the political management of culture and education. In the late 1920s, he was elected a member of the Guomindang Central Executive Committee, and was later named Guomindang secretary general in 1929. He went on to become president of the Central Politics College, a member of the Standing Committee of the Central Executive Committee, director of the Military

Affairs Department's Bureau of Statistics, and the minister of education, among many other posts.

Chen Lifu also invented a character retrieval system known as the "Five-Stroke" or *Wubi* retrieval system. Five-Stroke was born on the battle-fields of 1920s China, when the country was entering its second decade of political fragmentation. Vast territories of the former Qing empire were now governed by military elites who offered only paper allegiance, or none at all, to the central government of a defunct republic. The Guang-zhou-based Nationalists and their Communist allies launched a north-ward military campaign whose goal was to defeat or co-opt militarists and reunite the country under a single capital and regime.

Chen Lifu's Five-Stroke system was fashioned and field-tested during this campaign. Charged with managing confidential materials, includ-ing telegrams and papers, Chen and his team processed approximately 150 documents daily, always under an acute sense of time-sensitivity and urgency. "Time was short and Chiang was always impatient," Chen recalled in his memoir, referring to Nationalist leader Chiang Kai-shek. "Whenever he wanted a certain paper, he required us to find it for him immediately. This challenge forced me to conceive a classification method for Chinese characters to organize our files."[20] Specifically, Chen developed Five-Stroke retrieval as a way to categorize the Nationalist mili-tary's growing collection of enemy combatant telegraph code books, pre-sumably organizing them according to the Chinese character surname of the enemy combatant officer, or perhaps the name of their regime or alliance.[21] The Guomindang faced a multitude of military actors, each employing different numerical transformations to encrypt messages transmitted using the four-digit telegraph code we examined in chapter 2. Upon interception, decryption was time-sensitive and depended not only on identifying the code being used, but also on quickly finding the code books themselves from within the army's library.

The military campaign drew to a close in 1927, and had by its conclu-sion dramatically reconfigured the political landscape of China. In April 1927, the Chinese Communists—erstwhile partners of Chiang Kai-shek and the Nationalists, as part of the United Front alliance—became subject to Chiang's violent and enduring purge, known as the White Terror. The Nationalists established a new regime centered in Nanjing, seeking to transform its capital city into a gleaming beacon of rational, modern, scientific government and statehood, undertaking an immense urban development campaign. For his own part, Chen Lifu set about extending his retrieval system from military to broader state and civilian applications: to render searchable everything from schools to prisons, political parties to the country's taxation and communications infrastructures. Chen was better positioned than most in the "character retrieval problem" debates, drawing upon his professional and political connections to pilot his retrieval program in the Command Headquarters Confidential Work Section official correspondence archives, the Nanjing Household Registration Survey, the Guomindang Central Executive Committee, and the Membership Unit of the Organization Department, among others.[22] Five-Stroke retrieval enabled one to "search for one person out of 100,000" (*shiwan ren zhong qiu yi ren*), Chen boasted, whether in the context of household registration statistics, party member name lists, addresses and statistics regarding postal and telephone offices, name lists within schools and factories, the management of state archives, the organization of tax registers and receipts, land registration, prison inmate registers, or documents connected to the examination and evaluation of political and military personnel.

Chen also saw his retrieval method as a way to manage a modern capitalist economy. The diversification of consumer preferences, coupled with industrialized production, had already flooded the Chinese market with more commodities than current systems of information retrieval could handle, Chen argued. In large department stores and wholesale operations, he noted, commodity varieties could extend into the tens of millions.[23] The same was true of the accelerated tempo of consumer transactions in China's growing banking sector, Chen noted, with each transaction producing a record that the bank or client might need to access and review at a later date. If Chinese characters were to remain the semiotic substrate of the modern Chinese state and economy, a new character retrieval and organization system would be absolutely necessary. Some offices experimented with hybrid Chinese character/Latin alphabet systems—such as organizing Chinese characters alphabetically according to their phonetic values—but Chen Lifu considered this disgraceful. "To use Chinese characters, but to seek help from Western writing," he wrote, "is truly a great dishonor to the Chinese nation."[24] Using Wubi, one could maintain Chinese characters while at the same time operating a surveillance state capable of handling hundreds of thousands of party member dossiers, nationwide census statistics, and household registration files.

Chen saw himself as much more than the designer of a new information architecture for the modern Chinese state, though. Although his treatises on retrieval seem at first glance to be fixated on decidedly banal problems, his ambitions for Wubi extended to nothing less than the metaphysics of Chinese writing. Chen self-identified as a participant in a historical conversation with some rather impressive interlocutors—most notably, the renowned Jin dynasty calligrapher Wang Xizhi (303–361), whose theory of the "eight fundamental strokes of the character *yong*" we have encountered more than once in our study (figure 6.3).

Chen drew inspiration from Wang's calligraphic theory, and sought to transplant it to the world of taxonomy. If all characters could be said to have a highly limited number of fundamental strokes, then it stood to reason that one could transcend the radical-stroke system of the *Kangxi Dictionary* and its 214 categorical divisions. One needed no longer traffic in radicals at all, but could transfer one's energy to the much more economical world of strokes.

Sage of calligraphy though Wang Xizhi may have been, this Jin dynasty "predecessor" had failed to realize something, Chen Lifu argued. There existed something prior to the "stroke" itself, out of which all strokes were formed: the *point* or *dian*. All strokes, no matter their curvature, thickness, or direction, began their existence at the very moment when instrument first made contact with medium—when brush first made contact with paper. Every stroke, as Chen phrased it, "must begin as a point before becoming a stroke" (*bi shi yu dian er cheng hua*). Wang Xizhi had fallen victim to a basic misconception, so subtle that it escaped not only his attention, Chen argued, but indeed that of countless others down through the generations. The stroke was *not* the fundamental element of Chinese characters. Chen Lifu saw himself not merely as facilitating the interception of coded telegraphic transmissions, then, or creating more efficient telephone books, but as having righted the wrongs of the ancients. It was as if Chen had returned to the time before Wang Xizhi— before Wang had begun to cut his errant pathway—to rediscover the lost

6.3 Eight fundamental strokes of the character *yong* (eternity)

way or *dao* of Chinese writing, and renew its long-deferred exploration. Wang Xizhi had quite literally *missed the point*.[25]

Although seemingly a minor qualification—this focus on the point (*dian*) rather than the stroke (*bihua*)—this notion would have a profound impact on Chen Lifu's understanding of Chinese. What we know as strokes—in all their varieties—achieved their quality depending upon what followed the moment of origination: the direction the brush moved, and the quality of this movement. Having failed to grasp this essential truth of Chinese writing, Wang Xizhi had missed the mark by three strokes. There were not eight, but five essential transformations of the original *dian*, Chen claimed, all others being mere variations or combinations of these five: the *dian* could remain a *dian*; it could venture out laterally to become a horizontal stroke; it could advance downward to become a vertical stroke; it could move at a diagonal, in which direction did not matter; or it could move and bend, the direction of this bend being for Chen Lifu an insignificant distinction (figure 6.4).

If we look laterally, at Chen Lifu's contemporaries, a wide field of practitioners is found engaging in comparable pursuits. Indeed, one of the prerequisites of being the kind of person who would dedicate years of effort to the invention of an experimental Chinese character retrieval system was, it appeared, to be the kind of person who regarded himself as possessing astonishing, unprecedented insight into the essence of Chinese writing. Participants in the "character retrieval problem" saw themselves as linguistic argonauts, venturing out into a vast and unknown terrain to uncover still elusive fundamental truths that even the greatest minds of Chinese antiquity had failed to grasp. Beneath the banality and humility of such goals as improved filing cabinets, library shelves, and phone books resided a much deeper sense of historical purpose and self-importance. *I am not merely inventing a new card catalog case—I am righting the wrongs of the ancients. I am not merely creating a new phone book—I am discovering the truth of the order of Chinese and thereby bringing Chinese writing into a state of parity with other world scripts, in which the question of fundamental order had long ago been settled.*

第一筆	第二筆	舉　　　　　　　　例	應注意之點
丶	丶	江 穴 情 ⋯⋯⋯	
	一	文 高 方 ⋯⋯⋯	
	丨	⋯⋯⋯⋯⋯⋯⋯⋯	按無此類字
	丿	冷 馮 凌 ⋯⋯⋯	
	乀	姜 羊 火	
	乛	⋯⋯⋯⋯⋯⋯⋯⋯	按無此類字
	フ	房 祝 楠 ⋯⋯⋯	
一	丶	平 雲 泰	
	一	于 秦	
	丨	工 東 ⋯⋯⋯	
	丿	次 咨 資 ⋯⋯⋯	按此為正寫實則以⇒
	乀	原 右	
	乛	⋯⋯⋯⋯⋯⋯⋯⋯	按無此類字
	フ	丁 先 木 ⋯⋯⋯	按此類極少
丨	丶	當 ⋯⋯⋯⋯	按此類字極少
	一	步 豐 業	
	丨	對 業 ⋯⋯⋯	按此類字極少
	丿	小 蔣 ⋯⋯⋯	
	乀	⋯⋯⋯⋯⋯⋯⋯⋯	按無此類字
	乛	卜 (僅此一字)	按此為俗寫實則以反點
	フ	日 過	
丿	丶	美 谷 ⋯⋯⋯	
	一	和 朱 ⋯⋯⋯	
	丨	仁 白 ⋯⋯⋯	
	丿	⋯⋯⋯⋯⋯⋯⋯⋯	按無此類字
	乀	徐 須 ⋯⋯⋯	
	乛	人 公 ⋯⋯⋯	
	フ	包 周 ⋯⋯⋯	
フ	丶	小 桑 ⋯⋯⋯	
	一	屈 弓	
	丨	巴 胥 ⋯⋯⋯	
	丿	刁 (僅此一字)	
	乀	姚 賀 ⋯⋯⋯	
	乛	又 义 (僅此數字)	
	フ	陳 子	

6.4 Chen Lifu's outline of the relationship between the point and strokes

LOOKING FOR "LOVE" IN ALL THE WRONG PLACES: DU DINGYOU AND THE PSYCHOLOGY OF SEARCH

Distilling the orthographic essence of Chinese characters was not the only concern that motivated reformers in their attempts to create an ideal character retrieval system. The 1920s and 1930s were also the era of "the masses" and the "citizen," a time when these concepts were swiftly becoming a central focus of political, economic, and social thought in China. Whether in the domains of popular education or literacy campaigns, the mobilization of the Chinese citizenry was deemed essential for the survival of the nation; it was the masses that formed the new locus of Chinese sovereignty. In this context, multiple reformers came to the conclusion that the key to solving China's character retrieval problem required an *ethnographic* focus, not an orthographic one. While their systems and approaches varied greatly, a shared goal was the idea of developing a "transparent" system of organizing the Chinese language—that is, one that annihilated any and all ambiguity about how to find any given Chinese character, and thus could be readily used by "everyone," with as little training as possible. For the historian of technology, then, one finds in such debates over the "Chinese everyman" one of the earliest discussions within the Chinese context of design, human-machine interactionism (HCI), user experience (UX) analysis, and related subjects.

For a vivid example of the ethnographic claims that were central to the character retrieval crisis, we turn to another contemporary of Lin Yutang. Du Dingyou (1898–1967) was a pioneering figure in the history of modern Chinese library science. Born in Shanghai in 1898, Du undertook his undergraduate training at the University of the Philippines. Following the Japanese bombardment of Shanghai in 1932, during which the collections of the Oriental Library (*Dongfang tushuguan*) suffered severe losses, Du was instrumental in the foundation of the Shanghai Municipal Library, of which he became vice director. He would later serve as director of the Zhongshan University library, among other posts.

As part of his lifetime of service to library science, Du Dingyou was the developer of still another experimental Chinese retrieval system: the Shape-Position retrieval system (*Hanzi xingwei jianzifa*). In 1925, he published an article entitled "On the Psychology of How the Masses Search for

Characters," a lively and even humorous piece in which Du took aim at his competitors within the field of character retrieval. Du Dingyou regarded competing systems of character retrieval as failures on ethnographic grounds: a failure to "grasp the psychology of retrieval among the masses" (*zhuaju minzhong jianzi de xinli*).[26] Divorced from the objectivity of mass psychology, Du's competitors had yet to solve the problem of the Chinese *script* because they had yet to solve the problem of the Chinese *masses*.

Du Dingyou wove his own modern-day parable starring an over-worked mother and a young daughter looking for "love." "One day," the parable began, "a twelve-year-old girl suddenly asked her mother: 'What is love?'" "How frightening," Du inserted his editorial voice: "only twelve and already talking about love!" Thankfully for her mother, the "love" for which this young girl searched was neither the emotion nor the experience, but quite literally the word: the Chinese character *lian* (戀). "At that moment," Du Dingyou continued, "her mother was in the middle of doing housework, and couldn't leave her seat. So she explained that the character *lian* has the character for 'language' (言 *yan*) in the middle, flanked on each side by 'silk' (糸 *si*), with one 'heart' (心 *xin*) on the bottom." Du was referring to a practice with which all of his readers would no doubt have been familiar: what Joseph Allen refers to as the Chinese "graphology" or "metalanguage." In everyday practice, one can describe the structure of a character verbally by describing the components out of which it is formed. The Chinese surname *Li* (李), for example, might be referred to as "tree-child *li*" (*mu zi li* 木子李), so as to distinguish it from other characters with the same pronunciation.

"As soon as she heard it," Du's parable continued, "she wrote it straight-away, without the slightest error. This is the essence of the structure of our writing, and a point which character retrieval researchers should pay close attention to."[27] Du took his allegorical critique one step closer to the cradle, suggesting that mass understandings of Chinese characters were "akin to when a baby recognizes his mother for the first time, that she is generally tall or short, fat or skinny." He continued:

That which he recognizes is his entire mother. Before he has become aware, when his faculties of discrimination are still immature, he still cannot recognize if this pair of eyes is his mother's, or any other part of the body. But as for the overall concept of his mother, this he possesses. A character retrieval method … should pay close attention to the importance of the law of synaesthesia.[28]

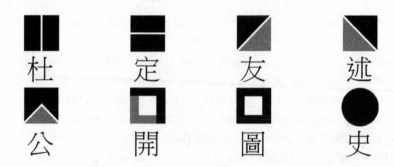

6.5 "Shape-Position" retrieval system by Du Dingyou

For Du, the average Chinese informational subject was a *Gestalt* pattern-finder: a mind for whom script was primordially meaningless and spatial. Highbrow questions of etymology had no place in a modern and mass-oriented character retrieval system, he felt, an ethnographic presupposition that manifested itself in the retrieval system he designed. Dispensing entirely with all etymological concerns, his system was premised on the contention that the best way to organize a writing system was, counterintuitively, to stop thinking of it as a writing system—that is, as a special class of object bound up with meaning. Instead, graphemes would be approached no differently than any other entity that occupied space (figure 6.5).

Moving through the corpus of Chinese characters, Du Dingyou isolated eight spatial archetypes: north–south/vertical (縱 *zong*); horizontal (橫 *heng*); oblique/diagonal (斜 *xie*); "carrying" (載 *zai*); "covering" (覆 *fu*); cornering/cornered (角 *jiao*); enclosed (方 *fang*); and whole/complete (整 *zheng*). To help potential users discern between these eight classes, he provided a handy eight-character phrase which, in deft entrepreneurial style, entailed his own name: *Du Dingyou shu gongkai tushuguan shi* ("The History of Public Libraries as Described by Du Dingyou").[29]

The groupings and classes of equivalence Du proposed in his system were unprecedented in Chinese history. Suddenly, the character 林 (*lin*, "forest") shared something in common, taxonomically speaking, with a host of characters with which it had never before been associated—characters such as 動 (*dong*, "movement") and 排 (*pai*, "to arrange"). These characters shared no radicals in common. They were neither composed of the same number of strokes nor did they share similar pronunciations. For Du, however, the salient property of *lin* was the side-by-sideness of its two component parts (木 *mu*)—that these components *mu* happened to signify "tree," from which the character *lin* derived its meaning "forest," concerned Du not in the slightest.

Du Dingyou's twelve-year-old girl and her harried mother would be difficult if not impossible to track down in the real world, however. They were not empirical subjects, decided upon through comparison and observation. They were the projections of Du's imagination: the invented Chinese masses that were constantly invoked by the linguistic elite of twentieth-century language reform efforts. Were we to extend our analysis to the dozens of other experimental retrieval systems, and to the world of twentieth-century Chinese language reform more broadly, we would gain many more illustrations of this field of ethnographic contestation. Each character retrieval system had embedded within it a presupposed, informational subject—a holographic Chinese user on whose behalf each system would function, and more precisely, a set of assumptions upon which the system *depended* if its claims to transparency and effortlessness were to bear truth.

Depending upon the system and inventor in question, this holographic *homme moyen* exhibited a distinct set of capacities, predilections, plasticities, and limitations. Du Dingyou's holographic Chinese was unsystematic and messy, and could not be relied upon to follow instructions of any complexity. Other designers put forth a holographic Chinese everyman of more optimistic design: possessing a high capacity for abstraction, pliable, attentive to detail, and able to reconceptualize Chinese script in ways no one ever had before. Who the Chinese everyman *actually* was was a different matter entirely.

FROM SEARCH TO SEARCH-WRITING

In the autumn of 1931, just as northeast China was falling to invading Japanese forces, Lin Yutang authored a letter in which he shared privately the work he was doing in his latest and boldest venture: a Chinese typewriter of his own novel design.[30] His continuing correspondence, extending into the late 1930s, would reveal his earliest thoughts on the subject.

In the first of his letters, Lin Yutang made three assertions about the history and prospect of typewriting for the Chinese language:

"No Chinese typewriter using a phonetic alphabet will ever have a real market."
"No Chinese typewriter can operate by a process of combination of strokes and dots."
"No Chinese typewriter can provide over 10,000 characters required in Chinese printing and correspondence."[31]

In three swift negations, Lin Yutang dismissed the entire history of Chinese typewriting that we have come to know, ruling out each of the three approaches that inventors had concentrated upon over the preceding half-century. With his first statement, Lin had refuted in one breath the entire model upon which Remington and others staked their hopes. With the second, he jettisoned divisible type, or combinatorial, approaches advanced by Qi Xuan and others. In his third pronouncement, his dissatisfaction with the limits of the common usage approach were made clear. Indeed, so damning were these statements by Lin that, at first glance, they seem indistinguishable from those of character abolitionists—a Qian Xuantong, perhaps, but not the would-be inventor of a Chinese typewriter. Nothing was left, it would seem, short of giving up or perhaps starting all over.[32]

As we delve more deeply into Lin's early thoughts on Chinese typewriting, however, we soon find that his goal was to *integrate* these three existing approaches, not to abandon them.[33] More accurately still, Lin wished to merge these three approaches and, in the process, create an entirely new *type* of typewriter—and, indeed, an entirely new mode of inscription. "First, the solution lies in reducing the number of characters to be provided for," he argued, in a clear invocation of the common usage approach to Chinese technolinguistic modernity. The machine he would set out to build would include a set of high-frequency Chinese characters, just like those machines we have examined thus far.

Combinatorialism would be essential as well. Nine-tenths of Chinese characters were composed in a combinatorial method, Lin explained in the course of his 1931 letter, with a left-side element called a "radical" and a right-side element called a "phonetic." Overall, Lin explained, there were approximately 1,300 right-side phonetic elements, while only around 80 left-side radicals. "By the combination of these elements,"

Lin wrote, perhaps unaware that he was invoking the spirits of Legrand, Beyerhaus, Qi Xuan, and others, "any number of Chinese characters can be formed, actually over 30,000 characters." To illustrate his approach, he provided an English-language analogy. Lin Yutang likened Chinese radicals and phonetics to English prefixes and suffixes, using the English-language "COM-" and "-BINE" to demonstrate how his system would work.[34] Like "com-," these "standardized left parts" would be concatenated "with any right part to form a perfect square beautiful character" (figure 6.6). Likewise, his "standardized right parts" could be combined "with any left part"[35] just like the English "-bine." Lin included in his letter a piece of folded paper, notated "2 pieces showing combination of right and left components into perfectly square and beautiful Chinese characters."[36]

Up to this point, the reader will notice, Lin's description of his type-writer differed not at all from many of the Chinese machines that came before. Indistinguishable thus far from Qi Xuan's combinatorial device from the 1910s, and sharing many of the same design principles as those manufactured by Commercial Press, it seems to have little more to examine. It was in Lin's third step, however, that something entirely new began to take shape. Rather than trying to fit all of the necessary characters and character components onto a standard Chinese typewriter tray bed, or on a cylindrical drum like those found on early prototypes by Qi Xuan and Zhou Houkun, Lin would steal a page from Chinese telegraphy and sequester all Chinese characters to the interior of his machine—away from the naked-eye view of the typist. Surrogacy, as we examined in chapter 2, would thus prove central to his typewriter as well. As with the Chinese telegraph code, the operator of Lin's imagined device would not manipulate or traffic in Chinese characters directly, but *indirectly*—in this case, by means of a keyboard-based control system. In one sense, then, Lin's machine would resemble Robert McKean Jones's "Chinese typewriter with no Chinese," with a keyboard that contained few if any Chinese characters. But unlike that of Jones, somehow Lin's machine *would* produce Chinese-character output.[37] The typist would use a keyboard not to type characters directly so much as to *instruct the machine as to which characters he or she wanted to type*. "The process of printing a word," Lin explained of his three-part process, "is therefore similar to printing an English word of three letters, like 'and' or 'the,' except that the first two

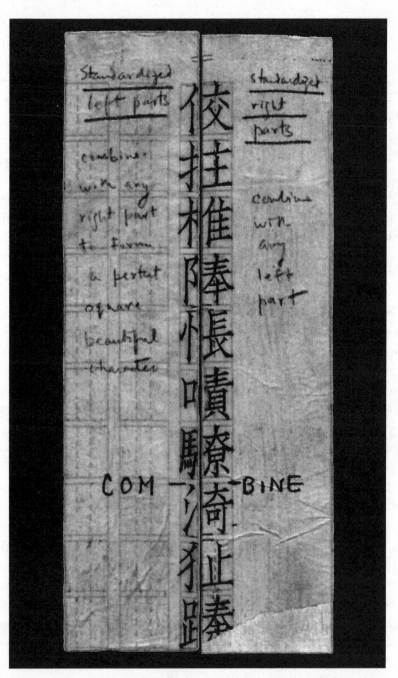

6.6 Lin Yutang's 1931 letter, demonstrating how characters would be formed on his Chinese typewriter

keys serve merely to bring the word into printing position, while with the third key, the whole word is printed complete."[38]

As part of his growing interest in typewriting, a concern laid out in these letters from the 1930s, Lin began to move away from the framework of *retrieval* and *search*—the domain of dictionaries, phone books, and card catalogs—to that of *inscription*. To make this transition from search to search-writing, however, would require Lin to transform the character retrieval frameworks he and others had worked on in the 1910s and 1920s. For the typewriter he had in mind, these symbolic systems would not suffice—a limitation we can only understand if we take apart Lin's MingKwai typewriter to grasp precisely how it functioned. When we do, we find the ways in which Lin's taxonomic thought, once restricted to the world of Chinese dictionaries and indexes, was refracted through the prism of the mechanical and material concerns of his device.

As outlined in his 1931 letter, Lin had set out to design a typewriter that, like those of both Zhou Houkun and Qi Xuan, would be outfitted with both full-body common usage Chinese characters, and divisible-type-style character components. Unlike these earlier systems, however, none of these characters or character components would be directly viewable or manipulated by the typist. They would be *inside* the machine. As a result, one of the first challenges Lin faced was how to fit these thousands of graphemes as tightly as possible inside the chassis of the MingKwai device, while also making them *accessible*. To achieve this, Lin departed from the dominant Chinese typewriter design of the era—the rectangular tray bed we have come to know well over the course of this study—and instead contemplated ways of wrapping or folding up his Chinese characters into a more condensed space.

The design he settled upon was something like a planetary system, with moons, planets, and a central star. The proverbial moons in Lin's system were a series of eight-sided metal bars, with his Chinese characters and radicals engraved upon each of their surfaces. With each of these octagonal bars long enough to fit 29 characters along its length, Lin could fit a total of 232 characters or character pieces on each bar (twenty-nine characters per side, multiplied by eight sides). Taking six of these bars, he fastened them to a circular, rotating gear—like six moons revolving around a shared planetary axis, as well as their own individual axes of rotation. Making six of these six-bar clusters in all, he in turn fastened these to a larger, circular, rotating drum—like six planets rotating about a central star. All told, then, Lin's system encompassed a total of forty-three separate axes of rotation: thirty-six metal bars rotating around their own lunar axes, six higher-order cylinders rotating around their own planetary axes, and finally one highest-order cylinder rotating around a singular, stellar axis (figure 6.7).[39] Owing to this ingenious design, one in which each of the eight faces on every one of Lin's metal bars could be brought into the printing position by means of a coordinated rotation process, Lin's machine boasted more than three times the capacity of the tray bed on a common usage Chinese typewriter, all in a smaller space. MingKwai offered up a total of 8,352 possible graphemes in all, with which it would be possible to compose *every Chinese character in existence.*

The second obvious challenge was that of layout and categorization: according to what taxonomic system would Lin's 8,352 Chinese characters and components be arrayed within this metallic hard drive? In answering

6.7 Mechanical design of MingKwai

this question, Lin would need his taxonomic system to function in ways that departed dramatically from the earlier character retrieval systems he and others had experimented with in the 1910s and 1920s. A dictionary organization system is one that does not require taxonomic evenness—that is, there is no need for the "A" section of a dictionary to contain the same number of words as the "G" or "Z" sections, or for the water-radical (氵) section of the *Kangxi Dictionary* to contain the same number of characters as the turtle-radical (龜) section. For Lin's typewriter to function, however, he needed to arrive at a categorical system in which each of his categories contained the same number: no more than eight characters in all, and ideally no fewer. This precise upper limit, we recall, was due to the "Magic Eye" viewfinder, through which the machine presented a total of up to, but no more than, eight character candidates from which the user could select. Even a ninth character in any one of the taxa would have sent Lin Yutang back to the drawing board, and back to his checkbook. There was a second, critical challenge, moreover. Lin also needed to fill each taxon as fully as possible. Underutilization of taxa—categories containing only three, four, or five characters, for example—ran the risk of severely diminishing the total capacity of the machine by hundreds if not thousands of characters, or requiring Lin Yutang to increase the total number of keys on the keyboard to handle a larger number of categories. Again, either eventuality would diminish the machine and cost money.

If all of these considerations were not challenging enough, Lin would also need to attend to questions of "user experience," to use the parlance of our own time—or as Du Dingyou had phrased it in 1925, the "psychology of how the masses search for characters." If we think back to the mechanical Chinese typewriters being used in offices across China at this moment in history, the typist could see all 2,500 characters on the tray bed, albeit in mirror image. They could scan, approximate, and rely upon a kind of metropolitan wayfinder's method of reaching a destination: setting out in the general direction of one's endpoint, using landmarks along the way, and asking for further directions while en route.

As for the typewriter Lin Yutang had in mind, however, Chinese characters would not be present and available in the same naked-eye sense. Characters would instead be sequestered to the interior of the machine, away from the direct observation and manipulation of the operator.

No longer navigating by sight and by stars, the typist on Lin Yutang's machine would need to rely exclusively on maps and coordinates: protocols and symbolic abstractions. Lin's machine would be a *zero tolerance* device in which every successive manipulation of the keys either worked, or not.

With this in mind, Lin could not afford to permit the material or machinic requirements of his device to result in a keyboard that confused or confounded potential users. Not only did Lin have to develop taxonomic categories that contained no more than (and ideally no fewer than) eight Chinese characters in all, but the symbols and protocols by which a user would "call up" each of these eight character clusters—that is, the symbols on the keyboard of his typewriter—would have to be intuitive and meaningful to the operator.

With the global debut of MingKwai in 1947, the fruits of Lin's labor were found on the keyboard of the machine. With the exception of six keys on which only one symbol appeared, the majority of keys featured a cluster of anywhere from two to five symbols, all as a means of spreading out and filling up each of his eight-character categories. Lin created novel groupings, moreover, clustering radicals together based on vernacular notions of resemblance. On one key, for example, the radicals *xin* (忄) and *mu* (木) appeared together, presumably because of their shared orthographic features: a pronounced vertical stroke, flanked by short accompanying strokes. A similar clustering was created for *mu* (目) and *ri* (日), which Lin assigned to a single key—radicals which bore no etymological relationship to one another, but which Lin grouped together presumably because of their shared rectangular shape.

Although groupings such as these might strike us as orthographically "natural," nevertheless each of Lin's clusters was unprecedented in Chinese linguistic practice. Detached entirely from etymological or semantic concerns, Lin created these groupings based upon what we might consider a pattern-finding *naiveté*—a kind of architectural classification in which "tall-thin" shapes were placed in one group, "rectangular shapes" in another, and so forth. The clustering of conventional shapes was clearly insufficient for the inventor, moreover. Lin also created his own proprietary "pseudo-radicals": uncanny shapes that, while they came ever so close to resembling conventional Chinese radicals and strokes, were utterly alien within the history of the Chinese script. One such symbol appeared along the bottom row of keys, fourth from the left: the Chinese character *horse* [*ma* 馬], cut in two and leaving only the bottom half.

PERFORMING MINGKWAI: LIN TAIYI AS THE CHINESE EVERYWOMAN

May 22, 1947, would live in the memory of the Lin family for a very long time: the day that Lin Yutang and his daughter Lin Taiyi brought the MingKwai machine home from the workshop, "as if bringing a newborn baby home from the hospital."[40] At eleven in the morning, the father-daughter duo carried the machine out from the workshop, and upon reaching their apartment, placed it on the table in their parlor. "When I sat in front of the machine to study typing on it," Lin Taiyi would later recall, "I could sense that it was a miracle."[41] Lin Yutang gestured to his daughter to try it out, typing whatever she pleased.[42] The experience was clearly a moving one for Taiyi: "Even though it cost 120,000 US dollars, even though it has saddled us with a lifetime of debt, this creation of my father that he had worked his entire life on, this newborn baby birthed with such difficulty, it was worth it."[43]

Baby metaphors appeared frequently in Lin family writings about MingKwai, alerting us to the profoundly personal nature of this project. In a private correspondence from April 2, 1947, Lin Yutang sent a photograph of the recently completed machine to his close associates and friends Pearl S. Buck and Richard Walsh. Upon the photograph, he penned: *To Dick-n-Pearl, This the first photo of the baby. Y.T.* (figure 6.8).

The summer of 1947 belonged to MingKwai. Lin Yutang set out upon an extensive promotional campaign, convening with journalists, contributing articles to the popular and technical press, and corresponding with cultural and political figures in China and the United States. He was in regular contact with his financial backers at the Mergenthaler Linotype company as well, along with executives at IBM and the Remington Typewriter Company, both of whom had expressed interest in the system. Lin cultivated endorsements from leading Chinese intellectuals and members of China's military, political, and financial circles as well. Lieutenant General Pang-Tsu Mow of the Chinese Air Force called MingKwai a "great

MINGKWAI TYPEWRITER invented by Lin Yutang,
size 14″ x 18″ x 9″; contains 8000 types.

6.8 Postcard to Richard Walsh and Pearl S. Buck

contribution to human society," while Tuh-Yueh Lee, manager of the New York offices of the Bank of China, remarked that he was "not prepared for anything so compact and at the same time comprehensive, so easy of operation and yet so adequate to even the most complicated characters." "It takes so little learning, whether for Chinese, or for Americans who don't know Chinese characters, to become familiar with the keyboard," linguist and Harvard professor of Chinese Zhao Yuanren (Yuen Ren Chao) commented. Zhao continued: "I think this is it."[44]

A defining moment in the MingKwai campaign was the demonstration of the device at the offices of the Remington Typewriter Company in Manhattan. If they saw in MingKwai the same sense of promise that Lin Yutang clearly did, they might place their formidable corporate weight behind the project, in cooperation with Mergenthaler. This would be a victory of immeasurable proportion for Lin Yutang, effectively winning over the two giants of modern information technology in the domains of both typewriting and Linotype. As Lin Taiyi describes in her biography of her father, the morning of the Remington demonstration was a stormy one. "Father and I wrapped up the machine in a wooden chest, took it from the apartment to the offices of Remington Typewriter Company in Manhattan," she reminisced. "Inside of the wooden box was our baby typewriter."[45] The responsibility of demonstrating the machine would fall to Lin Taiyi. Here in a quiet and severe boardroom sat representatives of the Remington company, with the typewriter placed on a small table.[46]

Lin Yutang began by painting a portrait for the Remington executives. One-third of the world's population used Chinese characters in one form or another—either completely, as in China, Taiwan, and Hong Kong, or partially, as in Japan and Korea. Engineers had thus far foundered in their attempts to produce a machine to serve this immense language community. The common usage Chinese typewriter developed by Commercial Press of Shanghai, or by Japanese competitors, offered no enduring solution to the puzzle of Chinese information technology, Lin emphasized. MingKwai was the answer. "After father had finished speaking," Lin Taiyi recalled, "he pointed to me to start typing."

Among the many bold claims made on behalf of MingKwai, none was bolder than when Lin Yutang called his machine "The Only Chinese Typewriter Designed for Everybody's Use." The promotional brochure phrased it even more succinctly: "Requires no previous training" (figure 6.9).

This insistence on effortlessness foisted heavy responsibility upon Lin Taiyi's shoulders, since she was the person who regularly found herself responsible for demonstrating MingKwai to journalists who visited the Lin family home. If they were to trust her father's claims, it certainly could not be Lin Yutang himself who presented the apparatus to skeptical observers—this linguist, this *New York Times* bestselling author, this creator of the machine. It had to be an "average" user, and she would need to make it look *easy*.

The person demonstrating the machine would also need to be a woman. We have already seen how, in China at this time, the ranks of Chinese typists were populated by a combination of young women and young men. The clerical worlds in the United States, Europe, Japan, and the greater part of the world, however, had long since become the nearly exclusive domain of women. By virtue of the decidedly international

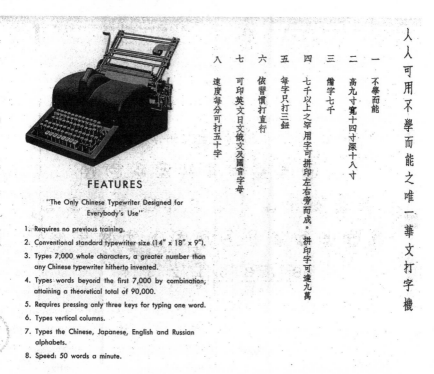

FEATURES

"The Only Chinese Typewriter Designed for Everybody's Use"

1. Requires no previous training.
2. Conventional standard typewriter size (14" x 18" x 9").
3. Types 7,000 whole characters, a greater number than any Chinese typewriter hitherto invented.
4. Types words beyond the first 7,000 by combination, attaining a theoretical total of 90,000.
5. Requires pressing only three keys for typing one word.
6. Types vertical columns.
7. Types the Chinese, Japanese, English and Russian alphabets.
8. Speed: 50 words a minute.

6.9 Promotional image: "The Only Chinese Typewriter Designed for Everybody's Use"

quality of Lin Yutang's promotional and financial efforts—efforts that found him appealing to business executives, cultural organizations, and media outlets primarily in the United States—this equation of "typist" and "young woman" would need to be observed carefully.[47]

Lin Taiyi recounted the event in vivid detail:

Under the crowd's watchful eyes, I turned on the machine. I pressed a key, but the typewriter did nothing. I pressed another key, but again nothing happened. I felt terribly embarrassed. My mouth went completely dry. I pressed another key, but it was no use. Father rushed over to my side, but the typewriter simply wouldn't move. The meeting room was dead quiet. The only thing you could hear was the sound of one key being pressed after the next. After a few minutes father had no choice but to apologize to the group, put it back in its oil rag, and slink out in embarrassment.[48]

Taiyi wondered to herself what was passing through the minds of the Remington executives—perhaps that her father was "a mad inventor."[49] Outside, the rain was still falling, and Lin Yutang contemplated whether it was best to cancel the event they had planned the following day for journalists—a potentially terrible embarrassment, but perhaps a necessary one. Upon returning home, Lin called the engineer, who promptly came over and had the MingKwai back up and running.[50] The press event was saved, yet the sting of the day's humiliation was still fresh.

Over the next three days, the Lin family home at 7 Gracie Square served as headquarters for the MingKwai press campaign. Journalists from the local and international Chinese press crowded about Taiyi, calling out "Miss Lin! Miss Lin!" (*"Lin xiaojie! Lin xiaojie!"*), one after the other.[51] With her father watching on, no doubt keenly interested in her performance, she was center stage. To convince newspapermen that the machine could handle "unscripted" texts, Lin Yutang invited reporters to "select any word," whereupon "Miss Lin typed it with speed and efficiency."[52] The gendered undertone of the event was perhaps nowhere clearer than in the photograph that later appeared in the *New York World Tribune* in which, evidently, the journalist did not even realize who this young woman was. The caption read: "Dr. Lin Yutang, author and philosopher, watches a secretary work a typewriter which types Chinese, English, Japanese and Russian" (figure 6.10).[53]

Lin Taiyi's performance was flawless by all accounts, moreover.[54] She made MingKwai look effortless, even going so far as to boast to the *Los Angeles Times* that the machine "took her two minutes to learn to operate."[55] Building upon this successful showing, the MingKwai media campaign bore fruit quickly. On August 22 alone, the *New York Times*, *Los Angeles Times*, *New York World Tribune*, *New York Herald Tribune*, *San Francisco Chronicle*, and *Chung Sai Yat Po* all featured articles on MingKwai, followed in subsequent days by the *Chicago Tribune*, *Christian Science Monitor*, *Business Week*, and *Newsweek*.[56] Articles began to appear in mainstream science and technical magazines in the fall and winter, beginning with *Popular Science* in November and *Popular Mechanics* in December. It would seem that MingKwai was, as Lin Yutang claimed, destined to become the first Chinese typewriter to achieve the same kind of widespread use and acclaim enjoyed by its Western counterpart.

6.10 Photograph of Lin Yutang and Lin Taiyi; from "Inventor Shows His Chinese Typewriter," Acme News Pictures—New York Bureau (August 21, 1947)

THE "FAILURE" OF MINGKWAI AND THE BIRTH OF INPUT

If MingKwai constituted a momentous break in the history of modern Chinese information technology, we might expect it to have swept the Chinese-language market to become the first widely celebrated Chinese typewriter in history. It did not. Instead, the one and only MingKwai prototype is gone, discarded some time during the 1960s with little fanfare by someone at the Mergenthaler Linotype corporation. Never massmanufactured, the machine perhaps lies in a landfill somewhere in New York or New Jersey, entombed in an unmarked grave beneath decades of accumulated garbage. Or perhaps it was scrapped for parts, or melted down. Why was the machine never mass-manufactured? How do we account for its failure, and what does its failure tell us—and *not* tell us—about the broader history of Chinese information technology at the midpoint of the twentieth century?

The anticlimactic conclusion of Lin's MingKwai dream played out in a private exchange between the inventor and his close friends, Pearl S. Buck and Richard Walsh. "Dear Y.T. and Hung, Dick and I have spent a sleepless night over your letter, thinking what we can do, in view of your financial situation." So began the May 1947 letter from Buck to Lin, coming in response to Lin's earlier, and no doubt humbling, request for financial assistance. After a tremendous outlay of capital for the MingKwai project, the writer-turned-inventor faced mounting and soon unbearable personal debt, and must have felt a certain embarrassment in seeking help from his longtime associates—the same friends to whom he had sent his personalized "birthday" card when MingKwai was but a newborn.

Buck recounted her own financial difficulties with her magazine, *ASIA*, and the farm her family now operated. "I am trying steadily to get the farm on a self-supporting basis, but it is like the typewriter—you have to put in before you can take out." "I shall have to say to you, dear friends, what I say to my own sister—I have shelter and food, and these I can share with you, but I have no money."[57]

If Lin's financial troubles were beginning to weigh heavily, making it less likely by the day that the MingKwai project would ever succeed, so too were the geopolitics of the late 1940s. On April 7, 1948, less than one year after Buck's letter, the city of Luoyang fell to Chinese Communist forces, followed by Kaifeng on June 19, Changchun on October 20, and Shenyang on November 1. The tide of the Civil War between Nationalist and Communist forces was turning rapidly, with the Communists launching their offensive for Beijing and Tianjin in mid-December. By January, Chiang Kai-shek had stepped down from office in an attempt to stop the bleeding, and the Chinese Communists established their headquarters in Beijing.

Watching the Chinese Civil War unfold from afar, executives at Mergenthaler Linotype and other US companies grew increasingly nervous about the fate of their patent rights in the event of a Communist victory:

"Little patent protection," an internal company report speculated, "could be looked for in a Communist dominated country"[58] and "import and currency restrictions might be imposed."[59] What is more, Mergenthaler saw on the horizon the possibility of the long-delayed arrival of the "Chinese alphabet," with the Chinese Cadmus coming this time, not in the form of the "Chinese phonetic alphabet" of the 1920s, but in the form of Mao Zedong and his calls for the full-scale Romanization of Chinese script (calls that would also be later abandoned). "It has been reported that the Communists favor the Romanization of the Chinese language," Martin Reed noted in his Mergenthaler report, "and such a program would of course lessen the demand for Lin's typewriter. ... If the educational system were based on Romanization, the need for such a typewriter would rapidly disappear."[60] "In view of the political and military developments in China," the report concluded, "it now appears that this program should be modified with a view to keeping expenditures of effort and money at a minimum until the situation in that country becomes clarified."[61]

Even after Mao Zedong proclaimed the founding of the People's Republic in October 1949, however, optimism prevailed that some sort of market might exist for MingKwai. Tests of the machine continued to be run on Chinese high school graduates in the United States, the experiments being directed by Professor Chung-yuan Chang of Columbia University, and consistently revealing positive results. As an internal report indicated, "Dr. Chang expressed the opinion that Lin's system of classification is the best yet devised."[62] Interest in the machine continued after 1949 as well, particularly on the part of the US State Department, the United Nations, and many of the period's leading scholars of Asia. MingKwai was conducive to other processes besides typewriting, it was felt, most notably telegraphic transmission and offset printing. MingKwai was "admirably adapted to the coding and decoding of Chinese for telegraphic coding," the report continued, and when coupled with offset printing technology, "the Lin machine forms a nucleus for a simple and inexpensive printing plant, which can be made mobile if desired by mounting on a medium sized truck."[63] MingKwai, it seemed, could be weaponized.[64]

The death knell of the MingKwai project came, however, with the escalation of hostilities on the Korean peninsula. Once China had committed its "volunteers" to the battle, and once it became clear that Communist Chinese forces were shedding the blood of US and Allied military personnel, all hope for the dream perished. In a fitting irony, we can read the demise of MingKwai in a grainy, black-and-white film produced by the Psychological Warfare Unit of the Eighth United States Army in Korea (EUSAK) (figure 6.11). Here we see a young female clerk—likely a Taiwanese woman—using a tray-bed, common usage Chinese typewriter to help produce leaflets to be scattered from Allied bombers flying near Communist-controlled sectors of the Korean peninsula. "Your leaders have deceived you," one leaflet reads, in an attempt to persuade Communist Chinese soldiers that their commanding officers were leading them down a "road to certain death."[65] MingKwai, it seemed, had failed.

In our attempts to explain MingKwai's failure, however, we risk overlooking a vitally important factor: *It did not fail*. Although created as a singular device—a prototype developed in the middle part of the twentieth century and debuted in the late 1940s—MingKwai was also something much broader: the instantiation of an entirely new human-machine relationship that, as examined at the outset of this chapter, is by now absolutely inseparable from all Chinese-language information technology. MingKwai marked the birth of "input." Central to the meaning of "input," we recall, is a technolinguistic condition in which the operator is not using the machine to *type* characters per se, but rather to *find* them. As distinct from the act of "typing," the act of "inputting" is one in which an operator uses a keyboard or alternate input system

6.11 United States Army propaganda film featuring the role of Chinese typewriters in the Korean War (selections)

to provide instructions or criteria to a protocol-governed, intermediary system, one that presents Chinese character candidates to the operators that fulfill said criteria. The specific characteristics of these criteria, be they phonetic or structural, are irrelevant to the core definition of input, as is the shape or design of the keyboard or device used in its operation. Just as calligraphy as a whole is not limited to any one brush, and just as movable type is not limited to any one particular typeface, input is not limited to any one particular *kind* of input system. Whether employing Lin's symbology, the symbolic system of Cangjie input, or indeed pinyin as employed by Sougou, Google, and others, input constitutes a new mode of human-machine interaction that encompasses a practically infinite variety of potential approaches, protocols, and symbolic systems. As a specific device developed in the 1930s and debuted in the 1940s, then, MingKwai as a typewriter may indeed have failed; but as a new mode of inscription and human-machine interaction, MingKwai marked the transformation of Chinese information technology in ways that Lin Yutang himself could scarcely have foreseen.

7

THE TYPING REBELLION

One person setting three or four thousand characters [an hour] doesn't amount to much. But if everyone were to set three or four thousand characters, now that'd be something.
—Zhang Jiying, Chinese typesetter, 1952

Long before I set out to write a history of the Chinese typewriter, I spent years staring at it without so much as realizing. In the research for my first book, a history of China's Ethnic Classification project (*minzu shibie*), I had pored over ethnological and linguistic reports authored by social scientists working in the southwestern province of Yunnan. In composing their studies on the categorization of ethnic identity in the region— reports I reread an untold number of times—researchers had often relied upon the very Chinese typewriters that would later constitute the focus of my research. The Chinese typewriter had been hiding in plain sight all those years.[1]

Alerting colleagues and friends in the field, I organized an informal "search party" for other purloined letters. I offered a brief crash course in how to identify typewritten documents (especially how to distinguish them from printed texts), and requested their assistance in surveying their own personal collections and archival holdings. Telltale signs included: occasionally handwritten characters tucked in amid typewritten ones (i.e., infrequent characters for which no character slug was available on the machine), the alternating faintness and darkness of different passages (an issue we discussed in chapter 4), an ever-so-slight zigzag of the text baseline, and the somewhat wider-than-normal spacing between characters. The hunt was on.

A new golden age in the history of Chinese typewriting rapidly came into view, with colleagues sighting the machine all the way from the metropolitan areas of Beijing, Shanghai, Harbin, and Kunming to remote regions of western China. No later than May 1950, the Political Protection Office of the Harbin Municipal Bureau of Public Security began to type up its surveillance reports, as shown in local investigations of Catholic communities.[2] In Beijing, typewritten reports from the Beijing Municipal Vice Food Products Industry Party appear no later than 1952.[3] Provincial-level Party secretaries in Hebei began producing typewritten documents no later than 1955.[4] Typewriting was no means limited to larger urban centers, moreover. In the county of Baoji in Shaanxi province, typewritten documents appear by 1957 in survey reports on local conditions.[5] Most telling of all, typewritten reports from 1956 and 1957 were produced in the remote pastoral region of Zeku county, Qinghai province.[6] Images of patriotic Chinese typists soon appeared as well, celebrated for producing the paperwork of state-building, postrevolutionary consolidation, economic planning, and class struggle. In March 1956, the Chinese typewriter made new strides with its first appearance within a Mao-era propaganda poster (figure 7.1).[7]

The 1950s marked a more vibrant period of Chinese typewriting than I had imagined, surpassing the late Republican period in extent and usage. During the Maoist period, an unrelenting series of sociopolitical and economic campaigns placed an unprecedented burden on Chinese typists, who found themselves tasked with the production of economic reports and low-run mimeographed materials for use in the ubiquitous "study sessions" taking place in work units across the country. So heavy was the burden on typists, indeed, that some work units resorted to outsourcing jobs to nonofficial "type-and-copy shops" (*dazi tengxieshe*), a phenomenon that raised concerns within the fledgling Communist state. With the proliferation of small-scale, independently operated typing shops, the Party-state's monopoly over the means of technolinguistic production

社会___的方向！

任何劳動,都是完成五年計劃不可缺少的劳動,都是光荣的劳動!

7.1 Propaganda poster featuring Chinese typist

was to a degree compromised, insofar as the same collection of typewriters, carbon paper, and mimeograph machines used to print and distribute state-commissioned speech transcripts, political study guides, and statistics was also being used to run a small-scale gray market publishing industry.

Chinese typewriters were used elsewhere to reproduce entire books, known as *dayinben* ("typed and mimeographed editions"), a mode of publication so prevalent that the term was later repurposed in the computer age as the Chinese translation of "to print out" (*dayin*) and "laser printer" (*jiguang dayinji*). In one type-and-mimeograph edition from 1968, a self-identified Red Guard typed and mimeographed the poetry of Chairman Mao, releasing the edition in time for International Workers' Day. In an even more profound act of devotion, members of the "Yunnan University Mao Zedong-ism Artillery Regiment Foreign Language Division Propaganda Group" transcribed Mao's speeches delivered during the years 1957 and 1958, copied from the *People's Daily*, the *China Youth Daily*, the *New China Bimonthly*, the *Henan Daily*, and others. Extending to over 280,000 characters in length, with page after page of densely packed typewritten text, this work would have taken between 100 and 200 hours—or four to eight full days—to type and mimeograph (figure 7.2).[8]

Occupying the vast terrain between handwriting and the printing press, type-and-mimeograph editions continued into the Reform Era (1978–1989). The nationally circulated Reform Era literary journal *Today!*, described by Liansu Meng as the "first unofficial journal in China since 1949," was composed using a Chinese typewriter and mimeograph stencils.[9]

The most fascinating dimension of Chinese typewriting in the second half of the twentieth century was not its prevalence or scope, however. Within this bustling world of typist activity, something else genuinely revolutionary was unfolding. Everyday clerks and secretaries in the Mao era spearheaded a series of innovations, centered around experiments with alternate ways of organizing the Chinese characters on their tray beds. Instead of adhering to radical-stroke organization—or indeed any of the experimental character retrieval systems designed by Chinese elites, as examined in the preceding chapter—these typists undertook a "radical departure," in both a figurative and literal sense. Specifically, they created their own idiosyncratic natural-language arrangements of Chinese characters designed to maximize the adjacency of characters that tended to appear together in actually written language, whether in the form of commonly used two-character compounds (known in Chinese as *ci*) or in key names and phrases within Communist nomenclature, such as "revolution" (*geming*), "socialism" (*shehui zhuyi*), "politics" (*zhengzhi*), and others.[10] Owing to the increased proximity of co-occurring characters, and the repetitiveness of Communist rhetoric, typists who used this experimental method boasted speeds of up to seventy characters per minute, or at least three times faster than the average typing speed in the Republican period.

In other words, among Mao-era typists we can trace the earliest known experimentation with and implementation of an information technology currently referred to as "predictive text"—now a common feature in Chinese search and input methods. Indeed, if "input methods" (*shurufa*) have been one of the pillars of modern Chinese information technology, as we examined in the preceding chapter, the second pillar is undoubtedly that of predictive text. It may come as a surprise that a technology so familiar to denizens of the digital age has deeply analog roots: Chinese

把我们的官僚主义什么东西吹掉，主观主义吹掉。我们以保护同志出发，从团结的愿望出发，经过适当的批评，达到新的团结。讲完了同志们。

在南京部队、江苏、安徽二省党员干部会议上的讲话
（一九五七年三月二十日）

我变成了一个游说先生，一路来到处讲一点话。现在这个时期，有些问题需要答复，就游说到你们这个地方来了，这个地方叫南京，从前也来过。南京这个地方，我看是个好地方，龙蟠虎踞。但有一位先生，他叫章太炎，他说龙蟠虎踞"古人之虚言"是古人讲的假话。看起来，这在国民党是一个虚言，国民党在这里搞了二十年，就被人民赶走了，现在在人民手里，我看南京还是个好地方。

各地方的问题都差不多，现在我们处在一个巨变的时期，就过去的一种斗争……………，基本上结束，基本上完毕了。对帝国主义的斗争是阶级斗争，对官僚资本主义、封建主义、国民党的斗争、抗美援朝、镇压反革命也是阶级斗争，后来呢，我们又搞社会主义运动、社会主义改造，它的性质也是阶级斗争的性质。

那么，合作化是不是阶级斗争呢？合作化当然不是一个阶级向一个阶级作斗争。但是合作化是由一种制度过渡到另一种制度，由个体的制度过渡到集体的制度，个体生产，它是资本主义的范畴，它是资本主义的地盘。资本主义发生在那个地方，而且经常发生着，合作化就把资本主义发生的地盘，发生的根据地去掉了。

所以从总的来说，过去我们几十年就干了个阶级斗争，改变了一个上层建筑，旧的政府、蒋介石的政府，我们把它打倒了，建立了人民政府，改变了生产关系，改变了社会经济制度，从经济制度和政治制度来说，我们的社会面貌改变了。你看，我们这个会场的人，不是国民党，而是共产党。从前，我们这些人，这个地方是不能来的，那一个大城市都不许我们去的。这样看来是改变了，而且改变了好几年了，这是上层建筑、政治制度；经济制度改为社会主义经济制度，就在最近几年了，现在也可以说是基本上成功了。这是过去我们几十年斗争的结果。拿共产党的历史来说，有三十几年；从鸦片战争反帝国主义算起，有一百多年，我们仅仅做了一件事——干了个阶级斗争。

同志们，阶级斗争改变上层建筑跟社会经济制度，这仅仅是为改变另外事开辟道路。现在，遇到了新的问题。过去那个斗争，就我们国内来说，现在基本上完结了，从国际上来说，还没有完结。为什么我们还要解放军呢？主要就是为了对付外国帝国主义，恐怕帝国主义来侵略，它是不怀好心的；国内也还有少数没有查出来的反革命残余分子，有一些过去被镇压过的，比如象地主阶级、国民党残余，如果我们没有解放军，它又会起来的。地主、富农、资本家现在守规矩了。资本家还不同些，我们把它当做人民内部的问题来处理。民族资产阶级接受社会主义，跟农民接受合作化不同，他们可以说是一种半强迫，就是说有些勉强，而且是在对他们相当有利的条件下接受改造的。所以，现在是处在这么一个巨变时期，………………由革命到建设，由过去我们反帝反封建的革命和后头的社会主义革

·109·

7.2 Type-and-mimeograph edition of *Long Live Chairman Mao Thought. Selections from 1957 and 1958* (c. 1958)

predictive text was invented, popularized, and refined in the context of mechanical Chinese typewriting before the advent of computing. What is more, this innovation cannot be ascribed to a single inventor, but rather came into being through a diffuse collective made up of largely anonymous typists.

CHINA'S FIRST "MODEL TYPIST"

In November 1956 a typist in the central Chinese city of Luoyang accomplished an astonishing feat. Using a mechanical Chinese typewriter, the very same we have come to know well over the course of this book, the operator typed 4,730 characters in one hour, setting a new record just shy of eighty characters per minute.[11] While seemingly unremarkable when considered within the more familiar context of alphabetic typewriting, the magnitude of the record becomes apparent when one considers the average speed of Chinese typists at the time: twenty to thirty characters per minute. This accomplishment thus represented a two- to fourfold acceleration of the apparatus. The typist had not achieved this record by means of electrical automation, moreover, or a new kind of typewriter. Instead, he had simply rearranged the Chinese characters on the machine's tray bed. Moving away from the longstanding taxonomic system of radical-stroke organization, and eschewing even the most experimental of character retrieval systems developed in the Republican period, the typist in 1956 reorganized the characters on the machine into natural-language clusters designed to maximize the adjacency and proximity of those characters in the Chinese language that tended to go together in actual writing.

Newspapers in the opening decades of the People's Republic of China (1949–present) were replete with stories of "model workers," proletarian champions who displayed feats of unprecedented production through the combined application of will and wisdom. While such exuberant claims cannot be accepted uncritically, in the case of the "model typist" from Luoyang, a diverse body of archival sources and material artifacts has led me to trust the accuracy—and perhaps even the quantitative claims—of this remarkable report.

To illustrate this typist's system briefly—we will return to it in greater detail later—we can examine a sample arrangement included as part of

ao	*de*	*huo*	*shu*	...	...	...
到	得	獲	述			
he	*da*	*ping*	*bian*	*shang*	...	...
何	達	評	邊	上		
uan	*gao*	*ru*	*xia*	*zhun*	*yu*	*yi*
轉	告	如	下	准	予	以
eng	*bao*	*ling*	*dao*	*pi*	*pan*	*suo*
呈	報	領	導	批	判	所
ui	*bing*	*qing*	*zhi*	*shi*	*fu*	*su*
滙	並	請	指	示	復	速

7.3 Sample of new character arrangement; from "Introduction to the 'New Typing Method' ('*Xin dazi caozuofa' jieshao*) ['新打字操作法'介紹]," *People's Daily* (*Renmin ribao*) (November 30, 1953), 3 (Romanizations included for purposes of illustration only)

a 1953 introduction for the public. The principles of this experimental organization can be seen in figure 7.3.

Beginning with the character *gao* (告), shaded in gray, we find located within the adjacent cells the characters *bao* (報) and *zhuan* (轉)—each of which can be combined with *gao* to form common two-character words: *baogao* (報告), meaning "report," and *zhuangao* (轉告), meaning "to pass on" or "transmit." Continuing our exploration from the vantage point of these two characters, we see further layers within this associative system. Adjacent to the character *bao*, we find three additional characters with which *bao* can be meaningfully combined: *cheng* (呈), *hui* (滙), and *zhuan* (轉), which can be paired to produce, respectively, *chengbao* (呈報 "to submit a report"), *huibao* (滙報 "to give an account of"), and *zhuanbao* (轉報 "message transfer"). Still other combinations surround the character *shang* (上), combinable with those of *xia* (下), *bian* (邊), and *shu* (述) to form "above and below," "top," and "above-mentioned," respectively. All told, the area pictured here contains no fewer than twenty-five perfectly adjacent compounds, as well as an additional set of proximate compounds. This density is impressive when we consider that there are only thirty individual characters in this sample area, out of the total number of 2,450 on the tray bed (all of which, as we will soon see, were likely organized in the same associative fashion), and that these characters would not have been adjacent if organized according to the conventional radical-stroke system. Using machines with this new method of character organization, typists in the early Maoist period set out on one of the most sweeping technolinguistic experiments in the modern age.

This transformation should be understood less as a leap of imagination and cognition than as the manifestation within a particular political environment of a longstanding, deeply corporeal relationship that obtained between machines and human bodies—or what Ingrid Richardson refers to as the "technosomatic" complex.[12] The sudden rearrangement of characters on Communist-era tray beds can be understood only if we see it as part of the much longer historical process we have examined thus far: an aggregation of tactile, decentralized, and largely unnoted experiences of thousands of typists and typesetters, individuals who interacted with their character racks and tray beds in embodied, nonverbal ways. It was within this hum of activity, and over the course of countless millions of fleeting, nanohistorical moments, that the leap to natural-language arrangement became conceivable, practicable, and, within the political milieu of the Maoist period, celebrated and incentivized—the countless acts of picking up type, setting it back down, moving from one character to the next on the typewriter, depressing the type lever, and so forth. To understand the history in question, it is essential that we engage with the *habitus* of Chinese technolinguistic practice, the "embodied history, internalized as a second nature and so forgotten as history."[13]

CHINA'S SECOND VERNACULARIZATION CAMPAIGN

The origins of this diffuse, decentralized, and grassroots movement are impossible to identify with certainty, and yet available evidence helps us trace out its contours with some confidence. The earliest and clearest example is found in the activities, not of a typist, but of a typesetter named Zhang Jiying, who became known to readers of the *People's Daily* in 1951 in an article entitled "Kaifeng Typesetter Zhang Jiying Diligently Improves Typesetting Method, Establishes New Record of 3,000-plus Characters per Hour."[14] Having worked as a typesetter for over a decade, first in the city of Zhengzhou and later in Kaifeng, Zhang was trained on both the older "24-tray character rack" and the newer "18-tray character rack," and posted very respectable typesetting speeds (ranging from 1,200 to 2,200 characters per hour over the course of his career thus far).[15] Only a few short months after the formation of the PRC, however, he experienced a reported surge of inspiration and began to engage in a sweeping, experimental reorganization of his character rack—an experiment that culminated in his 1951 feat.

What caught Zhang's attention in particular was his colleagues' practice of pairing characters together on the character rack that they employed frequently in the course of daily work. Three characters in particular had been clustered together, in clear violation of radical-stroke organization: *xin* (新), *hua* (華), and *she* (社), which together form the name "New China Press" or *Xinhuashe*. "I thought, if one were to put a group of related characters together," Zhang would later explain, "it would definitely be good for setting."[16]

Zhang set out to apply this principle across his entire character rack. His character rack would soon feature over 280 two-character compounds, eight three-character sequences, and even seven four-character sequences, a style of organization he termed *lianchuan*—meaning "series" or "chain."[17] He included terms and names such as "revolution" (*geming*), "American imperialist" (*Meidi*), "liberation army" (*jiefangjun*), "agriculture" (*nongye*), and many others employed in Communist parlance. These were *cliché*, in the original French meaning of the term, signifying a printer's "stereotype block" upon which is etched a commonly used phrase, rather than just a letter.[18] Derived from the past participle of the verb *clicher*, or *to click*, the term was connected to the sound made by a printing sort being set in place. Over time, *cliché* drifted semantically to its present-day meaning of a "trite, worn-out expression."

Zhang extended this organizational scheme to practically the entire character rack, not merely a small, specialized region thereof as his colleagues had done. Zhang's *clichés* were not consistent, moreover, but transformed depending upon the properties of the text under production and the overarching political environment.[19] If "materials on the workers' movement" constituted the operative theme of one period, Zhang explained, he prepared such compounds as "production" (*shengchan*), "experience" (*jingyan*), "labor" (*laodong*), and "record" (*jilu*). At other times, the demands of a specific propaganda campaign might dominate media attention, prompting Zhang to rearrange his character rack anew, prioritizing terms and phrases such as "Resist America, Aid Korea" (*kang Mei yuan Chao*) (the Korean War–era mass mobilization campaign).[20] In this way, Zhang set about transforming his body and his character rack into Chinese Communist Party (CCP) rhetoric incarnate, not in the sense of parroting certain key terms, but in the sense that his fingers, hands, wrists, elbows, eyes, peripheral vision, joints, movements, anticipatory reflexes—every part of his body—would be intimately attuned and maximally sensitized to the distinct cadences of CCP rhetoric.

Zhang pushed his new character arrangement system further, surpassing his own record: a record 4,778 characters set in one hour, or nearly eighty characters per minute, captured on film on July 29, 1952, by the South Central Film Team of the Central Film Production Company.[21] In the meantime, Party-state authorities saw in Zhang's accomplishment an opportunity to celebrate the kind of proletarian parable so vital at the time: the model worker who had used personal initiative and free time to push his industry beyond what others had imagined possible, overturning "tradition" broadly writ, and in doing so demonstrating to the masses both the possibilities of ingenuity and the impermissibility of complacency.[22] It was Zhang's departure from conventional practice—his *heteropraxy*—that helped him serve orthodoxy far more effectively than rote, conventional practice—or orthopraxy—ever could. "One person setting three or four thousand characters [an hour] doesn't amount to much," Zhang reflected. "But if everyone were to set three or four thousand characters, now that'd be something."[23] In quick succession, the Party extended an invitation to the typesetter to take part in the May Day celebration of 1952, helped him to co-author a book explaining his method in greater depth, sponsored his tour of publishing houses across the country, admitted him to the Party, and encouraged others to study and perhaps emulate his method (figure 7.4).[24]

The case of Zhang Jiying was soon followed by others. In 1952, Commercial Press in Shanghai undertook character tray reform, implementing the same *lianchuanzi* system and witnessing a clear increase in speed as a result.[25] The Jinggangshan Newspaper Printing House

7.4 Zhang Jiying

implemented the *lianchuanzi* arrangement as well, as part of their shift away from the industry standard twenty-four-part character tray to the so-called "'eight'-character-style" character rack (*ba zi shi*). And just as the city of Kaifeng had its model typesetter in Zhang Jiying, Jinggangshan could take pride in their own: local typesetter Wang Xinshun, who set an all-province hourly record on April 10, 1958, with 3,840 characters set. Wang broke his own record later that same year, setting 4,100.[26]

By 1958, *lianchuanzi* had become widespread enough to be spotlighted in a manual on page layout and typesetting published by the journalism research institute at People's University.[27] Here the method was referred to as the "Connected Language Tray Bed" (*lianyu zipan*) or the "Connected Character Tray Bed" (*lianchuanzi zipan*), and was explained as follows:

As much as possible, you want to place compounds together that are used together and are related—for example, by placing the four characters "jie" (解), "jue" (决), "wen" (问), "ti" (题) in four adjacent columns.[28] Another example would be putting characters together like "jian" (建), "she" (设), "zu" (祖), "guo" (国), and "ti" (提), "gao" (高), "chan" (产), "liang" (量), which would make setting up these characters a lot more convenient.[29] We could even consider using radiating style (*fangshexing*) or chainlink style (*liansuoxing*) to set up a common usage tray bed, for example "ren" (人), "min" (民), "gong" (公), "she" (社) —> "hui" (会), "zhu" (主), "yi" (义).[30] Radiating it out like this would resemble the word game "thimble linking" (*dingzhen xuxian*).[31]

A similar technique, the manual explained, was to pair linked words and place them in thematic zones. In one zone, the typesetter might arrange clusters such as "American imperialism" (*Meidi*), "to invade" (*qinlüe*), and "to destroy" (*pohuai*), calling this the "negative connotation terms tray" (*bianyi zipan*).[32] Another region could then be designated the "Social Structure Terms Tray" (*shehui zuzhi mingcheng de pan*), featuring, as the manual explained, "'Socialism,' 'cooperative,' 'Chairman Mao,' etc., etc." ("*shehui zhuyi*," "*hezuoshe*," "*Mao zhuxi*" *dengdeng*).[33]

To find "Chairman Mao" and "socialism" tucked inside quotation marks, followed by the term *dengdeng* ("so on and so forth" or "et cetera"), is deeply revealing, alerting us to the metacognitive distance that was central to this organizational method. In order to make use of natural-language arrangements, one could not be an unwitting parroter of routine political expressions, spouting Mao-era phraseology unreflectively. To the contrary, one had to be acutely aware of their roteness and regularity—to think of them precisely *as cliché*, in order to be able to do an effective job of preparing one's character rack or tray bed for maximum efficiency. The success of Zhang's method was a function of maximum sensitization to and anticipation of the distinct cadences of CCP rhetoric, not just with obvious cases such as "Mao Zedong" and "cadre" (*ganbu*), but also politically charged "neutral" terms such as "education" (*jiaoyu*), "to exist" (*cunzai*), "according to" (*genju*), and others. It was Zhang's ability to establish a deeply private, individualistic, even esoteric relationship with the public, authoritative, and standardized language of the era that increased his capacity to serve political orthodoxy. Radical individualism was, in this context, completely compatible with and conducive to state power.

CHINESE TYPEWRITING AND "MASS SCIENCE"

As illustrated in the case of "New China Press," Zhang Jiying was not the first to conceptualize a vernacular taxonomic approach to Chinese typesetting. He was, however, perhaps the first to pursue this possibility to its logical extremes, extending the *lianchuanzi* system across the entirety of the character rack, rather than a limited region thereof. The same distinction held true for Chinese typewriting, in which—as we recall from chapters 3 and 4—character slugs in the Chinese typewriter tray bed were not fixed, but rather removable and replaceable. Although the application of natural-language clusters achieved unprecedented stature and levels of experimentation beginning in the 1950s, in fact there were already limited "predictive" elements to the Chinese typewriters we examined from the Republican period. Two in particular merit consideration. First, we recall there existed one region on Republican-era typewriter tray beds reserved for "special usage characters" (*teyong wenzi*), in which characters were not organized according to radical classes or stroke count. Instead, within this narrow strip on the tray bed, measuring four columns wide and thirty-four rows deep, characters were clustered so as to form common two-character compounds and multicharacter sequences such as "the Republic of China" (*Zhonghua minguo*).[34] In particular, typists dedicated this small region to the characters that formed the names of Chinese provinces. For example, the characters *meng* (蒙) and *gu* (古) were located horizontally adjacent to one another, forming the name Menggu or "Mongolia." This arrangement became more complex and interesting in the case of characters that appeared in more than one toponym, such as *jiang* (江), *hu* (湖), *nan* (南), *xi* (西), *dong* (東), and *shan* (山) (figure 7.5). To the left of *jiang*, for example, was the character *zhe* (浙), forming the name Zhejiang; to its upper-left corner, the character *su* (蘇), forming Jiangsu; to its lower left was *long* (龍), beneath which was located a third character *hei* (黑)—together forming the province name Heilongjiang.[35]

The second example comes in a letter dated August 26, 1928, from newspaper reporter, editor, and language reformer Chen Guangyao to Wang Yunwu, editor in chief at Commercial Press and inventor of the "Four-Corner" retrieval system we saw in the previous chapter. Knowing of the company's Chinese typewriter division, Chen suggested to Wang the possibility of reorganizing the company's Chinese typewriter tray bed so as to maximize the adjacency of naturally co-occurring Chinese characters. By way of example, Chen cited the two- and more-character terms "then" (*ranze* 然則), "China" (*Zhongguo* 中國), "to invent/invention" (*faming* 發明), "Three People's Principles" (*Sanmin zhuyi* 三民主義), "world" (*shijie* 世界), and even longer idiomatic expressions like "to be unexpected" (*yiliao zhiwai* 意料之外).[36] When ordering the characters in this way, he explained, a compression of only a few characters would make possible many more compounds. Even a simple four-character sequence 發明達光 made it possible to write such common terms as "to invent/invention" (*faming* 發明), "to develop" (*fada* 發達), and "bright" (*guangming* 光明), all with much greater speed and ease than when using radical-stroke arrangement. Moreover, common particles could be organized more scientifically by placing them, not in accordance with dictionary order, but in proximity to those characters alongside which they normally appeared.

For all his encouragement, however, Chen Guangyao's experimental proposal to Commercial Press was never adopted by the company, and never took hold in other areas of Chinese typewriting. Indeed, within

su 蘇	*zhou* 洲
zhe 浙	*jiang* 江
long 龍	*bei* 北
hei 黑	*hu* 湖

7.5 Sample from "Special Character Region" of Huawen typewriter (pre-1928); from *Chinese Typewriter Character Arrangement Table* (*Huawen daziji wenzi pailie biao*) [華文打字機文字排列表], character table included with *Teaching Materials for the Chinese Typewriter* (*Huawen dazi jiangyi*) [華文打字講義], n.p., n.d. (produced pre-1928, circa 1917)

the rich archives on Republican-era typewriting, there is not a single further reference to the application of his system, whether in theory or in practice. To the contrary, all evidence suggests that, while elements of natural-language arrangement continued to be used on Republican-era machines, they remained entirely localized to the small "special usage" region of the typewriter.

The early PRC period marked a different era altogether. In the wake of the Zhang Jiying mini-phenomenon, natural-language organization was quickly appropriated from typesetting and applied to the field of Chinese typewriting. In November 1953, readers of the *People's Daily* encountered Shen Yunfen, a young woman who had joined the People's Liberation Army two years prior, at the age of seventeen.[37] A native of Shanghai, she was appointed to the North China Military Region Headquarters to serve as a typist in October 1951 during the Korean War. By her own account, her performance was slow at the outset, which led her to experience despondence—even to the point of losing weight from the anxieties involved. Under the guidance of a colleague, Shen reported increasing her speed to 2,113 characters per hour—a respectable improvement, but one that left her dissatisfied.

Shen decided to pursue the "New Typing Method" (*Xin dazi caozuofa*), a method she had learned of recently, attributed to one Wang Jialong.[38] Applying the same principle of adjacency as Zhang Jiying's "serial" method, Wang and soon others exploited the shape of the typewriter tray bed to extend Zhang's linear, one-dimensional organization into a two-dimensional, *x-y* matrix. The operative principle of the New Typing Method was termed "radiating compounds" (*fuci fangshe tuan*), explained as follows: by "selecting one character as the core and then radiating outward from it," the typist could populate the three to eight spaces around each character with as many related characters as possible.[39] Owing to this multidimensionality, the typist could push beyond the left-right sequencing of related characters on the compositor's character rack, and begin to experiment with both vertical and diagonal arrangements. It also made possible the stringing together of these mini-regions into widening associative networks.

In moving away from radical-stroke organization, a space of practically infinite possibility was opened up. Shen Yunfen's speeds steadily increased, reaching 3,012 characters per hour and then leading her to a first-place finish in the North China Military Region Typing Competition in 1953 with a record-setting 3,337 characters in one hour. On January 25, 1953, the young Shen was granted the title of "First-Class Hero" (*yideng gongcheng*) and "second-level model worker" (*erji mofan*). In September 1955, she was received by Mao Zedong at the National Conference of Youth Activists in Socialist Construction.[40]

An essential question emerges out of the stark contrast between Chen Guangyao's failure in the Republican period and the vibrant experimentation of the 1950s: why was it only during the Communist period that this already known technolinguistic technique—"new," "China," "press"— became the object of intense focus and exploration? How did the logic of *xinhuashe*, a region accounting for only a fraction of the overall character rack, become the organizational principle for the entire character rack? Likewise, how did the logic governing a thin strip on Republican-era tray beds, a "special" region that accounted for not even 6 percent of the total lexical space, eventually conquer the surrounding 94 percent? One potential answer that can be disqualified immediately is any suggestion that the Republican period was in some sense less innovative, experimental, or anti-traditional with regard to language reform overall. To the contrary, as we saw in the preceding chapter, the late Qing and Republican periods witnessed a practically uninterrupted exploration of alternate character organization and retrieval systems, and scores of reformers heaped criticism upon the organization of the *Kangxi Dictionary* and its attendant radical-stroke system. Why then was it not until the 1950s that predictive text strategies began to proliferate within typesetting offices and on Chinese typewriter tray beds? How do we account for the sudden rise of natural-language experimentation in the Maoist period?

To understand the emergence of natural-language experimentation in the early Maoist period, we must consider three key political changes that took place following the Communist revolution of 1949: the Chinese Communist endorsement and celebration of what have been termed "popular" or "mass" knowledges; an increasingly routinized if not "predictable" Chinese Communist rhetoric; and unprecedented time pressure experienced by typists during the immediate postrevolutionary period.[41] Just as Communist authorities made and endorsed radical pushes for nonelite participation in fields as diverse as paleontology, medicine, and seismology, so too did they endorse calls for a sweeping, bottom-up reorganization of the Chinese language—a mass taxonomy that would depart from earlier modes of categorizing Chinese characters, and better reflect the way in which "common people" organized their linguistic universes.[42] If it was possible to proletarianize medicine and the physical sciences, and to "challenge the notion that science was the province of elites," why not the systems by which script was organized?[43] The second transformation involved the development of a political discourse of unprecedented routinization—a "systemization of ideological categories and language" that came to influence, if not define, entire domains of textual production in the early PRC period.[44] This was true not merely for overtly Chinese Communist keywords such as "struggle" (*douzheng*) and "proletariat" (*wuchan jieji*), but also for an ever-expanding repertoire of what Franz Schurmann has aptly termed "seemingly conventional words with special significance"—terms such as "opinion" (*yijian*) and "discussion" (*taolun*).[45] A third factor was the unprecedented employment of Chinese typists in the conduct of government affairs, a topic broached at the beginning of this chapter. By the mid-to-late 1950s, Chinese typists and typewriters could be found throughout China, serving an increasingly regular role in the everyday functioning of government, as well as in the unrelenting series of Mao-era mobilization campaigns—campaigns for which typists were tasked with producing economic reports and low-run mimeographed materials. During the Great Leap Forward, in particular, this increasing pressure on typists developed into a metanarrative all its own, with typists beginning to adopt the conventional language of "quotas" and "output." Typists and work units competed with one another to outstrip target goals and to increase the efficiency of their "production"—that is, the total number of typewritten characters produced per month and per year.

These three novel conditions combined to catalyze the transformation of the tray bed into unprecedented linguistic configurations. The possibilities for such experimentation were practically boundless. With roughly 2,500 characters on the tray bed, typists had an unimaginably large number of different arrangements to try, making possible a total democratization of character organization in which each typist organized characters as he or she saw fit. More specifically, the possibility emerged of a system completely suited to two things simultaneously: one's own body, including its deeply personal and varied idiosyncrasies of movement; and the increasingly standardized Maoist discourse of the period.

DECENTRALIZATION, CENTRALIZED?

Having witnessed the celebration of Zhang Jiying, Shen Yunfen, and other model taxonomists in the popular press, we might be tempted to imagine that Party, state, and industry elites saw the virtue of this new, decentralized taxonomic experiment and encouraged the nation's typists to stride forth toward this brave new future. What transpired was something altogether different. Party and industry elites may have celebrated Zhang Jiying, but ultimately they had little faith that his methods could be adopted in any widespread manner by Chinese typists. Zhang and

Shen were models, to be sure, but not ones that Party and industry leaders believed it possible for all compositors and typists to emulate. Rather than lending support to further user-led experiments in vernacular taxonomy, the early Mao-era typewriter industry instead set off down a path common to other branches of media in the early PRC: centralization. Specifically, they set out to standardize and centralize this new phenomenon of vernacular taxonomy.

We gain insight into the minds of early PRC typewriting circles through the minutes of a 1953 meeting held in Tianjin. At the "Meeting for the Improvement of the Typewriter Character Chart" (*Xiugai daziji zibiao huiyi*), fifty-two representatives convened on August 30: representatives from manufacturers, typing schools, the Tianjin City Committee of the CCP, the Tianjin municipal government, and some thirty other work units. Here participants reflected upon the recent past of the Chinese typewriter industry, and their visions for its future.

The preeminent concern among participants was the problem of standardization. There were problems with the Chinese typewriter, one participant argued, "with regard to the old tray bed, as well as the selection and arrangement of characters. These have adversely affected the improvement of the efficiency of the typewriter."[46] More broadly, complaints were expressed about the general lack of standardization across typewriter manufacturers, noting that each manufacturer used a different character layout chart. In an age of standardization and rationalization, such a tendency could hardly be permitted to continue unchecked. Cited favorably at the meeting was the Party's campaign to simplify Chinese characters and to abolish "variant characters" (*yitizi*).[47] If a comparable level of standardization could be brought to bear within Chinese typewriting, participants agreed, the positive effects would be manifest. Once Chinese typewriter tray beds were set up in a unified way, for example, editors could in turn publish character indexes and typing textbooks to be used by one and all. "This would be extremely helpful for both study and usage," as one participant summarized.[48]

The typing reform committee in Tianjin was well aware of the experiments then underway with vernacular taxonomy, moreover. They just did not imagine them implementable across a wide community of practice. In their reports they referred to these as "radiating compounds" (*fangshe*

zituan), defined as "taking one character as the core, then arranging specially related compounds to the top, bottom, left, and right." Offering a
measure of praise for the system, the committee was also quick to outline
no fewer than three problems with it. First, it provided no "set sequence"
(*yiding de cixu*), relying instead upon "rote memorization and groping
around" (*qiangji mosuo*). Second, because Chinese characters connected
with one another to form an exceedingly large number of compounds,
the radiating method could hardly hope to achieve comprehensiveness.
The third and most critical problem, however, was its unmistakably idiosyncratic and individualistic quality. Were a typist using such a method
to depart her post, or were a clerk to fall ill, it would prove difficult to
replace her. This system would not suffice, it was decided—it was not
"absolutely good."[49] In sharp contrast to the stories of Zhang Jiying, Shen
Yunfen, and others, a core principle came to be shared by those in atten-
dance at the Tianjin meeting: Chinese typewriters should maintain the
industry standard radical-stroke system of organizing characters.

As if in consolation, however, the committee did put forth its own,
far more moderate proposal for the tray bed. Within a given radical class,
the sequence of characters need not adhere strictly to stroke count, they
conceded. If two characters within a given radical class tended to appear
together in Chinese words or phrases, it was reasonable to organize them
adjacently on the tray bed—even if this meant violating stroke count.[50]
The earliest manifestation we encounter of this "relaxed" radical-stroke
tray bed was the Wanneng ("All-Purpose") Chinese Typewriter, manufactured circa 1956—the machine once built by Japan, but now under
control of Chinese manufacturers.[51] Far more reserved than Zhang
Jiying's character rack, this tray bed featured what can be thought of as at
most an "adjustment" rather than abandonment of the radical-stroke system: radical classes were maintained, but within them characters could be
placed out of order in terms of stroke count.

A sign of the Wanneng's partial loosening of organization can be seen
when we look at the characters categorized within the "water" radical.
Directly to the left of the character *ze*, we find *mao*: the typist could thus
proceed directly between two of the three characters that formed the
name of the Great Helmsman, Mao Zedong. At the same time, the third
character in Mao's full name—*dong*, meaning "east"—was located where

it tended to be located on all typewriters to date: in the "special usage" region, positioned alongside the three other cardinal directions. With the Wanneng machine, then, we encounter the industry's first response to the bottom-up experiment with vernacular taxonomy. Unwilling to break with the radical-stroke system, typewriter manufacturers and industry leaders instead put forth a compromised vision in which typists would have to content themselves with refinements only slightly more suited to the rapid production of common compounds and names. At the same time, industry leaders cautioned that even this relaxation in the radical-stroke system needed to be undertaken slowly, and that it would take users time to learn. While a limited amount of individual personalization was "feasible" (*kexingde*),[52] the Tianjin committee noted, "it's best not to make any more big changes" (*zuihao buyao zai da gaizhuang*).

No doubt challenged and inspired by local-level experiments well underway by the mid-1950s, the typewriting industry decided to undertake a slightly more dramatic departure: the "Reformed" Chinese typewriter of 1956, which constituted the first machine to be outfitted with an "out-of-the-box" natural-language tray bed. In this tray bed arrangement, radical-stroke taxonomy was no longer obeyed, with the manufacturer instead developing its own vernacular arrangement based upon the same principles then circulating among everyday users.[53] In one sense, this move constituted an endorsement of local-level experimentation, while at the same time, it redoubled the industry's commitment to centralization and standardization. On this machine, the arrangement of Chinese characters would be subordinated not to the body of any individual typist per se, but instead to that of a hypothetical, average typist to be defined by the manufacturer itself—a kind of *homme moyen* not unlike those we encountered in chapter 6. Once this transition to vernacular arrangement had been made, presumably, textbooks could continue to be edited and published *en masse*, as could character tray bed charts. Once incorporated into typing institutes and programs, moreover, typists could be expected to memorize this new layout precisely as they had the earlier radical-stroke system. Phrased differently, the industry attempted to standardize the vernacular, exhibiting the same impulse as Chinese elites in the first half of the twentieth century during the country's first and more famous vernacularization movement.

THE TYPING REBELLION

Manufacturers may have been content to stop at the "reformed" tray bed, but typists were not. In a historical development that further highlights the importance of user-driven technological change, individual Chinese operators pushed nascent ideas of tray bed reorganization toward what in many ways was its logical extreme: a total democratization of character organization in which each typist organized his or her tray bed as he or she saw fit, in accordance with the many particularities of his or her own, individual body.[54] There would be no standards, no centralization, and, indeed, effectively infinite potential variation (with 2,500 characters being amenable to approximately 1.6288×10^{7528} different arrangements).[55]

In this sense, Chinese typists took centrally issued propaganda about "model typists" and "model typesetters" more seriously and literally than central authorities had anticipated, creating machines that were at once deeper extensions of their bodies into Chinese Communist rhetoric, and deeper ingestions of this rhetoric into their bodies. Departing entirely from the radical-stroke system, and from the dictates of any centrally authorized taxonomic "starting point" to the tray bed, the goal became the development of organizational systems completely suited to one's own body and to the discourse of the Maoist period. What ensued was the development of a practically infinite number of deeply personal pathways to an increasingly rote, standardized political discourse. By means of this comprehensive subordination of the machinery of language to the body—not to one centrally determined, hypothetical body, but to all bodies, democratically, empirically, and privately determined—what became possible was a more perfect and ever more personal connection and commitment to the rhetorical apparatus of Maoism.

Confronted by dizzying possibility, typists hardly engaged in blind or random rearrangements of characters. There was an emerging logic to vernacular taxonomy, as well as an emerging community of practice in which one could share and learn principles. Indeed, decentralized, user-driven reorganization became so important within Chinese typewriting that, beginning in the 1960s, we begin to see a formalization of natural-language experimentation—an attempt not to centralize it, but to set down certain "best practices" in writing. In a fascinating explanation of natural-language tray bed arrangements from 1960, authors Wang Gui-hua and Lin Gensheng drilled down into the question of how one should go about setting up a vernacular tray bed.[56] They outlined for their readers the different factors that influenced when one should undertake such a renovation, as well as certain spatial-linguistic factors that should be kept in mind during the process.

The authors contrasted two strategies for setting up a vernacular machine: "gradual improvement" (*zhuri gaijin*) and "all-at-once rearrangement" (*yici gaipai*). Gradual improvements of the tray bed, they explained, involved making incremental changes each day, taking careful notes about which characters one used more and less frequently, selectively moving these higher- and lower-frequency characters around the tray bed, and taking detailed notes about any changes one made. "Gradual improvement" was ideally suited to work units with only one typist, Wang and Lin suggested, because it would minimize disturbance to the unit's workflow, and because the redistributed characters would be simpler to remember. There were disadvantages to the method, however. With more than two thousand characters on the tray bed, it could take an exceedingly long time to complete the process—as long as a year if one changed between six and seven characters every day. Perhaps most importantly, the gradual method was unsystematic, since it was carried out in piecemeal fashion. The typist in this method did not engage in extensive preparation or consideration, thereby raising the risk of making poor taxonomic decisions that, while seemingly appropriate at first, could later prove detrimental and difficult to remedy.

The "all-at-once" method was a more extreme alternative. After extensive mapping and planning, the typist would use his or her spare time to empty the tray bed completely, and then build it back up, cell by cell, in accordance with a carefully determined lexical blueprint. In a single, concerted exertion of mental and physical labor, the process could be completed in its entirety. An all-at-once transformation incurred obvious risks, however. First, it placed a tremendous onus on the typist's memory, requiring him or her to memorize an entirely new organizational layout right away (a challenging task, even if we are speaking of a system that the typist would have developed personally). To cope, Wang and Lin explained, it was advisable for the typist to spend his or her spare time

memorizing the new layout as rigorously as possible, once the new layout was established. Either the whole arrangement had to be memorized
from day one—an improbable feat—or the typist's productivity and speed
would necessarily suffer for a period of time. As such, this method was
less advisable than the "gradual method" for work units that relied upon
a single typist. "Rearranging a tray bed must be done with care and attention, and assiduously," Wang and Lin summarized. "You must not engage
in such a thing carelessly. But at the same time, you must overcome all
kinds of conservative thinking."[57]

Wang and Lin also provided a detailed overview of the best systems
one could use, focusing on five spatial-linguistic approaches in particular: *association style, radiating style, fortress style, connecting verse style*, and
repeating character style (*chongfuzi shi*).[58] Within association style (*jituan-
shi*), the typist started by placing a character within the tray bed matrix,
and then surrounded it on all sides with related characters—building further associative clusters from there, using each of those characters as a
new starting point. The example given by the author was that of *shi* (时),
indicating "time" in a general sense. With this character as epicenter,
a typist using association style would then surround it with characters
such as *ping* (平), *tong* (同), *ji* (及), *zan* (暂), *xiao* (小), *sui* (随), *lin* (临), and
so forth. Each of these characters, when concatenated with *shi*, formed
common two-character Chinese words: "normally" (*pingshi*), "at the
same time" (*tongshi*), "timely" (*jishi*), "hour" (*xiaoshi*), "at a time of one's
choosing" (*suishi*), and "temporarily" (*linshi*). In fortress style (*baoleishi*),
by comparison, the typist combined place names, personal names, or
technical terms in one dedicated zone of the tray bed, even if these terms
could not be combined with one another to any significant degree. One
simply knew in this technique that all country names, for example, were
to be found in the lower right zone of the tray bed, whereas personal
names were to be found in the lower left.[59]

The principles outlined by Wang, Lin, and others were adopted and
elaborated upon by typists throughout the Maoist period, and indeed
beyond. From the evidence of two machines manufactured in mainland China, and used by mainland Chinese typists during the 1970s
and 1980s in two separate locations—the United Nations in Geneva, and
the offices of UNESCO in Paris—we discover both shared patterns and

idiosyncratic differences that reveal the factors and strategies guiding this vernacularization movement.[60] As these examples illustrate, these typists were engaged in ongoing, situated processes of "tinkering and making ad hoc arrangements" and "reconfigurations" not unlike those described in other user-machine contexts examined by Adele Clarke, Joan Fujimura, and Lucy Suchman.[61] Starting with the character *mao* (as in "Mao Zedong"), we can juxtapose each of these two machines against one another and against a machine from the pre-Communist period configured according to radical-stroke arrangement (figure 7.6).

As shown here, the character *mao* was arranged according to the radical-stroke system on the Republican-era machine, just above the character *hao* (毫) built out of the same "hair radical" component (毛). On the UNESCO and UN machines, by contrast, the placement of the character *mao* makes clear that these two typists were weighing an entirely different and new set of considerations. Rather than placing the character where it belonged within the radical-stroke system, the UNESCO typist placed *mao* within a specifically political configuration: directly above the characters *ze* (泽) and *dong* (东) (forming Mao's full name); and directly to the left of the characters *zhu* (主) and *xi* (席) (forming "*Mao zhuxi*" or

"Shu-style" typewriter (1930s)	UNESCO/Paris typewriter (1970s/80s)[1]			United Nations/Geneva typewriter (ca. 1983)[2]		
mu 母	*wei* 委	*she* 社	*jian* 建	*dong* 东	*ze* 泽	*mao* 毛
mei 每	*yuan* 员	*hui* 会	*yi* 议	*hei* 黑	*ma* 马	*zhu* 主
bi 比	*mao* 毛	*zhu* 主	*xi* 席	*ge* 哥		*xi* 席
mao 毛	*ze* 泽	*yi* 义	*tan* 谈	*he* 河	*lie* 列	*zhu* 著
hao 毫	*dong* 东	*zhen* 阵	*pan* 判			

[1]Other notable features of this region include the terms "committee" (*weiyuanhui*), "socialism" (*shehuizhuyi*), and "construction" (*jianshe*).
[2]Other prominent features of this region include the names "Hegel" (*Heige*) and the first two characters of "Marxism-Leninism" (*Ma Lie*).

7.6 The location of the character *mao* (毛) on three Chinese typewriters

"Chairman Mao"). Owing to this rearrangement, the production of the name "Mao Zedong" on the UNESCO machine now required the operator to traverse only two units of space. By comparison, the same three characters typed on the popular Commercial Press machine of the 1930s would have required the user to traverse some 57.66 units of space: from *mao* at (34,37) to *ze* at (54,5), and finally to *dong* at (35,11). A survey of the tray beds reveals hundreds of other similar examples, including "Chairman Mao," "committee member" (*weiyuan*), "independent" (*duli*), "plan" (*jihua*), "to attack" (*gongji*), and "nationality" (*minzu*).[62]

When considered collectively, all of these small changes added up to something revolutionary. If we visualize two tray beds as heat maps—one tray bed from the Republican period, and one from after the move to natural-language arrangements—we can begin to appreciate the consequence of this new form of classification (figure 7.7).[63] In these heat maps, the color of each cell is a scaled chromatic representation of *the number of adjacent characters with which a given character can be combined to form a real, two-character word*, with white equaling 0 (indicating that a character cannot be meaningfully combined with any adjacent characters) and shades of light to dark gray corresponding to the range of values between 1 and 8 (8 indicating that a character can be combined meaningfully with all of its adjacent characters). Looking at them, we witness the "predictive turn" in Chinese information technology—a revolutionary vernacularization of taxonomy that formed the conceptual and practical foundations of what we now refer to as "predictive text."

As illustrated in this visualization, Mao-era experimentation with the Chinese typewriter resulted in a significantly "hotter" tray bed, one in which only a very small proportion of characters were not located next to at least one other character with which they tended to appear in natural language.

A further comparison of the UNESCO tray bed visualization with that of the UN machine—that is, between two machines which both employed predictive text organization—is equally revealing, alerting us to the dramatically decentralized and democratic dimension of this experimental movement. Although the goal was to produce more perfectly the rote and repetitive nature of phraseology within Maoist China, each tray bed was utterly individual and personal (figure 7.8).

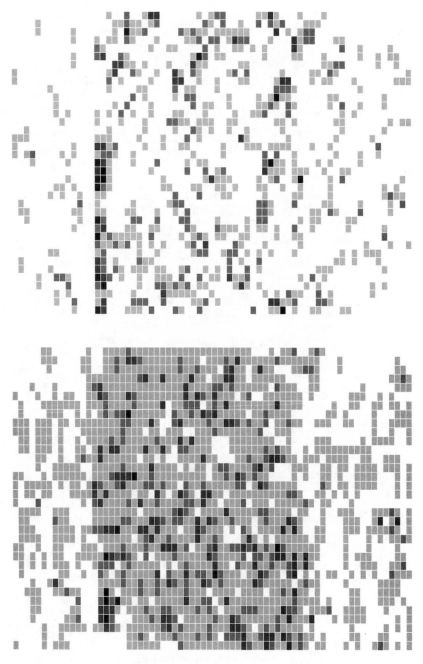

7.7 Heat map comparison of typewriters from before and after the predictive turn

7.8 Heat map comparison of two natural-language tray beds

There was an immense space for individualization within this practice, that is to say, with many factors to consider beyond those of vocabulary. To create a predictive text tray bed, one had to determine which characters to include on the tray bed; which two-, three-, and four-character sequences to make adjacent; where and how to create these adjacencies; where on the tray bed to place the centermost character (so as to avoid crowding or bunching up); how to place certain "dead-end" characters that were limited to only very specific two-character pairings (such as the *jin* of Tianjin, which pairs with few other characters); and how to shape the directionality of these pairings, among many others. There was also quite likely a mnemonic dimension to predictive text tray beds—that is, the ways in which typists would have used associative clusters not only to accelerate the speed of typing, but also as an *aide-mémoire* for the location of specific characters within specific clusters (for example, remembering the location of the character *mei*—美 "beauty"—by remembering that it formed part of the cluster *Mei diguo*—美帝国 "American imperialist"). To produce a predictive text tray bed was no simplistic matter of regurgitating rote phraseology, but a profoundly subtle "memory practice."[64]

By the close of the 1980s, natural-language arrangements had become so popular among typists that typewriter manufacturers began providing consumers with *blank* tray bed tables. Rather than printing conventional tray bed guides—guides that, since the 1910s, had mapped out the precise location of every one of the roughly 2,500 characters on the machine—these new manuals included a tray bed table left *purposefully blank*, providing users with nothing more than "suggestions" as to general principles of natural-language organization (figure 7.9).[65] Still other manuals provided more detailed recommendations on how to conceptualize a natural-language arrangement, but again left it up to individual typists to determine exactly how they would implement the system on their devices (figure 7.10).[66]

Once eager to standardize and centralize these new experimental efforts in the domains of typesetting and typewriting, publishers and companies were now capitulating to local-level, user-led changes, spending the waning years of the Chinese typewriting industry trying to catch up along a path that had already been forged by the "masses."

7.9 Predictive text tray bed organization chart (1988)

图2-2　放射式排字

7.10 Explanation of "arrow style" organization in 1989 typewriting manual

CONCLUSION: TOWARD A HISTORY OF CHINESE COMPUTING AND THE AGE OF INPUT

The forthcoming sequel to *The Chinese Typewriter* will be the first history in any language of Chinese computing, tracing this history from its inception in the immediate postwar period to its efflorescence in a burgeoning network of Chinese, Taiwanese, and Japanese computer scientists from the 1970s onward. The pathway of this history will take readers through subjects as diverse as machine translation, computer graphics, the rise of computer programming, the software revolution, the feminization of Chinese intellectual labor, and the growth of personal computing. At the center of the story will be a cast of eccentric personalities, drawn from the ranks at IBM, RCA, MIT, the CIA, the US Air Force, the US Army, the Pentagon, the RAND Corporation, the British telecommunications giant Cable and Wireless, Silicon Valley, the Graphic Arts Research Foundation, the Taiwanese military, the Soviet military, Japanese industrial circles, and the highest rungs of mainland Chinese intellectual, industrial, and military establishments.

As we will examine, continued experiments with mechanical Chinese typewriting became the doorway into new domains of information technology more properly called word processing and early computing. Using a variety of input devices—custom-made keyboards, QWERTY keyboards, pressure-sensitive surfaces, and even early pen tablets—inventors and companies from China, Taiwan, the United States, Western Europe, and the Soviet bloc engaged in ever more sophisticated explorations into Chinese technolinguistic modernity. With the dawn of Chinese computing, the already porous borders between common usage, combinatorialism, and surrogacy eroded completely, with the strategies and materialities of these once relatively distinct modes rushing together, swirling and eddying into new technolinguistic configurations. What is more, the border that once separated inscription from retrieval—the border that Lin Yutang first crossed in his work on MingKwai—disappeared entirely. Approaches to Chinese typewriting, typesetting, dictionary organization, telegraphy, and other domains and practices were brought into single devices in various combinations.

Most importantly, it will provide readers with the first-ever examination of *input*, a new technolinguistic condition that is quietly transforming the relationship between humans, machines, and language—not only in China but worldwide. Input constitutes the core technolinguistic condition of a new era in the history of Chinese, a condition that no longer adheres to the assumption that took hold more than a century ago as part of the collapsing imaginary of the alphabetic world: the all-pervasive fiction of Tap-Key and his monstrous Chinese keyboard. This fiction told the world that for every symbol there is a key, and by consequence, that alphabetic scripts were blessed with a kind of technolinguistic efficiency and immediacy that character-based Chinese writing could never hope to achieve.

If the appearance of Tap-Key marked the formation of this fiction, MingKwai and input marked its annihilation, rescuing Chinese script—and perhaps script itself—from the powerful myth that some writing systems are more immediate than others, and deeper still, that any script is immediate, ever. And so, if our story began in a time when the very notion of a "Chinese keyboard" was an oxymoron, our story now ventures into a time when the keyboard is ubiquitous in China, and yet when *typing itself no longer exists*—a time when QWERTY is everywhere in China, and yet when the keyboard as we know it is dead. For the keyboard and Chinese script alike, in order for everything to stay the same, everything needed to change.

The rise of input was by no means inevitable, nor is its history one easily celebrated. Input took shape as the result of a 150-year, sleep-deprived, torment-ridden history in which human beings functioning within the Chinese-character information environment were never permitted to drift off into the comfortable dream of immediacy like their counterparts in the alphabetic world. While the alphabetic world fell deeper and deeper into the comforting depths of Q-W-E-R-T-Y, A-Z-E-R-T-Y, and so forth, the world of Chinese characters was pierced day and night by near-constant alarm bells: Morse code, braille, typewriting, Linotype, Monotype, punched-card memory, text encoding, dot matrix, word processing, personal computing, optical character recognition, and the alphabetic order of the Olympic Parade of Nations, each expelling Chinese characters yet again from the bounds of the "universal." Each expulsion, in turn, buoyed up to the surface of consciousness fundamental truths about the inherent arbitrariness of Chinese script—of *all* script—in a recapitulation of that most basic of principles within Semiotics 101: that there is no inherent, invariant, or natural relationship between the signifiers we use and those concepts we wish to signify. At the aggregate level, this longstanding condition of technolinguistic sleep deprivation translated into more than a century in which a transnational cast of engineers, linguists, telegraphers, education reformers, phone book compilers, library scientists, typists, and more were left with little choice but to exploit the vast spaces of possibility that open up at precisely the moment when one gives up on the myth of immediacy—when one accepts as one's starting point a condition of *nonidentity* between keys and screens—and then begins to *play*.[1]

The rise of input is also not easy to grasp, particularly for those of us whose entire life has been conditioned by the *what-you-type-is-what-you-get* framework. As we venture into the next volume, three analogies help us distinguish between input and type: stenography, telecommunications, and MIDI. On a stenotype machine, such as those used within court stenography and elsewhere, only a fraction of the letters of the Latin alphabet is present. To produce letters that are absent from the machine—such as *b*, *d*, *f*, *g*, and many others—the typist must use the letters that are present to represent or stand in for them: the letter "f," for example, is not found on the keyboard, but must be symbolized by simultaneously typing two of the letters that are: in this case, "t" and "p." Meanwhile, the letter "b" is produced by simultaneously striking the "chord" composed of "p" and "w." When the typist reads back over the transcript, and sees the letter "p" by itself, he or she knows that—in this case—"p" equals "p." But when "p" appears alongside "w," the stenotypist knows that—in this case—these letters are being used to represent *another* letter, rather than themselves. By reviewing these particular letter chords, he or she knows which letters are being indicated and can thereupon translate this coded primary transcript into a legible secondary transcript.

In one sense, computer-age China is a country of stenotypists. The symbols one desires to see on the page or screen are not present on the keyboard in any naked-eye sense; instead, everything that is typed upon the "primary transcript" is but a temporary and disposable set of instructions that subsequently need to be translated into a secondary, "plaintext" transcript in accordance with a set of protocols. In the context of input, it is this "secondary transcript"—the characters that appear on the page—that is endowed with preeminent value, while the original transcript—the keystrokes that are intercepted by the input method editor (IME)—is discarded immediately after the translation is complete, without ever being viewed by human eyes.

Telecommunication provides a second analogy. In China, all text input—even text input that takes place within ostensibly "non-transmitting" programs like Microsoft Word—is in fact a form of *communication at a distance with oneself*—or *auto-telecommunication*. Even though this human-computer interaction seems to be taking place entirely "locally," confined to one person's relationship with his or her machine, nevertheless the model at play is classically telecommunicative: the operator is not sending telegrams to another party, of course, nor signaling from ship to shore, but is sending out coded transmissions to the IME, which are then translated and retransmitted back to the operator in the form of Chinese "plaintext." Taking place along the way is a process of retrieval: Chinese script is being summoned from elsewhere. In this way, Chinese input is a form of *retrieval-composition*, rather than the *inscription-composition* of typing.

The third and perhaps most evocative analogy is drawn from the domain of electronic music. With the advent of MIDI, or the Musical Instrument Digital Interface, entirely new modes of performing and composing musical scores became possible in the second half of the twentieth century. While patterned after and often indistinguishable from conventional instruments, MIDI pianos, guitars, drum pads, and woodwinds, among many others, are in fact instrument-agnostic controllers with which a performer can use the technosomatic format of one device (e.g., piano playing) to perform another (e.g., a violin). One can use a piano-shaped MIDI controller to play cello, a woodwind-shaped device to play the drums, a guitar-shaped MIDI device to play piano, and so on. The relation between a MIDI controller and its output is so plastic, in fact, that the controller need not resemble any conventional instrument at all. One could just as readily manipulate sound—*any* sound—through actuators embedded in the fabric of one's clothing (as in choreographer Gerry Girouard's *Songs for the Body Electric*), in a banana (as in the Sonic Banana project), or even the structure of an entire building (as in David Byrne's *Playing the Building* installation piece mounted in Stockholm, New York, London, and Minneapolis).[2]

Whatever analogy we choose, one fact remains clear: in China, the QWERTY keyboard and typing as we know them are dead, and have been reborn in the service of something altogether different: input. In the Chinese context, the QWERTY keyboard was long ago transformed into a "smart" peripheral, in contemporary parlance, that derives speed, power, and accuracy in direct proportion to the increasing processing power, algorithmic sophistication, and memory richness of consumer-grade computers, tablets, smart phones, and more. Meanwhile, QWERTY in the alphabetic world has remained largely unchanged since the era of typewriting since, as we recall, "computer designers were happy not to reinvent text-based input and output."

Chinese input is becoming more sophisticated with each passing year, moreover, with the steady advancement of predictive text, autocompletion, and most recently, a Wi-Fi-augmented input framework known as "cloud input" (*yun shuru*). In cloud input, IMEs use the cloud to provide smarter and smarter "suggestions" to users by comparing their QWERTY keystrokes against those inputted by Chinese computer users elsewhere in the network. Not unlike autosuggestions encountered in the Google search bar, Chinese cloud input has one all-important difference: this Wi-Fi-augmented process is not limited to the web, but is becoming a core part of all text input—even input that takes place in ostensibly "local" or "non-transmitting" programs. Whether one is searching Baidu or composing a Word document, one's keystrokes are being intercepted by and ingested into third-party cloud servers, with Chinese character suggestions being retransmitted back. As Chinese input moves further in the direction of the cloud, then, there is increasingly no such thing as "entirely local" text, contained exclusively on one's own computer or device.[3]

As we venture more deeply into the history and practice of Chinese input, however, it bears noting that our imaginations will continue to be stalked by the imagined Chinese monstrosity of Tap-Key. Long after the Chinese typewriter will have been displaced by Chinese computing and word processing, visions of immense, antimodern machines will continue to make regular appearances. "The Chinese typewriter is a longstanding joke in the West," a 1973 piece for the UK-based *Times* will read, "where it is almost synonymous with the paradoxical or impossible." Likening the machine's tray bed to an immense, strange, lunar surface (when, in fact, it measures only eighteen inches by nine inches in "terrestrial" dimensions), it will describe the process of Chinese typing as an "operation similar to landing on the Moon."[4] Another British journalist, this one writing in 1978, will describe the Chinese typewriter as a "cumbersome form of miniature dive-bombing, on which a really proficient typist can achieve only ten characters a minute."[5] Another Chinese typewriter will appear at the Museum of the History of Science in Oxford in a special exhibit tellingly entitled "Eccentricity: Unexpected Objects and Irregular Behavior."[6] Still another will form part of the permanent exhibit in Sweden at the Tekniska Museet.[7] Here in Stockholm, the machine will seem at first to occupy a privileged position, stationed in a display case at the very beginning of an exhibition dedicated to printing and other forms of writing technology. The accompanying description, however, will reveal an altogether different motive:

Hieroglyphics are picture characters, pictograms. ... Today's traffic signs are a type of pictogram. The Chinese alphabet is also pictographic and contains tens of thousands of characters. ... As opposed to pictographic script, alphabetic script has very few characters, or "letters." It is much easier to develop printing techniques if you use an alphabetic writing system.[8]

Even *The Simpsons* will enter the fray in 2001. In his new job writing copy for a fortune cookie manufacturer, Homer Simpson will be shown extemporizing terse jewels of wisdom to his daughter, as she takes dictation on a Chinese typewriter. "You will invent a humorous toilet lid"; "You will find true love on Flag Day"; "Your store is being robbed, Apu." He will pause for a moment to confirm that she is keeping up. "Are you getting all this, Lisa?" The frame will switch to Lisa, poised tentatively in front of yet another absurdly complex machine, pressing buttons cautiously and with great hesitation. In elongated syllables, she will respond: "I don't knowwww."[9]

The monstrous Chinese typewriter will not only persist, moreover—it will have reinforcements in the form of monstrous, imaginary Chinese *computers*. As exemplified in an online Q&A exchange from 1995, many of the same tropes of Chinese typewriting will soon migrate unconsciously from one arena of IT to another:

Dear Cecil, How on earth can the Chinese and Japanese use computers, given that their writing uses thousands of different characters? The keyboard must look like something off a Wurlitzer pipe organ.
Dear Nora, Nah, it looks pretty much like any keyboard, and using it is a piece of cake. All you have to do is adhere closely to the following six hundred steps. You might want to pack a lunch.[10]

As we continue our examination of Chinese and global information technology in the age of computing and new media, then, one of our biggest challenges remains: to liberate our imaginations from a past that never actually existed.

TABLE OF ARCHIVES

ASIA

AS	Academia Sinica (Taipei, Taiwan)
BMA	Beijing Municipal Archives (Beijing, China)
NDL	National Diet Library (Tokyo, Japan)
SMA	Shanghai Municipal Archives (Shanghai, China)
TMA	Tianjin Municipal Archives (Tianjin, China)

UNITED STATES

BL	Bancroft Library (University of California, Berkeley)
DZS	Devello Z. Sheffield Letters & Photographs (Ruth S. Johnson Family Collection)
GKP	George Kennedy Papers (Yale University, New Haven, CT)
HI	Hoover Institute (Stanford University, Stanford, CA)
HML	Hagley Museum and Library (Wilmington, DE)
HUNT	Huntington Library (San Marino, CA)
IBM	International Business Machines Corporate Archives
LOC	Library of Congress (Washington, DC)
MBHT	Museum of Business History and Technology (Wilmington, DE)
MIT	Massachusetts Institute of Technology (Cambridge, MA)
MPH	Museum of Printing (North Andover, MA)
NARA-CP	National Archives and Records Administration (College Park, MD)

NARA-SB	National Archives and Records Administration (San Bruno, CA)
NCM	National Cryptologic Museum (Annapolis Junction, MD)
PCA	Philadelphia City Archives (Philadelphia, PA)
PENN	University of Pennsylvania Archives (Philadelphia, PA)
PSBI	Pearl S. Buck International (Perkasie, PA)
RBML	Columbia University Rare Books and Manuscripts Library (New York, NY)
SI	Smithsonian Institute (Washington, DC)
SMML	Seeley Mudd Manuscript Library (Princeton University, Princeton, NJ)
STAN	Stanford University Special Collections (Stanford University, Stanford, CA)
TSM	Private Collection of Thomas S. Mullaney (Stanford University, Stanford, CA)
UD	University of Delaware Special Collections (Wilmington, DE)
UH	University of Hawai'i at Manoa Special Collections (Honolulu, HI)
UN	United Nations Archives (New York, NY)
WES	Wesleyan University Special Collections (Middletown, CT)
YULMA	Yale University Library Manuscripts and Archives (New Haven, CT)

EUROPE

BNF	Bibliothèque Nationale de France (Paris, France)
CFV	Dipartimento di Studi sull'Asia Orientale, Università Ca' Foscari di Venezia (Venice, Italy)
CW	Cable and Wireless Archives (Porthcurno, United Kingdom)
DNA	Danish National Archives (Copenhagen, Denmark)
HH	Private Collection of Henning Hansen (Copenhagen, Denmark)
MAM	Musée des Arts et Métiers (Paris, France)
MME	Musée de la Machine à Écrire (Lausanne, Switzerland)
MS	Mitterhofer Schreibmaschinenmuseum (Partschins/Parcines, Italy)
NI	Needham Institute (Cambridge, United Kingdom)
OHA	Archivio Storico Olivetti (Ivrea, Italy)
PTM	Porthcurno Telegraph Museum (Porthcurno, United Kingdom)
RH	Private Collection of Rolf Heinen (Drolshagen, Germany)
TM	Tekniska Museet (Stockholm, Sweden)
UDD	Private Collection of Umberto Di Donato (Milan, Italy)

BIOGRAPHIES OF KEY HISTORICAL PERSONS (ALPHABETIC BY SURNAME)

AUGUSTE BEYERHAUS (?–?), printer in Berlin and developer of divisible type "Berlin Font" for Chinese

BI SHENG [畢生] (990–1051), inventor of movable type

CHEN LIFU [陳立夫] (1900–2001), former Guomindang secretary general, minister of education, and inventor of five-stroke (*wubi*) Chinese character retrieval system

DU DINGYOU [杜定友] (1898–1967), library scientist and inventor of shape-position (*xingwei*) Chinese character retrieval system

SAMUEL DYER (1804–1843), British Protestant missionary and printer in China, and developer of steel typeface for Chinese

PIERRE HENRI STANISLAS D'ESCAYRAC DE LAUTURE (1826–1868), developer of experimental Chinese telegraph code

WILLIAM GAMBLE (1830–1886), Irish-born American printer, oversaw operations at the Presbyterian Mission Press in Ningbo, China, beginning in 1858

THOMAS HALL (1834–?), inventor of Hall index typewriter

CARLOS HOLLY (1838–1919), engineer, inventor, and assistant in manufacturing Chinese typewriter by Develle Sheffield

HU SHI [胡適] [Hu Shih] (1891–1962), Chinese language reformer, philosopher, former president of Peking University, and diplomat

JIN JIAN [金簡] (?–1795), bannerman from Shenjing, former Grand Minister of the Household Department, and Supervisor of the Wuying dian

ROBERT MCKEAN JONES (1855–1933), typographer, keyboard designer, Remington employee, and inventor of Phonetic Chinese typewriter

KATAOKA KOTARŌ (?–?), inventor of Otani Japanese Typewriter

MARCELLIN LEGRAND (?–?), type designer and codeveloper of divisible type Chinese font

LIN TAIYI [林太乙] (1926–2003), daughter of Lin Yutang

LIN YUTANG [林語堂] (1895–1976), linguist, author, and inventor of MingKwai Chinese typewriter

EDWIN MCFARLAND (1864–1895), inventor of first Siamese typewriter

GEORGE MCFARLAND (1866–1942), brother of Edwin McFarland, and proprietor of McFarland typewriter store in Siam

SAMUEL MORSE (1791–1872), inventor of telegraph and codeveloper of Morse code

JEAN-PIERRE GUILLAUME PAUTHIER (1801–1873), Orientalist, translator, and codeveloper of divisible type Chinese font

QI XUAN [祁暄] (1890–?), inventor of combinatorial Chinese typewriter

QIAN XUANTONG [錢玄同] (1887–1939), philologist, language reformer, and advocate of abolishing Chinese characters

DEVELLO ZELOTES SHEFFIELD (1841–1913), American missionary and inventor of first Chinese typewriter

SHIMADA MINOKICHI (?–?), inventor of "Oriental Typewriter"

SHU ZHENDONG [舒震東] (?–?), codeveloper of Shu Zhendong Chinese typewriter [舒震東華文打字機] (first mass-manufactured Chinese typewriter)

GEORGE STAUNTON (1781–1859), sinologist and first translator of Qing legal code into English

SUGIMOTO KYŌTA [杉本京太] (1882–1972), inventor of Japanese typewriter

TAP-KEY (1900–present), apocryphal inventor of Chinese typewriter

SEPTIME AUGUSTE VIGUIER (1837–1899), codeveloper of four-digit Chinese telegraph code

WANG JINGCHUN [王景春] [Chin-Chun Wang, C.C. Wang] (1882–1956), former editor of *Chinese Students' Monthly*, codirector of Peking-Mukden and Peking-Hankow Railways, and inventor of experimental telegraph code for Chinese

WANG YUNWU [王雲五] [Wang Yun-wu] (1888–1979), inventor of four-corner numerical Chinese character retrieval system (*sijiao haoma*)

YASUJIRO SAKAI (?–?), inventor of early kanji typewriter

Y.C. JAMES YEN [晏阳初] (1893–1990), educationalist, literacy activist, and cofounder of National Association of Mass Education Movements

YU BINQI [俞斌祺] (1901–?), athlete and inventor of Yu-style Chinese typewriter

ZHANG JIYING [張繼英] (?–?), developer of natural-language Chinese character rack for printers

ZHAO YUANREN [趙元任] [Y.R. Chao, Yuen Ren Chao] (1892–1925), linguist

ZHOU HOUKUN [周厚坤] [Hou Kun Chow, H.K. Chow] (1889–?), codeveloper of Shu Zhendong Chinese typewriter (first mass-manufactured Chinese typewriter)

CHARACTER GLOSSARY

aikoku taipisuto	愛国タイピスト	patriotic Japanese typist
baihua	白話	vernacular language
baoleishi	堡壘式	fortress style
ba zi shi	八字式	"eight"-character-style character rack
bianyi zipan	貶義字盤	negative connotation terms tray
biaozhun heng zhi shi Zhong Ri wen daziji	標準橫直式中日文打字機	Standard Horizontal-Vertical-style Chinese-Japanese typewriter
biaozhun shi Huawen daziji	標準式華文打字機	Standard-style Chinese typewriter
bihua	筆畫	stroke
bimu moxi	閉目默習	blind (typing) practice
Bi Sheng	畢生	
bi shi yu dian er cheng hua	必始於點而成畫	"must begin as a point before becoming a stroke"
bitan	筆談	brushtalk
bushou	部首	radical/classifier
chan	产	to produce/production
Chen Lifu	陳立夫	
chinmun	真文	truth script
chokugo	勅語	imperial rescript

chongfuzi shi	重複字式	repeating character style
cunzai	存在	to exist
dao	道	way
Da Qing lüli	大清律例	Great Qing Legal Code
dayin	打印	to print out
dayinben	打印本	typed and mimeographed edition
daziji	打字機	typewriter
dazinü	打字女	typewriter girl
dazi rencai	打字人才	typing talent
dazi tengxieshe	打字謄寫社	type-and-copy shop
daziyuan	打字員	typist
dengdeng	等等	et cetera
denshin jigō	電信字号	telegraph signals
dian	點	dot/point
dianbao	電報	telegram
dianma	電碼	telegraph code
diannao	電腦	computer
dianxin	電信	telegraphy
dianxin zhuquan	電信主權	telegraph sovereignty
dingzhen xuxian	顶针续线	thimble linking
Dishu	地书	*Book from the Ground*
dōbun dōshū	同文同種	same script, same race
Dongfang tushuguan	東方圖書館	Oriental Library
douzheng	鬥爭	struggle
Du Dingyou	杜定友	
duli	獨立	independent
erji mofan	二級模範	second-level model (worker)
fada	發達	to develop
faming	發明	to invent/invention
fangshexing	放射性	radiating style
fuci fangshe zituan	複詞放射字團	radiating compounds
gada gada gada	嘎哒嘎哒嘎哒	onomatopoeia describing sound of Chinese typewriter
gao	高	high

geming	革命	revolution
genben fangfa	根本方法	fundamental method
genju	根據	according to
gexin	革新	to reform/reformed, to innovate/innovation
gong	公	public
gongji	攻擊	to attack
gou	勾	hook
guangde Huawen dazi buxi xueshe	廣德華文打字補習學社	Guangde Chinese Typing Supplementary School
guibing bushou fa	歸併部首法	Merged Radicals Method
guo	国	country
Guoyu	國語	national language
guoyu yuekan	國語月刊	*National Language Monthly*
hanzi jianfa	漢字檢法	Chinese Character Retrieval Method
hanzi jianzi he paidie fa	漢字檢字和拍疊法	Chinese Character Retrieval and Stacking Method
hanzi paizi fa	漢字排字法	Chinese Character Arrangement Method
hanzi suoyin zhi	漢字索引制	Index System for Chinese Characters
hanzi wu zui	漢字無罪	Chinese characters are innocent
hanzi xingwei jianzifa	漢字型位檢字法	Shape-Position Retrieval System for Chinese Characters
haoma jianzifa	號碼檢字法	Numerical Retrieval System
hezuoshe	合作社	cooperative
Hiraga Gennai	平賀源内	
hitsudan	筆談	brushtalk
hōbun taipisuto kyōkai	邦文タイピスト協会	Japanese Typist Association
hōbun taipuraitā tokuhon	邦文タイプライター讀本	*Japanese Typewriter Textbook*
hongxing daziji chang	紅星打字機廠	Red Star Typewriter Factory
Hua Mei dazi zhuanxiao	華美打字轉校	Chinese-American Typing Institute
huangdi yong zhi gongwen	皇帝用之公文	clerical documents for use by the emperor

Huanqiu Huawen daziji zhizaochang	環球華文打字機製造廠	Huanqiu Chinese Typewriter Manufacturing Company
Huawen daziji	華文打字機	Chinese typewriter
Hui shi Hua-Ying wen dazi zhuanxiao	惠氏華英打字專校	Mr. Hui's Chinese-English Typing Institute
Hu Shi	胡適	
iroha	いろは	Phonetic system for organizing Japanese kanji
jian	建	to construct/construction
jiancha zi fa	檢查字法	character retrieval method
Jianzhaoshi Huawen daziji	菅沼式華文打字機	Suganuma-style Chinese typewriter
jianzifa wenti	檢字法問題	character retrieval problem
jiaoyu	教育	education
jiatian quezi	加添缺字	to add missing characters
jiefangjun	解放軍	People's Liberation Army
jiguang dayinji	激光打印機	laser printer
jihua	計畫	plan
jilu	記錄	record
jingyan	經驗	experience
Jin Jian	金簡	
jituanshi	集團式	association style
jōyō kanji hyō	常用漢字表	List of Characters for General Use
jue	決	to decide
kaku yubi no bundan	各指の分担	allocation [of keys] to each finger
kang Mei yuan Chao	抗美援朝	Resist America, Aid Korea
Kangxi zidian	康熙字典	*Kangxi Dictionary*
kanji seigen o teishō su	漢字制限を提唱す	Advocating the Restriction of the Number of Kanji
Keimōsha	啓蒙社	
kexingde	可行的	feasible
laodong	勞動	labor
lianchuan	連串	series/chain
lianchuanzi zipan	連串字字盤	connected character tray bed

liang	量	amount/quantity
liansuoxing	連鎖性	chainlink style
lianxi fenzhi	練習分指	practicing fingerwork
lianyu zipan	聯語字盤	connected language tray bed
Lin Taiyi	林太乙	
Lin Yutang	林語堂	
Li Zhaofeng	李兆丰	
Mainichi hiragana shinbun	毎日ひらがな新聞	*Daily Hiragana News*
Mao zhuxi	毛主席	Chairman Mao
Ma Shoujian	馬守劍	
Meidi	美帝	American imperialist
mianji	面積	surface area
min	民	the people
Mingji daziji hang	銘記打字機行	Ming Kee Typewriter Company
minzu	民族	nationality/ethnic group
minzu shibie	民族識別	Ethnic Classification
mobi jianzifa	末筆檢字法	Last-Stroke Character Retrieval Method
moji no oshie	文字の教え	*The Teaching of Words*
mumi wuse	目迷五色	bewildered
Nipponshiki	日本式	Japanese-style
nongye	農業	agriculture
nüdaziyuan	女打字員	female typist/typewriter girl
pildam	筆談	brushtalk
pohuai	破壊	to destroy
qiangji mosuo	強記摸索	rote memorization and groping around
Qian Xuantong	錢玄同	
Qian zi jing	千字經	*Thousand Character Classic*
qinlüe	侵略	to invade
Qi Xuan	祁暄	
ren	人	person
richang yingyong jichu erqian zi	日常應用基礎二千字	*2000 Fundamental Everyday Usage Characters*
rinji kokugo chōsakai	臨時国語調査会	Interim Committee on the National Language

sanmin zhuyi	三民主義	Three People's Principles
sanzenji jibiki	三千字字引	*The Three Thousand Character Dictionary*
San zi jing	三字經	Trimetrical Classic
Shanghai Zhongwen daziji zhizaochang lianyingsuo	上海中文打字機製造廠聯營所	Shanghai Chinese Typewriter Manufacturers Association
shehui zhuyi	社會主義	socialism
shehui zuzhi mingcheng de pan	社會組織名稱的盤	social structure terms tray
shengchan	生產	production
Sheng Jiping	盛濟平	
Sheng Lianzhen	盛練貞	
Sheng Yaozhang	盛耀章	
shi	式	style
shiwan ren zhong qiu yi ren	十萬人中求一人	search for one person out of 100,000
shuangge pai Zhongwen daziji	雙鴿牌中文打字機	Double Pigeon-brand Chinese typewriter
shuru	輸入	input
shurufa	輸入法	input method
Shu Zhendong	舒震東	
Shu Zhendong Chinese Typewriter	舒震東華文打字機	
sijiao haoma	四角號碼	Four-Corner Code
siku quanshu	四庫全書	Four Treasuries
songti	宋體	Song-style character
Sougou Pinyin	搜狗拼音	Sougou pinyin input
Sugimoto Kyōta	杉本京太	
Tadahisa Sakurada	櫻田常久	
tadashii shushi no keitai	正しい手指の形態	correct form for the fingers
tadashii taipuraichingu no shisei	正しいタイプライチングの姿勢	correct typewriting posture
taipisuto netsu	タイピスト熱	typist fever
taipuraitā	タイプライター	typewriter

Tang Chongli	唐崇李	
taolun	討論	discussion
Ta Tsing Leu Lee.	大清律例	Great Qing Legal Code
teyong wenzi	特用文字	special usage characters
ti	提	to lift/raise
ti	題	topic
Tianshu	天书	*Book from the Sky*
tongsu dazipan	通俗打字盤	popular typewriter tray bed
tongsu jiaokeshu	通俗教科書	popular textbook
tongsu jiaoyu	通俗教育	popular education
tongsu tushuguan	通俗圖書館	popular library
tongsu yanjiang	通俗演講	popular lecture
tongwen	同文	shared civilization
Wang Jingchun	王景春	
Wang Yunwu	王雲五	
wanneng shi Huawen daziji	萬能式華文打字機	Wanneng-style Chinese typewriter
Wei Geng	魏賡	
weiyuan	委員	committee member
wen	问	to question/question
wuchan jieji	無產階級	proletariat
Wuying dian	武英殿	Imperial Printing Office
Xin dazi caozuofa	新打字操作法	New Typing Method
xing	形	shape
xin nüxing	新女性	new woman
xinxing gongye	新興工業	contribute to emerging enterprise
xin yun suoyin fa	新韻所引發	New Rhyme-Based Indexing System
xiugai daziji zibiao huiyi	修改打字機字表會議	Meeting for the Improvement of the Typewriter Character Chart
Xu Bing	徐冰	
xuesheng	學生	student
Y.C. James Yen	晏阳初	
yi	義	meaning
yici gaipai	一次改排	all-at-once rearrangement
yideng gongchen	一等功臣	First-Class Hero

yijian	意見	opinion
yin	音	sound
yinshuaji	印刷機	printing machine
yinwei	因為	because
yin-yi-xing	音義形	sound-meaning-shape
yitizi	異體字	variant characters
yong shou chaolu	用手抄錄	copy and record by hand
yong zi bafa	用字八法	eight fundamental strokes of the character *yong*
Yu Binqi	俞斌祺	
Yu Binqi Zhongwen dazi zhiye xuexiao	俞斌祺中文打字職業學校	Yu Binqi Chinese Typing Professional School
yun shuru	雲輸入法	cloud input
Zhang Jiying	張繼英	
Zhang Xianglin	張祥麟	
Zhang Xieji daziji gongsi	張協記打字機公司	Chang Yah Kee Typewriter Company
Zhang Yuanji	張元濟	
zhaoshu	詔書	imperial edict
Zhao Yuanren	趙元任	
zhengzhi	政治	politics
Zhongguo	中國	China
Zhonghua minguo	中華民國	Republic of China
zhongshi wenzi	重視文字	place importance on writing
Zhou Houkun	周厚坤	
zhuaju minzhong jianzi de xinli	抓住民眾檢字的心理	to grasp the psychology of retrieval among the masses
zhuri gaijin	逐日改進	gradual improvement
zhuyin zimu	注音字母	Chinese phonetic alphabet
zibiao	字標	character positioning device
zipan	字盤	tray bed
zi ru qiren	字如其人	writing is like the person
zongheng shi Huawen daziji	縱橫式華文打字機	Vertical-Horizontal-style Chinese typewriter
zu	祖	ancestor

NOTES

INTRODUCTION: THERE IS NO ALPHABET HERE

1. This figure is ten times that of the 2004 games in Athens.

2. In competing estimates, a new milestone in torch relay duration was set in 2010 as part of the Vancouver Olympics. See Yvonne Zacharias, "Longest Olympic Torch Relay Ends in Vancouver," *Vancouver Sun* (February 12, 2010); Thomas K. Grose, "London Admits It Can't Top Lavish Beijing Olympics When It Hosts 2012 Games," *U.S. News* (August 22, 2008).

3. Eric A. Havelock, *Origins of Western Literacy* (Toronto: Ontario Institute for Studies in Education, 1976), 28, 44.

4. Walter J. Ong, *Orality and Literacy* (New York: Routledge, 2002 [1982]), 89.

5. Leonard Shlain, *The Alphabet Versus the Goddess: The Conflict between Word and Image* (New York: Penguin, 1999).

6. The Parade of Nations itself dates back to 1906, although I have been unable to locate any evidence of written regulations before 1921. Now considered "intercalated" games, the 1906 games are no longer regarded as one of the official Olympic games.

7. "Chaque contingent en tenue de sport doit être précédé, par une enseigne portant le nom du pays correspondant et accompagné de son drapeau national (les pays figurent par ordre alphabétique)." See "Cérémonie d'ouverture des jeux olympiques" in "Règlements et Protocole de la Célébration des Olympiades Modernes et des Jeux Olympiques Quadriennaux," 1921, 10.

8. "Les nations défilent dans l'ordre alphabétique de la langue du pays qui organise les Jeux" (rule 33, "Ceremonie d'ouverture des Jeux Olympiques," CIO, Régles Olympiques [Lausanne: Comité International Olympique, 1949]), 14. "Les

délégations défilent dans l'ordre alphabétique de la langue du pays hôte, sauf celle de la Grèce, qui ouvre la marche, et celle du pays hôte qui la clôt" (rule 69, "Cérémonies d'ouverture et de clôture," *Charte Olympique* [1991], n.p.).

9. When Germany hosted in 1972, the Greek national team was followed by Egypt (*Ägypten*) and Ethiopia (*Äthiopien*) in an order that would have been altogether familiar to anyone in the Latin alphabetic world. The Moscow Olympics of 1980 introduced the curious wrinkle of Cyrillic, wherein the sequence of Greece, Australia, and Afghanistan perhaps sparked curiosity as to how *Au* should precede *Af*. Explanation was readily available, however, by the simple fact that the 3rd letter of the Russian alphabet "*в*" (in "*Австралия*") precedes its 22nd letter "*ф*" (in "*Афганистан*").

10. On the topic of Chinese ethnic diversity, and the PRC state's role in shaping contemporary understandings thereof, see Thomas S. Mullaney, *Coming to Terms with the Nation: Ethnic Classification in Modern China* (Berkeley: University of California Press, 2010).

11. "Did NBC Alter the Olympics Opening Ceremony?," *Slashdot* (August 9, 2008), http://news.slashdot.org/story/08/08/09/2231231/did-nbc-alter-the-olympics-opening-ceremony (accessed March 1, 2012).

12. Had Chinese organizers selected pinyin as the basis for the parade, the evening would have unfolded very differently. Sailor Ciara Peelo would have enjoyed the honor of leading Aìěrlán (Ireland) on the field directly after Greece, instead of waiting for what must have felt like an eternity in position 159. Following Ireland would have come Āijí (Egypt) and Āīsāiébǐyà (Ethiopia), whose athletes instead were required to wait until positions 146 and 147. Pinyin was far from the only alternative, moreover. As we will see later in this book, there have existed many dozens of organizational schemes by which to arrange Chinese characters.

13. Zhou Houkun and Chen Tingrui, "A Newly Invented Typewriter for China (*Xin faming Zhongguo zhi daziji*) [新發明中國之打字機]," *Zhonghua xuesheng jie* 1, no. 9 (September 25, 1915): 6.

14. "On Literary Revolution (*Wenxue geming lun*) [文學革命論]," *Xin qingnian* [新青年] 2, no. 6 (February 1917).

15. Qian Xuantong, "China's Script Problem from Now On (*Zhongguo jinhou zhi wenzi wenti*) [中國今後之文字問題]," *Xin qingnian* 4, no. 4 (1918): 70–77.

16. Lu Xun, "Reply to an Interview from My Sickbed (*Bingzhong da jiumang qingbao fangyuan*) [病中答救亡情報訪員]" (1938), in *Complete Works of Lu Xun* (*Lu Xun Quanji*) [鲁迅全集], vol. 6 (Beijing: Renmin Wenxue, 1981), 160.

17. Statement by Chen Duxiu quoted in Qian Xuantong, "China's Script Problem from Now On (*Zhongguo jinhou zhi wenzi wenti*) [中國今後之文字問題]," *Xin qingnian* 4, no. 4 (1918): 76.

18. "A Few Methods for the Creation of Indexes for Chinese Books (*Bianzhi Zhongwen shuji mulu de ji ge fangfa*) [編織中文書籍目錄的幾個方法]," *Eastern Miscellany* (*Dongfang zazhi*) [東方雜誌] 20, no. 23 (1923): 86–103.

19. Simon Leung, Janet A. Kaplan, Wenda Gu, Xu Bing, and Jonathan Hay. "Pseudo-Languages: A Conversation with Wenda Gu, Xu Bing, and Jonathan Hay," *Art Journal* 58, no. 3 (Autumn 1999): 86–99.

20. Michael Lackner, Iwo Amelung, and Joachim Kurtz, eds., *New Terms for New Ideas: Western Knowledge and Lexical Change in Late Imperial China* (Leiden: Brill, 2001).

21. Harry Carter, *A View of Early Typography up to About 1600* (Oxford: Oxford University Press, 2002 [1969]), 5. Harry Carter (1901–1982) was a typographer and type historian who read law at Oxford, worked in the drawing office at the Monotype Corporation, and who served as Archivist at Oxford University Press.

22. Xu Bing [徐冰], "From 'Book from the Sky' to 'Book from the Ground' (*Cong Tianshu dao Dishu*) [从天书到地书]," manuscript provided by Xu Bing to author via email, May 15, 2013. For our discussion of the technolinguistic, the most compelling passage is as follows, in the original: 我决定造四千多个假字,因为出现在日常读物上的字是四千左右,也就是说,谁掌握四千以上的字,就可以阅读,就是知识分子。我要求这些字最大限度地像汉字而又不是汉字,这就必须在构字内在规律上符合汉字的规律。为了让这些字在笔画疏密、出现频率上更像一页真的文字,我依照《康熙字典》笔画从少到多的序例关系,平行对位地编造我的字。…字体,我考虑用宋体。宋体也叫"官体,"通常用于重要文件和严肃的事情,是最没有个人情绪指向的、最正派的字体。

23. Carter, *A View of Early Typography*, 5. As I venture into questions of Chinese technolinguistic modernity, I draw inspiration from the scholarship of Bernard Siegert, Delphine Gardey, Markus Krajewski, Ben Kafka, Miyako Inoue, Mara Mills, Matthew Hull, and others who, as Kafka has phrased it, work "to put the *bureau* back in *bureaucracy*." As Hull reminds us, documents "are not simply instruments of bureaucratic organizations, but rather are constitutive of bureaucratic rules, ideologies, knowledge, practices, subjectivities, objects, outcomes, even the organizations themselves." Meanwhile, Kafka insists, even at the extremes of historical experience—such as in Nazi-controlled Europe—it was the quotidian and even tedious processes like indexing that "made possible the 'banality' of the banality of evil" of which Hannah Arendt wrote. See Ben Kafka, "The State of the Discipline," *Book History* 12 (2009): 340–353, 341; Delphine Gardey, "Mécaniser l'écriture et photographier la parole: Utopies, monde du bureau et histoires de genre et de techniques," *Annales: Histoire, Sciences Sociales* 3, no. 54 (May–June 1999): 587–614; Ben Kafka, *The Demon of Writing: Powers and Failures of Paperwork* (Cambridge, MA: MIT Press, 2012); Markus Krajewski, *Paper Machines: About Cards & Catalogs, 1548–1929* (Cambridge, MA: MIT Press, 2011); Matthew S. Hull, "Documents and Bureaucracy," *Annual Review of Anthropology* 41 (2012): 251–267; Hull, "Documents and Bureaucracy," 251; Kafka, "The State of the Discipline," 341.

24. There are, of course, important and noteworthy exceptions to this observation, drawn primarily from work on technical communication in premodern China. See, in particular, Francesca Bray et al., eds., *Graphics and Text in the Production of Technical Knowledge in China: The Warp and the Weft* (Leiden: Brill, 2007). Comparable work on the modern period is limited, among the most compelling being that of Christopher Reed, Michael Hill, Elizabeth Kaske, Robert Culp, and Milena Dolezelová-Velingerová.

25. Christopher Rea and Nicolai Volland, eds., *The Business of Culture: Cultural Entre-preneurs in China and Southeast Asia, 1900–65* (Vancouver: University of British Columbia Press, 2015).

26. JoAnne Yates, *Control through Communication: The Rise of System in American Management* (Baltimore: Johns Hopkins University Press, 1993 [1989]), 41.

27. Bruce Robbins, "Commodity Histories," *PMLA* 120, no. 2 (2005): 456.

28. José Goldemberg, "Technological Leapfrogging in the Developing World," *Georgetown Journal of International Affairs* 12, no. 1 (Winter/Spring 2011): 135–141.

29. Havelock, *Origins of Western Literacy*, 15.

30. Li Gui [李圭], *New Records of My Travels Around the World* (*Huanyou diqiu xinlu*) [環游地球新錄]. Citation from Charles Desnoyers, *A Journey to the East: Li Gui's A New Account of a Trip Around the Globe* (Ann Arbor: University of Michigan Press, 2004), 121. My heartfelt thanks to Tobie Meyer-Fong for bringing this terrific source to my attention.

31. Daniel Headrick, *The Tentacles of Progress: Technology Transfer in the Age of Imperialism, 1850–1940* (Oxford: Oxford University Press, 1988); Everett M. Rogers, *Diffusion of Innovations* (New York: Free Press, 2003 [1962]).

32. Friedrich A. Kittler, *Gramophone, Film, Typewriter*, trans. Geoffrey Winthrop-Young and Michael Wautz (Stanford: Stanford University Press, 1999), 190–191. For a more recent introduction to the central importance of aurality within the history of Western typewriting, see "The History of the Typewriter Recited by Michael Winslow," http://www.filmjunk.com/2010/06/20/the-history-of-the-typewriter-recited-by-michael-winslow/ (accessed September 5, 2010).

33. Sequence from Merchant Ivory's film *Bombay Talkie* (1970), music by Shankar Jaikishan, lyrics by Hasrat Jaipuri, "Typewriter tip tip tip/Tip tip tip karata hai/Zindagi ki har kahaani likhata hai." My thanks to Andrew Elmore for alerting me to this film.

34. For "China-centered history," the *locus classicus* is Paul Cohen, *Discovering History in China: American Historical Writing on the Recent Chinese Past* (New York: Columbia University Press, 1984).

CHAPTER 1: INCOMPATIBLE WITH MODERNITY

1. "A Chinese Typewriter," *San Francisco Examiner* (January 22, 1900).

2. Ibid. The final string of exclamations is meant to parody the numerals *1, 2, 3, 4, 8,* and *9* as spoken in Cantonese.

3. *St. Louis Globe-Democrat* (January 11, 1901), 2–3.

4. Louis John Stellman, *Said the Observer* (San Francisco: Whitaker & Ray Co., 1903).

5. *The Chinese Typewriter*, written by Stephen J. Cannell, directed by Lou Antonio, starring Tom Selleck and James Whitmore, Jr., 78 mins., 1979, Universal City Studios.

6. Bill Bryson, *Mother Tongue: The English Language* (New York: Penguin, 1999), 110.

7. Walter J. Ong, *Orality and Literacy* (New York: Routledge, 2013 [1982]), 86.

8. Specimens of these machines can be found at the Mitterhofer Schreibmaschinen-museum in Partschins (Parcines), Italy, along with a number of other public and private collections.

9. Edwin Hunter McFarland (1864–1895); George Bradley McFarland (1866–1942).

10. Samuel Gamble McFarland (1830–1897); Jane Hays McFarland (?–1908). Samuel Gamble and his wife traveled to Bangkok from New York via Singapore and the Cape of Good Hope, and there they joined the small Presbyterian mission in Bangkok. The family soon moved to Petchaburi under the patronage of the local governor, where the McFarlands would spend the next seventeen years, shuttling back and forth between there and Bangkok via a three-day, two-night rowboat journey. George B. McFarland, *Reminiscences of Twelve Decades of Service to Siam, 1860–1936*, Bancroft Library, BANCMSS 2007/104, box 4, folder 14, George Bradley McFarland, 1866–1942, 2.

11. The McFarland children—William, Edwin (also known as Samuel), George, and Mary—were all born in Siam, having one of their first opportunities to visit the United States in 1873. When they returned home to Siam in August 1875, William and Edwin stayed behind to attend school.

12. Tej Bunnag, *The Provincial Administration of Siam, 1892–1915: The Ministry of the Interior under Prince Damrong Rajanubhab* (Kuala Lumpur: Oxford University Press, 1977).

13. McFarland, *Reminiscences*, 5. Having reached its third edition, the reference work had fallen out of print for at least five years when his father's passing inspired the creation of a fourth. He would continue to release further editions in 1916, 1930, and 1932. For these and other activities, George McFarland was the recipient of multiple honors. Under King Chulalongkorn (r. 1868–1910—Rama V), he was bequeathed the 4th Order of the White Elephant—and by King Vajiravudh (r. 1910–1925—Rama VI) the 3rd Order of the Crown of Siam and title of First Councilor, Phra Ach Vidyagama. He was also named emeritus professor in the Faculty of Medicine of Chulalongkorn University. McFarland, *Reminiscences*, 13.

14. G. Tilghman Richards, *The History and Development of Typewriters: Handbook of the Collection Illustrating Typewriters* (London: His Majesty's Stationery Office, 1938), 13.

15. "The Hall Typewriter," *Scientific American* (July 10, 1886), 24.

16. See english.stackexchange.com/questions/43563/what-percentage-of-characters-in-normal-english-literature-is-written-in-capital (accessed October 26, 2015).

17. Richards, *The History and Development of Typewriters*, 41.

18. "Accuracy: The First Requirement of a Typewriter," *Dun's Review* 5 (1905): 119; "The Shrewd Buyer Investigates," *New Metropolitan* 21, no. 5 (1905): 662.

19. "A Siamese Typewriter," *School Journal* (July 3, 1897), 12.

20. McFarland, *Reminiscences*, 9.

21. He would later reign as King Rama VI until his death in 1925. See Walter Francis Vella, *Chaiyo! King Vajiravudh and the Development of Thai Nationalism* (Honolulu: University of Hawai'i Press, 1978); Stephen Lyon Wakeman Greene, *Absolute Dreams: Thai Government Under Rama VI, 1910–1925* (Bangkok: White Lotus, 1999).

22. King Rama V presented to George McFarland the 4th Order of the White Elephant, an accolade to be followed by the 3rd Order of the Crown of Siam and the title of first councilor, *Phra Ach Vidyagama*, both conferred upon him by King Rama VI. In 1902, George was ordained elder of the Second Church of Bangkok, and later served as president of the Conference of Christian Workers until 1914.

23. For the denizens of Syracuse, this relationship inspired curiosity. In July 1897, *The School Journal* expressed excitement over the impending visit of the king and queen of Siam to the United States. "Some time ago," the article proudly recounted, "the Prince of Siam was sent to this country to have made a number of typewriters fitted with Siamese letters." "The Smith Premier Typewriter Company was selected to build the machines, the work being superintended by the secretary." "A Siamese Typewriter," 12. Smith Premier would later host the new crown prince of Siam, the young man who would later accede to the throne to become King Rama VI. *Phonetic Journal* (May 15, 1897), 306–307; "Highlights of Syracuse Decade by Decade," *Syracuse Journal* (March 20, 1939), E2; "Siam's Future King Guest in Syracuse," *Syracuse Post-Standard* (November 4, 1902), 5.

24. McFarland, *Reminiscences*, 12.

25. Ibid., 13–14.

26. Ibid., 13.

27. Ibid. There was an undoubtedly personal dimension to George's lament at play, his late brother having deliberately chosen the double-keyboard over its shift keyboard counterpart. As George made a point of emphasizing in his memoir, the deceased Edwin had "chosen the Smith Premier as best adapted to his purpose because of its large number of keys." McFarland, *Reminiscences*, 9.

28. Ibid., 13.

29. Ibid.

30. Underwood was the pioneer of visible, front-strike typewriting, the Underwood Number 1 (1897). Front-strike, visible typewriting was developed by Franz X. Wagner, originally of the Yost Caligraph company. John T. Underwood took up his design, and in the process established a company that offered serious competition to the giant Remington. Richards, *The History and Development of Typewriters*, 43; A.J.C. Cousin, "Typewriting Machine," United States Patent no. 1794152 (filed July 13, 1928; patented February 24, 1931).

31. George had left the world of typewriting behind by this point, concentrating his efforts on other parts of the family's legacy. He went on to rebuild the Phetchaburi Church first constructed by his father, and at the same time oversaw the republication and expansion of his father's Siamese-language dictionary. McFarland, *Reminiscences*, 14.

32. Photographs, October 23, 1938, George Bradley McFarland Papers, box 3, folder 15, Bancroft Library, University of California, Berkeley.

33. In 1886, the firm purchased the Remington interests, with the manufacturing division being reconstituted in 1903 as the Remington Typewriter Company (and subsequently as Remington Rand Inc. in 1927). Wyckoff, Seamans & Benedict, *The Remington Standard Typewriter* (Boston: Wyckoff, Seamans & Benedict [Remington Typewriter Co.]), 1897, 7.

34. Wyckoff, Seamans, and Benedict, *The Remington Standard Typewriter*, 33–34.

35. Ibid., 16–17.

36. *Remington Notes* 3, no. 10 (1915); Richards, *The History and Development of Typewriters*, 72. At the Columbia Exposition of 1893, Remington circulated postcards featuring Missipi, or Edna Eagle Feather, named the first Indian to learn shorthand and typewriting. Missipi was of the Osage Nation.

37. The most celebrated index typewriter is the Mignon, first built in 1904 by the A.E.G. Company in Berlin, with a reported speed of 250 to 300 characters per minute. See Richards, *The History and Development of Typewriters*, 45.

38. McFarland Papers, Bancroft Library, University of California, Berkeley, box 3, folder 14.

39. "Typewriters to Orient: Remington Rand Sends Consignment of 500 in the Mongolian Language," *Wall Street Journal* (April 26, 1930), 3.

40. "Ce n'est donc pas sans fierté, que la Maison Olivetti contribue à leur marche en avant par son apport de machines à écrire." "La Olivetti au Viet-Nam, au Cambodge et au Laos," *Rivista Olivetti* 5 (November 1950): 70–72, 71.

41. "Le Clavier Arabe," *Rivista Olivetti* 2 (July 1948): 26–28, 26.

42. For a fuller picture of the machines housed at this and other museums, please see the "Machines" section of the Bibliography.

43. Samuel A. Harrison, "Oriental Type-Writer," United States Patent no. 977448 (filed December 15, 1909; patented December 6, 1910).

44. Ibid.

45. Richard A. Spurgin, "Type Writer," United States Patent no. 1055679 (filed August 11, 1911; patented March 11, 1913). Remington followed suit again in 1921, when Elbert S. Dodge filed a patent to adjust their machines for "Hebrew and similar languages." "A specific object of the invention," Dodge explained, "is to adapt the Remington typewriting machine with very slight constructional changes to reverse the usual direction of feeding movement." Elbert S. Dodge, "Typewriting Machine," United States Patent no. 1411238 (filed August 19, 1921; patented March 28, 1922).

46. Selim S. Haddad, "Types for Type-Writers or Printing-Presses," United States Patent no. 637109 (filed October 13, 1899; patented November 14, 1899).

47. Ibid.

48. Ibid.

49. Ibid.

50. A comparable idea is advanced by Vassaf Kadry, assignor to the Underwood Typewriter Company residing in Constantinople. See Vassaf Kadry, "Type Writing Machine," United States Patent no. 1212880 (filed January 15, 1914; patented January 30, 1917). Assignor to Underwood Typewriter Company. Kadry identifies himself as a subject of the Sultan of Turkey.

51. Baron Paul Tcherkassov and Robert Erwin Hill, "Type for Type Writing or Printing," United States Patent no. 714621 (filed November 21, 1900; patented November 25, 1902).

52. In 1910, Herbert H. Steele of Marcellus, New York, submitted a patent for an Arabic typewriter, assigned to the Monarch Typewriter Company. While bearing this general name—"Arabic Typewriter"—the focus of the patent is far more precise, focused on the carriage mechanism of the Monarch machine and its adjustment for the purposes of Arabic typewriting. "My invention has for its principal object," Steele explained, "to produce an efficient carriage feed mechanism for use in a typewriting machine designed for writing the Arabic and similar languages wherein the paper carriage is required to have its step-by-step movement from left to right instead of in the other direction and wherein some characters require a greater extent of carriage movement than others." H.H. Steele, "Arabic Typewriter," United States Patent no. 1044285 (filed October 24, 1910; patented November 12, 1912). In 1917, Seyed Khalil (1891–1974), a self-described inventor and "freelancer" born in Kashan, Persia, filed for a patent assigned to the Underwood typewriter company and issued in January 1922. Khalil had immigrated to the United States in 1916, and shortly following his twenty-sixth birthday, offered up his vision of Arabic typewriting. He positioned his work somewhat in opposition to that of Tcherkassov and Hill. Their use of a nonsignifying graphic element served its function in reconciling Arabic writing to the typewriter, Khalil argued, but resulted in a method that was too slow and placed too onerous a demand on the typist. Moreover, Khalil argued, this approach to Arabic typewriting led to an erroneous belief "that certain letters of such languages must be distorted to be typewriting." Each letter on Khalil's machine had only two forms: a terminal form and a nonterminal form. Khalil saw broader application of his machine as well, looking beyond Arabic to Persian, Hindustani, and Turkish. See World War I Draft Registration Card (United States Selective Service System, World War I Selective Service System Draft Registration Cards, 1917–1918, National Archives and Records Administration, Washington, DC, M1509); World War Two Draft Registration Card (United States Selective Service System, Selective Service Registration Cards, World War II: Fourth Registration, National Archives and Records Administration Branch locations: National Archives and Records Administration Region Branches); Seyed Khalil, "Typewriting Machine," United States Patent no. 1403329 (filed April 14, 1917; patented January 10, 1922); Fourteenth Census of the United States, 1920 (National Archives and Records Administration, Washington, DC, Records of the Bureau of the Census, record group 29, NARA microfilm publication T625).

53. H.H. Steele, "Arabic Typewriter."

54. In 1917, John H. Barr and Arthur W. Smith submitted a patent application for an Arabic typewriter, in which the spirit of minimal modification was once again expressed. "A further object of our invention," Barr and Smith wrote as assignors to Remington, "is to readily adapt an ordinary typewriting machine employed for writing English or other European languages to use as a so-called Arabic machine of the character specified above without modifying, or materially modifying, the structural features of such ordinary machines as they now exist." What is more, the mechanical modifications they were proposing brought into reach not only Arabic but by extension Turkish, Persian, Urdu, Malayan, "and many others in addition to Arabic itself." John Henry Barr was Associate Professor of Machine Design at Cornell University, part of their faculty of Mechanical Engineering and Mechanic Arts. See *The Cornell University Register 1897–1898,* 2nd ed. (Ithaca: University Press of Andrus and Church, 1897–1898), 18; John H. Barr and Arthur W. Smith, "Type-Writing Machine," United States Patent no. 1250416 (filed August 4, 1917, patented December 18, 1917).

55. Georg Wilhelm Friedrich Hegel, *The Philosophy of History,* trans. John Sibree (New York: Wiley Book Co., 1900 [1857]), 134.

56. Edward W. Said, *Orientalism* (New York: Vintage Books, 1979); Rey Chow, "How (the) Inscrutable Chinese Led to Globalized Theory," *PMLA* 116, no. 1 (2001): 69–74; John Peter Maher, "More on the History of the Comparative Methods: The Tradition of Darwinism in August Schleicher's work," *Anthropological Linguistics* 8 (1966): 1–12; Lydia H. Liu, *The Clash of Empires: The Invention of China in Modern World Making* (Cambridge, MA: Harvard University Press, 2004: 181–209). See also, for example, August Schleicher, "Darwinism Tested by the Science of Language," trans. Max Müller, *Nature* 1, no. 10 (1870): 256–259.

57. Samuel Wells Williams, "Draft of General Article on the Chinese Language," n.d., Samuel Wells Williams family papers, YULMA, MS 547 location LSF, series II, box 13, 3.

58. Samuel Wells Williams family papers, YULMA, MS 547 location LSF, series II, box 13, 3.

59. Peter S. Du Ponceau, *A Dissertation on the Nature and Character of the Chinese System of Writing, Transactions of the Historical and Literary Committee of the American Philosophical Society,* vol. 2 (1838).

60. "Du Ponceau on the Chinese System of Writing," *North American Review* 48 (1848): 306.

61. Ibid., 272–273.

62. Henry Noel Humphrey, *The Origin and Progress of the Art of Writing: A Connected Narrative of the Development of the Art, Its Primeval Phases in Egypt, China, Mexico, etc.* (London: Ingram, Cooke, and Co., 1853). Creel cites the second edition, published in 1885. Herrlee Glessner Creel, "On the Nature of Chinese Ideography," *T'oung Pao* 32 (2nd series), no. 2/3 (1936): 85–161, 85.

63. *China As It Really Is* (London: Eveleigh Nash, 1912), 154.

64. Ibid., 160.

65. W.A. Martin, *The History of the Art of Writing* (New York: Macmillan, 1920), 13. Cited in Creel, "On the Nature of Chinese Ideography," 85–161, 85.

66. Bernhard Karlgren, *Philology and Ancient China* (Cambridge, MA: Harvard University Press, 1926), 152.

67. T.T. Waterman and W.H. Mitchell, Jr., "An Alphabet for China," *Mid-Pacific Magazine* 43, no. 4 (April 1932): 353.

68. Creel, "On the Nature of Chinese Ideography," 85.

69. Ibid., 86.

70. Ibid., 160.

71. Geoffrey Sampson, *Writing Systems* (Stanford: Stanford University Press, 1985).

72. Jack Goody, *The Interface between the Written and the Oral* (Cambridge: Cambridge University Press, 1987), xvii–xviii.

73. Ibid., 64.

74. Jack Goody, "Technologies of the Intellect: Writing and the Written Word," in *The Power of the Written Tradition* (Washington: Smithsonian Institution Press, 2000), 138.

75. Havelock, *Origins of Western Literacy*, 18; Robert Logan, *The Alphabet Effect* (New York: William Morrow, 1986), 57.

76. Goody, "Technologies of the Intellect," 138.

77. Goody, *The Interface between the Written and the Oral*, 37. Joseph Needham underwent his own transformation. In the second volume of *Science and Civilisation*, he voiced suspicions about the Chinese language, citing it as one potential answer to the question of why China had failed to achieve a scientific revolution on par with that of Western Europe. "At a later date," he wrote, "we shall enquire how far the difference of linguistic structure between Chinese and the Indo-European languages had influence on the differences between Chinese and Western logical formations." When that later date arrived, Needham announced his conclusion: the "inhibiting influence of the ideographic language has been grossly over-rated." "It has proved possible to draw up large glossaries of definable technical terms used in ancient and medieval times for all kinds of things and ideas in science and its applications. At the present day, the language is no impediment to contemporary scientists. If the social and economic factors in Chinese society had permitted or facilitated the rise of modern science there as well as in Europe, then already 300 years ago the language would have been made suitable for scientific expression." Joseph Needham, *Science and Civilisation*, vol. 2 (Cambridge: Cambridge University Press, 1956), 199; Joseph Needham, "Poverties and Triumphs of the Chinese Scientific Tradition," in *Scientific Change (Report of History of Science Symposium, Oxford, 1961)*, ed. A.C. Crombie (London: Heinemann, 1963).

78. Alfred H. Bloom, "The Impact of Chinese Linguistic Structure on Cognitive Style," *Current Anthropology* 20, no. 3 (1979): 585–601.

79. Derk Bodde, *Chinese Thought, Society, and Science: The Intellectual and Social Background of Science and Technology in Pre-Modern China* (Honolulu: University of Hawai'i Press, 1991), 95–96.

80. William C. Hannas, *Asia's Orthographic Dilemma* (Honolulu: University of Hawai'i Press, 1996); William C. Hannas, *The Writing on the Wall: How Asian Orthography Curbs Creativity* (Philadelphia: University of Pennsylvania Press, 2003).

81. William G. Boltz, "Logic, Language, and Grammar in Early China," *Journal of the American Oriental Society* 120, no. 2 (April–June 2000): 218–229, 221.

82. "Le macchine Olivetti scrivono in tutte le lingue," *Notizie Olivetti* 55 (March 1958): 1–4.

83. Allen Ginsberg, *Howl and Other Poems* (San Francisco: City Lights Publishers, 2001).

CHAPTER 2: PUZZLING CHINESE

1. William Gamble, *List of Chinese Characters in the New Testament and Other Books*, 1861, Library of Congress, G/C175.1/G15.

2. In the original, the two names are rendered by Gamble as "Tsiang sin san" and "Cü sin san."

3. In addition to "anti-reading," another apt term might be "distant reading," to invoke the concept and practice made famous by Franco Moretti. See Franco Moretti, *Distant Reading* (New York: Verso, 2013).

4. In literary Chinese, the character 之 (*zhi*) is typically used as a possessive particle and as the substitute pronoun "it"; 而 (*er*) is typically used as a conjunction, a particle to convey "as well as" or "and yet," or a particle to convey a change in state or situation, among other potential meanings; while 不 is used to negate verbs. For Gamble's character frequency analysis, see William Gamble, *Two Lists of Selected Characters Containing All in the Bible and Twenty Seven Other Books* (Shanghai: Presbyterian Mission Press, 1861), ii, in which William Gamble published the results of his research. As Gamble argued in the preface, movable type printers of Chinese needed to "form a scale, as printers call it, of the characters in the written language."

5. See Endymion Wilkinson, *Chinese History: A New Manual* (Cambridge, MA: Harvard University Asia Center, 2012), 78.

6. "List of Chinese Characters Formed by the Combination of the Divisible Type of the Berlin Font Used at the Shanghai Mission Press of the Board of Foreign Missions of the Presbyterian Church in the United States" (Shanghai: n.p., 1862), 1. In 1863, William Gamble published *Liangbian pin xiaozi* [兩邊拼小字] (*List of 1878 Chinese characters which can be formed by divisible type*), December 22, 1863, manuscript Library of Congress G/C175.1/G18.

7. Gamble, *Two Lists of Selected Characters Containing All in the Bible and Twenty Seven Other Books*, ii.

8. See Cynthia J. Brokaw, "Book History in Premodern China: The State of the Discipline," *Book History* 10 (2007): 254.

9. The number of works was later expanded to 146. See Wilkinson, *Chinese History*, 951.

10. Jin Jian, *A Chinese Printing Manual*, trans. Richard S. Rudolph (Los Angeles: Ward Ritchie Press, 1954), xix. Jin Jian (?–1795) was a bannerman from Shenjing, whose ancestors emigrated from Korea in the waning years of the Ming dynasty. Having served as grand minister of the household department, Jin Jian was later appointed supervisor of the Wuying dian (1772–1774). See Wilkinson, *Chinese History*, 912.

11. The system of 214 radicals originated in the late Ming dictionary the *Lexicon* (*Zihui*), edited by Mei Yingzuo and published in 1615. It later formed the taxonomic basis of the *Kangxi Dictionary*, compiled and completed by Zhang Yushu [張玉書] (1642–1711) and Chen Tingjing [陳廷敬] (1639–1712) and published in 1716.

12. The first of the 214 radicals is composed using a single stroke (*yi* [一]), while the last is composed by using seventeen (*yue* [龠]).

13. The radical-stroke system further stipulates that characters within the same radical class and with the same number of strokes be organized according to the specific types of strokes out of which they are composed. Strokes themselves are categorized into eight types, which are then placed in a sequence. If two characters have the same radical and stroke characteristics, their dictionary sequence depends on the first (and sometimes second or third) stroke used to create the character.

14. "Label and arrange twelve wooden cabinets according to the names of the twelve divisions of the *Imperial Kangxi Dictionary*," Jin wrote in his printing manual, *Collectanea Printed by the Imperial Printing Office Movable Type*. "When selecting type, first examine the make-up of the character for its corresponding classifier, and then you will know in which case it is stored. Next, count the number of strokes, and then you will know in which drawer it is. If one is experienced in this method, the hand will not err in its movements." Jin Jian's first proposal to the emperor was to organize characters into a phonetic "rhyme" system. Ultimately, however, Jin centered upon the system of character organization employed in the *Kangxi Dictionary*. For a translation of Jin Jian's account of the printing process, see Jin Jian, *A Chinese Printing Manual*.

15. Claude-Marie Ferrier and Sir Hugh Owen, *Exhibition of the Works of Industry of All Nations: 1851 Report of the Juries* (London: William Clowes and Sons, 1852), 452.

16. For foreign printers of Chinese, the particular spatial characteristics of Chinese movable type had not always been considered vexing or "puzzling"—a point that deserves emphasis. Over the course of the nineteenth century and earlier, for example, a number of foreign printing establishments created or employed their own Chinese fonts, such as the College of Saint Joseph in Macao and the Imprimerie Royale in Paris—a font containing an estimated 126,000 sorts in all, and thus one that would certainly have "surrounded" the foreign typesetters just as characters in the *Wuying dian* had surrounded Jin Jian. Over the course of the nineteenth century, however, the desire to surround and sedentarize Chinese characters became an

increasingly powerful one within printing, typographic, and pedagogical circles. See Walter Henry Medhurst, *China: Its State and Prospects, with Especial Reference to the Spread of the Gospel* (London: John Snow, 1838), 554–556; J. Steward, *The Stranger's Guide to Paris* (Paris: Baudry's European Library, 1837), 185.

17. Li Chen, *Chinese Law in Imperial Eyes: Sovereignty, Justice, and Transcultural Politics* (New York: Columbia University Press, 2015). In particular, see chapter 2, "Translation of the Qing Code and Origins of Comparative Chinese Law."

18. See also Joshua Marshman, *Elements of Chinese Grammar: with a preliminary dissertation on the characters, and the colloquial medium of the Chinese, and an appendix containing the Tahyoh of Confucius with a translation* (printed at the Mission Press, 1814).

19. William Gamble went on to compile his own dictionary, arranging these 5,150 characters according to the radical-stroke arrangement system of the *Kangxi Dictionary*, and then indicating alongside each character the number of times each appeared in his survey. His dictionary also included the same list of characters, grouping them into fifteen categories according to their frequency of appearance, and then internally organizing each of these fifteen categories according to the Kangxi radical-stroke system. Christian missionary educators in the nineteenth century were also concerned with the Chinese lexicon, but more perhaps for the goal of introducing new concepts (and characters) than reproducing the Chinese language as they found it. Returning to William Gamble and other missionary printers introduced above, their lexical studies of the Chinese language were centered, not merely on the Chinese classics, but also on Chinese translations of the Bible. Gamble, *Two Lists of Selected Characters Containing All in the Bible*, v–vi.

20. Ibid. Another influential scholar in this regard was Samuel Dyer (1804–1843). Dyer was born in Greenwich in 1804, the fourth son of John Dyer, the secretary of the Royal Hospital for Seamen. As an undergraduate at Cambridge, he read classics, mathematics, and law at Trinity Hall, but later reached out to the London Missionary Society in the summer of 1824, offering himself up for the cause of missionary work in "heathen lands." That summer, he entered the seminary and commenced his studies under the tutelage of David Bogue—including the study of Chinese. Dyer was ordained in February 1827, and in less than one month was dispatched to Malacca with his wife, Maria. Arriving in Penang in August, he promptly set about studying the Hokkien dialect. At a time when the printing of Chinese was almost exclusively lithographic and xylographic, with certain notable exceptions (Morrison's dictionary was printed in Macao using metal movable types), Dyer was committed to the creation of a movable metal type font for Chinese. See Ibrahim bin Ismail, "Samuel Dyer and His Contributions to Chinese Typography," *Library Quarterly* 54, no. 2 (April 1984): 157–169.

21. Jean-Pierre Guillaume Pauthier, *Foe Koue Ki ou Relation des royaumes bouddhiques* (Paris: Imprimerie Royale, 1836); Jean-Pierre Guillaume Pauthier, *Le Ta-Hio ou la Grande Étude, ouvrage de Confucius et de ses disciples, en chinois, en latin et en français, avec le commentaire de Tchou-hi* (Paris: n.p., 1837).

22. Jean-Pierre Guillaume Pauthier, *Chine ou Description historique, géographique et littéraire de ce vaste empire, d'après des documents chinois*, first part (Paris: Firmin Didot Frères, Fils, et Cie, 1838).

23. *Les caractères de l'Imprimerie Nationale* (Paris: Imprimerie Nationale, 1990), 114–117.

24. See *Châh Nameh, sous le titre: Le livre des rois par Aboul'kasim Firdousi, publié, traduit et commenté par M. Jules Mohl*, 7 vols. (Paris: Jean Maisonneuve, 1838–1878); Arthur Christian, *Débuts de l'Imprimerie en France* (Paris: G. Roustan and H. Champion, 1905).

25. Medhurst, *China*, 557.

26. Ibid., 558.

27. Marcellin Legrand, "Tableau des 214 clefs et leurs variants" (Paris: Plon Frères, 1845); L. Léon de Rosny, *Table des principales phonétiques chinoises disposée suivant une nouvelle méthode permettant de trouver immédiatement le son des caractères quelles que soient les variations de prononciation, et adaptée spécialement au Kouan-hoa ou dialecte mandarinique*, 2nd ed. (Paris: Maisonneuve et Cie, Libraire-Éditeurs pour les Langues Orientales, Étrangères et Comparées, 1857).

28. Medhurst, *China*, 556.

29. Ibid., 558.

30. Robert E. Harrist and Wen Fong, *The Embodied Image: Chinese Calligraphy from the John B. Elliott Collection* (Princeton: Art Museum, Princeton University in association with Harry N. Abrams, 1999), 4.

31. Ibid., 152.

32. Ibid.

33. Adapted from Yee Chiang, *Chinese Calligraphy: An Introduction to Its Aesthetics and Techniques* (Cambridge, MA: Harvard University Press, 1973 [1938]), 163–164.

34. See John Hay, "The Human Body as a Microcosmic Source of Macrocosmic Values in Calligraphy," in *Self as Body in Asian Theory and Practice*, ed. Thomas Kasulis, Roger Ames, and Wimal Dissanayake (Albany: State University of New York Press, 1993), 179–212; Amy McNair, "Engraved Calligraphy in China: Recension and Reception," *Art Bulletin* 77, no. 1 (March 1995): 106–114; Craig Clunas on categorization in Maxwell Hearn and Judith Smith, eds., *Arts of the Sung and Yuan* (New York: Metropolitan Museum of Art, 1996); Richard Curt Kraus, *Brushes with Power: Modern Politics and the Chinese Art of Calligraphy* (Berkeley: University of California Press, 1991).

35. *Twelfth Annual Report of the American Tract Society* (Boston: Perkins and Marvin, 1837), 63.

36. *Chuang Tzu: The Basic Writings*, trans. Burton Watson (New York: Columbia University Press, 1964), 47.

37. *Twelfth Annual Report of the American Tract Society*, 62–63. *Characters Formed by the Divisible Type Belonging to the Chinese Mission of the Board of Foreign Missions of the*

Presbyterian Church in the United States of America (Macao: Presbyterian Press, 1844). As with Pauthier and Legrand's font, it contained both whole and partial characters, the latter kind being further subcategorized into the same two groups as in the Frenchmen's work: "horizontally divided" partials and "vertically divided" partials. "Chinese Divisible Type," *Chinese Repository* 14 (March 1845): 129.

38. This number might seem large on its own, and yet it fell well within the range of other, whole-character Chinese fonts from the era. Indeed, if we reflect back on the work of William Gamble and others, Legrand and Pauthier's proposal was modest by comparison. More significantly, unlike whole-character fonts, divisible type would not be confined to the one-to-one, sort-to-character ratio inherent in conventional printing. With these three thousand full and partial characters, Legrand and Pauthier's font could compose an estimated 22,841 characters in total. "Chinese Divisible Type," 124–129, 129. Unlike Legrand's divisible type, which included both horizontally and vertically divided characters, Beyerhaus opted instead to focus exclusively on vertically divisible characters, presumably to simplify the system. This font was sufficient to prepare full Chinese translations of the Old and New Testaments, which Beyerhaus prepared for the American Missionary Society. See George Dodd, *The Curiosities of Industry and the Applied Sciences* (London: George Routledge and Co., 1858), 4.

39. *Annales des empereurs du Japon* was begun by the Dutch merchant and Orientalist Isaac Titsingh (1745–1812), and continued by the French Orientalist Abel Rémusat (1788–1832) (*Nihon Ōdai Ichiran* 日本王代一覧). *Relation des royaumes bouddhiques* was the French translation of the *Foguo ji* by Rémusat. In 1837, further news was reported of Walter Lowrie's intention to purchase a complete set of Legrand's Chinese font in his station as secretary of the Western Foreign Missionary Society. In 1844, Legrand went on to exhibit his Chinese punches in Paris, 4,600 in all, capable of producing all Chinese characters. By 1859, the Presbyterian Board of Missions in New York adopted Beyerhaus's "beautiful font" for some of its Chinese-language publications. See *Twelfth Annual Report of the American Tract Society*, 62–33; Ferrier and Owen, *Exhibition of the Works of Industry of All Nations*, 409; and Samuel Wells Williams, *The Middle Kingdom: A Survey of the Chinese Empire and Its Inhabitants* (New York: Wiley & Putnam, 1848), 604.

40. Ferrier and Owen, *Exhibition of the Works of Industry of All Nations*, 452.

41. Medhurst, *China*, 557. In 1854, Legrand's Chinese font appeared in Stanislas Hernisz's curious primer, *A Guide to Conversation in the English and Chinese Languages for the Use of Americans and Chinese in California and Elsewhere*. Hernisz paid thanks to Legrand in his preface, adding: "It is perhaps not the least interesting fact of our times that Chinese books should be printed in China with types manufactured 'beyond the ocean,' by an 'outside barbarian'!" See Stanislas Hernisz, *A Guide to Conversation in the English and Chinese Languages for the Use of Americans and Chinese in California and Elsewhere* (Boston: John P. Jewett and Co., 1854). For Pauthier's part, evidently he was pleased, returning to work with Legrand again in 1858 for his *L'Inscription syro-chinoise*.

42. Comte d'Escayrac de Lauture, *On the Telegraphic Transmission of Chinese Characters* (Paris: E. Brière, 1862).

43. Ibid., 6.

44. Also inspired by this trip, Escayrac de Lauture would later publish his multivolume *Mémoires sur la Chine* (1864–1865).

45. "Le télégraphe veut une langue plus brève, intelligible à tous les peuples. Je vais montrer que cette langue n'est point une utopie; que non-seulement son emploi est possible, mais encore qu'il est facile, indiqué, nécessaire." Comte d'Escayrac de Lauture, *Grammaire du télégraphe: Histoire et lois du langage, hypothèse d'une langue analytique et méthodique, grammaire analytique universelle des signaux* (Paris: J. Best, 1862 [August]), 9.

46. Yakup Bektas, "The Sultan's Messenger: Cultural Constructions of Ottoman Telegraphy, 1847–1880," *Technology and Culture* 41 (2000): 206.

47. Ibid., 669–696.

48. Between a 4- and 5-unit pulse sequence, one faced an increase of anywhere from 25 to 75 percent in the time required for transmission. Because of its inefficiency, this five-unit code area—which contained 2^5 or 32 additional spaces—was originally limited to numerals and special symbols (including punctuation). Connections to the English language went further than this, however, and included subtle considerations of the particular types of ambiguities that might surface during the transmission of letters. Within Morse code, the allocation of pulse patterns also took into account the co-occurrence of letters, and the ambiguities that could arise if a sequence of two originally distinct patterns was partitioned incorrectly, thereby leading to erroneous transcription. Confronted with the most common two-letter sequences ("digrams"), then, the pulse patterns of such letters had to be sufficiently distinct so as to prevent misreading, even though this meant allocating pulse patterns whose energetic efficiency fell short of what a given letter's frequency might otherwise demand.

49. *Convention télégraphique internationale de Paris, révisée à Vienne (1868) et Règlement de service international (1868)—Extraits de la publication: Documents de la Conférence Télégraphique Internationale de Vienne* (Vienna: Imprimerie Impériale et Royale de la Cour et de l'Etat, 1868), 58.

50. *Convention télégraphique internationale de Saint-Pétersbourg et Règlement et tarifs y annexés (1875). Extraits de la publication—Documents de la Conférence Télégraphique Internationale de St-Pétersbourg: Publiés par le Bureau International des Administrations Télégraphiques* (Bern: Imprimerie Rieder & Simmen, 1876), 22; *Convention télégraphique internationale de Berlin (1885): Publiés par le Bureau International des Administrations Télégraphiques* (Bern: Imprimerie Rieder & Simmen, 1886, 15.

51. These would not be available, however, to those using "Hughes Signals," an alternate system to Morse.

52. *Convention télégraphique internationale et règlement et tarifs y annexés révision de Londres (1903)* (London: The Electrician Printing and Publishing Co., 1903), 16.

53. Daniel Headrick, *The Tentacles of Progress: Technology Transfer in the Age of Imperialism, 1850–1940* (Oxford: Oxford University Press, 1988). See chapter 11.

54. W. Bull, "A Short History of the Shanghai Station" (Shanghai: n.p., 1893) [handwritten manuscript], Cable and Wireless Archive DOC/EEACTC/12/10, 4.

55. For a fascinating study of telegraphy in its early years, and the radical visions that accompanied its emergence, see Carolyn Marvin, *When Old Technologies Were New: Thinking about Electric Communication in the Late Nineteenth Century* (Oxford: Oxford University Press, 1998).

56. Escayrac de Lauture, *Grammaire du Télégraphe*, 4.

57. The original reads, "langue des faits et des chiffres, langue sans poésie, planant cependant au-dessus des vulgarités de la vie commune." "Le discours est comme un calcul avec des mots: il faut trouver l'algèbre de ce calcul, imparfait comme chaque idiome; il faut trouver la commune mesure de la pensée et des discours humains." Ibid., 4, 8.

58. The original reads: "le catalogue des idées principales constitue la nomenclature: c'est comme la matière et le corps du discours." Ibid., 4, 15.

59. Actions too would be categorized into fundamental and secondary terms. As Escayrac de Lauture captured it: "Now that we have given form to the body of language, we may now give it life." Beginning with the principal idea of "movement," one could modify it to achieve directionality (to go, to come, to circulate, to traverse), instrumentality (to carry, to strike, to divide), and any number of other variations. Neither tense nor person would be required in Escayrac de Lauture's system, moreover, with conjugation and declension being replaced with a further set of modifying protocols. Ibid., 12. The natural world could be classified in comparable ways, Escayrac de Lauture contended. Whether vegetables, chemicals, mammals, reptiles, mollusks, fish, or otherwise, all entities within the natural world could be captured using abbreviated letter codes. Each letter within the system would carry significance. An initial consonant in a sequence could be used to indicate Linnaean classes, with the subsequent vowel then indicating Linnaean orders and suborders. For geographic locations, latitude and longitude could be used, with individual mountains demarcated by their summits, rivers by their sources, and oceans by their central points. See Escayrac de Lauture, *Grammaire du Télégraphe*, 11–12.

60. "Sans connaître un seul de ces mots, on pourrait, à l'aide d'un simple vocabulaire, établir avec certitude le sens d'une phrase." Ibid., 15.

61. "… serait plus propre qu'aucune langue connue aux communications internationales d'un certain ordre." Ibid., 15.

62. Bull, "A Short History of the Shanghai Station," 7–10.

63. Zhu Jiahua [Chu Chia-hua], *China's Postal and Other Communications Services* (Shanghai: China United Press, 1937), 149.

64. Kurt Jacobsen, "A Danish Watchmaker Created the Chinese Morse System," *NIASnytt (Nordic Institute of Asian Studies) Nordic Newsletter* 2 (July 2001): 17–21.

65. Septime Auguste Viguier [威基謁] (*Weijiye*), *Dianbao xinshu* [電報新書], in "Extension Selskabet—Kinesisk Telegrafordbog," 1871; Arkiv nr. 10.619, in "Love og vedtægter med anordninger," GN Store Nord A/S SN China and Japan Extension Telegraf. Rigsarkivet [Danish National Archives], Copenhagen, Denmark.

66. No mention is made in Chinese, Danish, or British sources as to whether Chinese telegraphers made use of the "abbreviated numerals" in Continental Morse (also known as International Morse Code). If so, it is possible that telegraphers were able to bypass the inefficiency of standard Morse codes for numerals, while still facing the limitations inherent to using only numerals and never letters.

67. Tom Standage, *The Victorian Internet: The Remarkable Story of the Telegraph and the Nineteenth Century's On-line Pioneers* (New York: Berkeley Books, 1999).

68. Steve Bellovin, "Compression, Correction, Confidentiality, and Comprehension: A Modern Look at Commercial Telegraph Codes," paper presented at the Cryptologic History Symposium, 2009, Laurel, MD. See also chapter 5 of N. Katherine Hayles, *How We Think: Digital Media and Contemporary Technogenesis* (Chicago: University of Chicago Press, 2012).

69. Edward Benjamin Scott, *Sixpenny Telegrams: Scott's Concise Commercial Code of General Business Phrases* (London: published by the author, 1885), 18, 35. Among many other possible examples, see in particular: Frank Shay, *Cipher Book for the Use of Merchants, Stock Operators, Stock Brokers, Miners, Mining Men, Railroad Men, Real Estate Dealers, and Business Men Generally* (Chicago: Rand McNally and Co., 1922).

70. Viguier's original list of characters was later slightly adjusted by De Mingzai (德明在). See Erik Baark, *Lightning Wires: The Telegraph and China's Technological Modernization, 1860–1890* (Westport, CT: Greenwood Press, 1997), 85. Following the 1949 revolution, we witness the further creation of two versions of the code, one in the mainland and one in Taiwan, which both made use of four-digit codes but with different code assignments. Even when accounting for these changes, we see that the basic model of Viguier's system became the industry standard for Chinese for over a century.

71. For a recent study of China's telegraphic infrastructure see Roger R. Thompson, "The Wire: Progress, Paradox, and Disaster in the Strategic Networking of China, 1881–1901," *Frontiers in the History of China* 10, no. 3 (2015): 395–427.

72. *Documents de la Conférence Télégraphique Internationale de Berlin (1909)*, 482.

73. Over the course of the 1920s and 1930s, radio transmission bore a heavier responsibility for telegraphic transmission. Observing Shanghai alone, thirteen lines connected the metropole to destinations in Europe, the United States, and Southeast Asia through intermediaries such as the Radio Corporation of America, Telefunken Telegraph and Wireless Company, Compagnie Générale de T.S.F., the Soviet Commission of Communications, the Directorate-General of Telegraphs and Posts of Annam, and other outfits. When radio transmission began in 1931, it carried approximately 10 percent of international traffic from China. Just four years later in 1935, radio accounted for approximately 40 percent of all international traffic. Statistics according to Zhu Jiahua, minister of communications from 1932 to 1935. Zhu Jiahua [Chu Chia-hua], *China's Postal and Other Communications Services*.

74. Bull, "A Short History of the Shanghai Station," 115. Even victories such as these carried unforeseen and sometimes negative repercussions, however. In this case, the

general lack of familiarity with the Chinese language, and thus inability to authenticate a given transmission as "bona fide," meant that China would be required to "deposit" official copies of the code book in multiple foreign countries, to serve as the proverbial gold standard. In an act not unlike the ritual interment of the "standard meter" and the "standard kilogram" in the French town of Breteuil in 1889—an act designed to preserve and render sacred an otherwise artificial system of weights and measures for eternity—the standard Chinese telegraph code would have to be "buried" in sites across the globe to ensure uniformity and temporal persistence. Such perpetuity, while it was to be desired in the case of weights and measures, could in the case of Chinese telegraphy only restrict future efforts to redesign the code. *Dictionnaire télégraphique officiel chinois en français (Fawen yi Huayu dianma zihui)* [法文譯華語電碼字彙] (Shanghai: Dianhouzhai [點后齋], n.d); Peter Galison, *Einstein's Clocks, Poincaré's Maps: Empires of Time* (New York: W.W. Norton and Co., 2003), 91; *Documents de la Conférence Télégraphique Internationale de Madrid (1932)— Tome I* (Bern: Bureau International de l'Union Télégraphique, 1933), 429. Owing to the complex, variegated structure of the conventions and amendments that governed telegraphic tariffs, moreover, the goal of ameliorating the position of Chinese could not be achieved through any single victory. Each gain was hard-won, incremental, and in need of constant maintenance lest it be repealed by subsequent amendments. Every time a new subtechnology or practice was added to the expanding telegraphic repertoire—such as preferential pricing for holiday telegrams, novel methods of addressing telegrams, the emerging coordination of telegraphy and surface post in the form of "letter-telegrams," and so forth—China could be sure that any new amendment would undoubtedly include at least one article pertaining to restrictions on encoded and numbered language. Christmas telegrams were to be assessed at a lower rate than regular telegrams, and yet such telegrams had to be written in "plaintext"—something Chinese did not technically possess. Likewise, it was stipulated in multiple agreements that telegram addresses had to be written in "plaintext" as well, and could not be written in either codes or ciphers. In the 1932 conference in Madrid, the congress outlined more detailed rules governing the practice of both "télégrammes de félicitations" and "letter-telegrams," stipulating that such transmissions had to be written entirely in "clear language." If regulations pertaining to discounted Christmas telegrams might appear utterly unrelated to the concerns of China, that is because they were—and it is precisely this unrelatedness that lies at the center of our ongoing consideration of "semiotic sovereignty." China's peculiar disadvantage within the international telegraphic community manifested itself, not in terms of targeted, conscious infringements on China's interests by callous and self-interested European and American powers, but through processes far more aloof and unaware. In far-off convention halls in Lisbon, London, and the like, a distant and largely indifferent Euro-American community passed one regulation after the next, which, despite having nothing to do with China per se, in fact exerted measurable influence upon China's interests, pricking, bumping, and bruising China practically each time. Each time that "clear language," "plain language," or analogous terms were invoked, representatives of China once again found themselves having to push for adjustments, lest China find itself once again excluded from one or another part of the ever-expanding and diversifying complex of global telegraphy.

75. The system was further mediated, moreover, through what appears to have been one or another Chinese dialect—perhaps Cantonese. That is, while table 2.1 lists only the pinyin pronunciation of each character in *putonghua*, a close consideration of particular letter-character combinations indicates that the pairings were quite likely made with a Chinese dialect in mind. The pairing of *zai* and "j" only makes sense, for example, when we consider its Cantonese pronunciation *joi*. The same is true for the "w," whose accompanying character is pronounced *wu* in Cantonese; and the letter "y" whose character is pronounced *wai*. Other elements within the list suggest a dialect other than Cantonese, however, such as the character used to represent "k" (凱—which is pronounced *hoi* in Cantonese). My thanks to Roy Chan for bringing this wrinkle to my attention. See Ministry of Communications (*Jiaotong bu*) [交通部], *Plaintext and Secret Telegraph Code—New Edition* (*Mingmi dianma xinbian*) [明密碼電報新編] (Shanghai: n.p., 1916); Ministry of Communications (*Jiaotong bu*) [交通部], *Plaintext and Secret Telegraph Code—New Edition* (*Mingmi dianma xinbian*) [明密電碼新編] (Nanjing: Jinghua yinshu guan [南京印書館], 1933), Rigsarkivet [Danish National Archives], Copenhagen, Denmark, 10619 GN Store Nord A/S. 1870–1969 Kode- og telegrafbøger, Kodebøger 1924–1969; Ministry of Communications (*Jiaotong bu*) [交通部], *Plaintext and Secret Telegraph Code—New Edition* (*Mingmi dianma xinbian*) [明密電碼新編] (n.p.: Jiaotongbu kanxing [交通部刊行], 1946).

76. Changing regulations governing international telegraphic communication often closed pathways of mediation as well. Trigraph encryption was eventually limited by the very same corporate and statist watchdogs that policed money-saving and/or clandestine communication. By the early nineteenth century, all telegraphic transmissions had to be "pronounceable"—a somewhat vague designation which translated more concretely into the mandatory inclusion of a minimum number of vowels within any coded transmission. Long sequences of consonants would no longer be permitted, thereby greatly reducing the number of code spaces in the original trigraph system.

77. Ministry of Communications (*Jiaotong bu*) [交通部], *Plaintext and Secret Telegraph Code—New Edition* (*Mingmi dianma xinbian*) [明密電碼新編] (n.p.: Jiaotongbu kanxing [交通部刊行], 1946).

78. Geoffrey C. Bowker, *Memory Practices in the Sciences* (Cambridge, MA: MIT Press, 2005).

CHAPTER 3: RADICAL MACHINES

1. "No Chinese Typewriters," *Gregg Writer* 15 (1912): 382.

2. "Judging Eastern Things from Western Point of View," *Chinese Students' Monthly* 8, no. 3 (1913): 154.

3. "Judging Eastern Things from Western Point of View," 154.

4. O.D. Flox, "That Chinese Type-Writer: An Open Letter to the Hon. Henry C. Newcomb, Agent of the Faroe Islands' Syndicate for the Promotion of Useful Knowledge," *Chinese Times* (March 31, 1888), 199.

5. Ibid.

6. "A Chinese Type-Writer," *Chinese Times* (January 7, 1888), 6.

7. Henry C. Newcomb, "Letter to the Editor: That Chinese Type-writer," *Chinese Times* [Tianjin] (March 17, 1888), 171–172.

8. Ibid.

9. Passport Applications January 2, 1906–March 31, 1925, National Archives and Records Administration, Washington, DC, ARC identifier 583830, MLR, number A1534, NARA Series M1490, roll 109.

10. A.H. Smith, "In Memoriam, Dr. Devello Z. Sheffield," *Chinese Recorder* (September 1913), 564–568, 565. See also Stephan P. Clarke, "The Remarkable Sheffield Family of North Gainesville" (n.p., n.d.), 3, manuscript provided by Stephan Clarke to the author.

11. "Missionaries of the American Board," *Congregationalist* (September 26, 1872), 3. See also Roberto Paterno, "Devello Z. Sheffield and the Founding of the North China College," in *American Missionaries in China,* ed. Kwang-ching Liu (Cambridge, MA: Harvard East Asian Monographs, 1966), 42–92. Eleanor Sherrill Sheffield was educated at Pike Seminary. See Clarke, "The Remarkable Sheffield Family of North Gainesville."

12. "Child of the Quarantine: One More Passenger on the Nippon Maru List—Baby Born During Angel Island Stay," *San Francisco Chronicle* (July 11, 1899), 12; Smith, "In Memoriam, Dr. Devello Z. Sheffield," 568.

13. Flox, "That Chinese Type-Writer," 199.

14. The treaty also expanded foreign navigation rights along the Yangtze River, allowed four Western powers to station legations in Beijing, and prohibited the use of the term *yi*—glossed in English as "barbarian"—in Chinese governmental records and correspondence in reference to the British and other foreign nationals. Devello and Eleanor raised five children, the first four of whom were born in Tongzhou: Alfred (1871–1961), John (1873–1874), Mary (1875–1961), Flora (1877–1975), and Carolyn (1880–1962). See Clarke, "The Remarkable Sheffield Family of North Gainesville," 14.

15. Devello Z. Sheffield, "The Chinese Type-writer, Its Practicability and Value," in *Actes du onzième Congrès International des Orientalistes,* vol. 2 (Paris: Imprimerie Nationale, 1898), 51.

16. Flox, "That Chinese Type-Writer," 199.

17. Letter from Devello Sheffield to his parents, January 27, 1886. Devello Z. Sheffield and Family Letters & Photographs (Ruth S. Johnson Family Collection).

18. Devello Z. Sheffield [謝衛樓], *Shendao yaolun* 神道要論 [*Important Doctrines of Theology*] (Tongzhou: Tongzhou wenkui qikan yin 通州文魁齊刊印, 1894).

19. Devello Z. Sheffield [謝衛樓], "Di'er zhang minshou youhuo weibei huangdi [第二章民受誘惑違背皇帝]," *Xiao hai yuebao* [小孩月報] 4, no. 3 (1878): 5; Devello Z. Sheffield [謝衛樓], "Diba zhang taizi duanding liangmin de baoying [第八章太子斷定良民的報應]," *Xiao hai yuebao* 5, no. 2 (1879): 2–3; Devello Z. Sheffield [謝衛樓], "Shangfa yuyan diliu zhang liangmin quanren fangzhan huigai [賞罰喻言第六章良民勸人放瞻悔改]," *Xiao hai yuebao* [小孩月報] 4, no. 10 (1879): 5; Devello Z. Sheffield [謝衛樓], "Shangfa yuyan disan zhang minshou youhuo fanzui geng shen [賞罰喻言第三章島民受誘惑犯罪更深]," *Xiao hai yuebao* [小孩月報] 4, no. 4 (1878): 6–7; Devello Z. Sheffield, [謝衛樓], "Shangfa yuyan diyi zhang daomin shou bawang xiahai [賞罰喻言第一章島民受霸王轄害]," *Xiao hai yuebao* [小孩月報] 4, no. 2 (1878): 3.

20. For nineteen years, from 1890 to 1909, he served as head of North China College. After a furlough in the United States, Sheffield returned to China in the fall of 1900, in the wake of the Boxer Rebellion. He assisted in the reconstruction of North China College. A new committee had formed, charged with overseeing the revision of the New Testament in literary Chinese. The Shanghai Conference appointed Sheffield as chairman of the committee, in which role he would oversee the project through its completion seventeen years later, in 1907. Shortly thereafter, Sheffield's tenure was extended, this time as chairman to oversee revision of the Old Testament in literary Chinese.

21. Sheffield, "The Chinese Type-writer, Its Practicability and Value," 63.

22. Ibid., 62.

23. Ibid., 62–63.

24. Ibid., 63. For clerks and businesspeople, as well, such an amanuensis machine would offer no opinions or interpretations of its own, and would permit foreigners to compose and issue Chinese-language communications "with the assurance that private matters would not be divulged by his writer." For a fascinating examination of parallel anxieties in the British imperial context, see Christopher Bayly, *Empire and Information: Intelligence Gathering and Social Communication in India, 1780–1870* (Cambridge: Cambridge University Press, 1999 [1996]).

25. Sheffield, "The Chinese Type-writer, Its Practicability and Value," 62–63.

26. Ibid., 63.

27. Ibid., 51.

28. Ibid., 50.

29. Ibid.

30. Ibid.

31. Ibid.

32. Ibid., 51.

33. Letter from Devello Sheffield to his family, December 3, 1888; letter from Devello Sheffield to his parents, January 27, 1886; Devello Z. Sheffield and Family Letters & Photographs (Ruth S. Johnson Family Collection).

34. Joseph Needham, *Science and Civilisation in China*, vol. 5, part 1 (Cambridge: Cambridge University Press, 1985), 206–207.

35. Sheffield, "The Chinese Type-writer, Its Practicability and Value," 51.

36. In the case of extremely rare characters, movable type printers were sometimes required to punch certain characters on demand. Sheffield's machine encompassed far fewer characters than did the average Chinese printing office, however, making such recourses to extra-systemic and/or alternate modes of inscription far more common.

37. Devello Z. Sheffield, *Selected Lists of Chinese Characters, Arranged According of Frequency of their Recurrence* (Shanghai: American Presbyterian Mission Press, 1903). Sheffield evidently did not count "untabulated characters" as part of his final calculation of 4,662.

38. Sheffield, *Selected Lists of Chinese Characters*. In a brief passage, Sheffield reflects as well on the wheel of his machine, which could be adjusted to include alternate Chinese characters, pertinent to other industries. Sheffield, "The Chinese Typewriter, Its Practicability and Value," 63.

39. This was not the case for Sheffield, however, who included each of these character forms on his machine, and thus clearly was treating them as separate entities.

40. "Science and Industry," *Arkansas Democrat* (October 10, 1898), 7; "China," *Atchison Daily Globe* (April 11, 1898), 1; *Daily Picayune-New Orleans* (April 9, 1898), 4; "Our Benevolent Causes," *Southwestern Christian Advocate* (July 8, 1897), 6; "Will Typewrite Chinese," *Atchison Daily Globe* (June 1, 1897), 3; "Typewriter in Chinese," *Denver Evening Post* (May 29, 1897), 1; "Salmis Journalier," *Milwaukee Journal* (May 3, 1897), 4.

41. *Daily Picayune-New Orleans* (April 19, 1898), 4.

42. "Science and Industry," 7.

43. "A Chinese Typewriter," *Semi-Weekly Tribune* (June 22, 1897), 16.

44. Sheffield, "The Chinese Type-writer, Its Practicability and Value," 60.

45. Smith, "In Memoriam, Dr. Devello Z. Sheffield," 565.

46. Sheffield arrived in Honolulu on March 24, 1909, aboard the *S.S. Siberia*. His destination was listed as Detroit, Michigan. See "Passenger Lists of Vessels Arriving or Departing at Honolulu, Hawaii, 1900–1954." National Archives and Records Administration, Washington, DC, Records of the Immigration and Naturalization Service, record group 85, series/roll no. m1412:6.

47. "Child of the Quarantine,"12.

48. "Passenger Lists of Vessels Arriving or Departing at Honolulu, Hawaii, 1900–1954," National Archives and Records Administration, Washington, DC, Records of the Immigration and Naturalization Service, record group 85, series/roll no. m1412:6.

49. Nanyang gongxue. Elsewhere, Zhou Houkun is listed as being from Wuxi in Jiangsu province. See "MIT China (*Meiguo Masheng ligong xuexiao Zhongguo*) [美國麻

省理工學校中國]," *Shenbao* (July 19, 1915), 6; University of Illinois Urbana-Champaign, ed., *University of Illinois Directory: Listing the 35,000 Persons Who Have Ever Been Connected with the Urbana-Champaign Departments, Including Officers of Instruction and Administration and 1397 Deceased* (Urbana-Champaign, 1916), 118. Nanyang College (*Nanyang Gongxue*), built in 1896–1898, was founded by order of the Qing's office for the development of foreign investments and telegraphic offices. Later, in 1921, it became Jiaotong Engineering University (交通大學).

50. Other notable shipmates and fellow travelers included Zhou Ren (周仁) and Zhang Pengchun (張彭春). Hongshan Li, US-China Educational Exchange, 62–63, 65–67, 70; "Name List of Students Selected to Travel to America (*Qu ding you Mei xuesheng mingdan*) [取定遊美學生名單]," *Shenbao* (August 9, 1910), 5; "Draft Proposal regarding Students Taking Exam to Study Abroad in the United States (*Kaoshi liu Mei xuesheng cao'an*) [考試留美學生草案]," *Shenbao* (August 8, 1910), 5–6.

51. "New Invention: A Chinese Typewriter (*Zhongguo daziji zhi xin faming*) [中國打字機之新發明]," *Shenbao* (August 16, 1915), 10; University of Illinois Urbana-Champaign, ed., *University of Illinois Directory: Listing the 35,000 Persons Who Have Ever Been Connected with the Urbana-Champaign Departments, Including Officers of Instruction and Administration and 1397 Deceased* (Urbana-Champaign: University of Illinois Press, 1916), 118.

52. His thesis was entitled "Experimental Determination of Damping Coefficients in the Stability of Aeroplanes." See Lauren Clark and Eric Feron, "Development of and Contribution to Aerospace Engineering at MIT," *40th AIAA Aerospace Sciences Meeting and Exhibit* (January 14–17, 2002), 2; "New Invention: A Chinese Typewriter (*Zhongguo daziji zhi xin faming*) [中國打字機之新發明]," *Shenbao* (August 16, 1915), 10; "MIT China (*Meiguo Masheng ligong daxue Zhongguo*) [美國麻省理工大學中國]," *Shenbao* (July 15, 1914), 6.

53. Zhou Houkun, "Explanation of Newly Created Chinese Typewriter (*Chuangzhi Zhongguo daziji tushuo*) [創制中國打字機圖說]," trans. Wang Ruding, *Eastern Miscellany (Dongfang zazhi)* [東方雜誌] 12, no. 10 (October 1915): 28.

54. A Monotype machine is a form of "hot metal" composing machine, a technique of typesetting in which operators use a keyboard to set a series of type matrices, hollow molds of letters into which the machine injects molten metal which cools into type slugs. Unlike earlier techniques of movable type printing, in which type setting and type founding constituted two separate processes, hot metal composing machines merged these processes into single devices. Monotype machines created injection-mold letters one by one, hence the name *mono*-type; while Linotype machines cast letters in the form of bars or *lines of type*. Hot metal composing marked an epoch in the history of print, replacing and bringing to an end the age of industrial-scale movable type printing made famous by Johannes Gutenberg and his descendants.

55. Here we see the continuation of late Qing reformers examined by Tze-ki Hon and others. As Hon phrases it, these late Qing participants in Chinese pedagogical reforms "realized that they were the 'educator of citizens,' participating in

nation-building by forging a collective identity among young Chinese." See Tze-Ki Hon, "Educating the Citizens: Visions of China in Late Qing History Textbooks," in *The Politics of Historical Production in Late Qing and Republican China,* ed. Tze-ki Hon and Robert Culp (Leiden: Brill, 2007), 81.

56. Charles W. Hayford, *To the People: James Yen and Village China* (New York: Columbia University Press, 1990), 40–41.

57. George Kennedy, "A Minimum Vocabulary in Modern Chinese," *Modern Language Journal* 21, no. 8 (May 1937): 587–592, 590. The precise size of Chen's corpus was 554,478 Chinese characters.

58. To be precise, the 177 most frequent characters accounted for 57 percent of the entire corpus. As we recall from chapter 2, William Gamble had determined in 1861 that the thirteen most common characters in his corpus accounted for one-sixth or 16.67 percent of the whole—a number that maps on to Chen Heqin's findings rather well. As we move out along both scholars' frequency curves, the fit becomes even tighter. In Gamble's analysis, the first 521 characters accounted for nine-elevenths (or 81.8 percent) of the whole. In Chen's analysis, the first 569 characters accounted for 80 percent of the whole.

59. Hayford, *To the People,* 60.

60. Ibid., 44.

61. Ibid., 50. Westerners were also concerned with this, but we will not focus on this community of practitioners. See, for example, William Edward Soothill, *Student's Four Thousand 字 and General Pocket Dictionary,* 6th ed. (Shanghai: American Presbyterian Mission Press, 1908), v. See also Courtenay Hughes Fenn, *The Five Thousand Dictionary,* rev. American ed. (Cambridge, MA: Harvard University Press, 1940), based on 5th Peking ed., which included additions and revisions by George D. Wilder, B.A., D.D., and Mr. Chin Hsien Tseng, eleventh printing.

62. Hayford, *To the People,* 48.

63. Ibid., 45. With the growth of mass education movements, algorithmic reading became a kind of pedagogical cottage industry. Competing parties put forth their own proprietary visions of "minimum Chinese." The National Association for the Advancement of Mass Education later published another list, this one targeted at farming communities. See National Association for the Advancement of Mass Education (*Zhonghua pingmin jiaoyu cujinhui*) [中華平民教育促進會], ed., *Farmer's One Thousand Character Textbook* (*Nongmin qianzi ke*) [農民千字課] (n.p., 1933), Rare Book and Manuscript Library, Columbia University, Papers of the International Institute of Rural Reconstruction, MS COLL/IIRR; and *Universal Word List for the Average Citizen* (*Ping guomin tongyong cibiao*) [平國民通用詞表], Rare Book and Manuscript Library, Columbia University, Papers of the International Institute of Rural Reconstruction, MS COLL/IIRR, n.d.

64. Kennedy, "A Minimum Vocabulary in Modern Chinese," 589.

65. Ibid., 588–591. Li extended the corpus to analyze 1,497,182 Chinese characters. Lu-Ho Rural Service Bureau (*Luhe xiangcun fuwubu*) [潞河鄉村服務部], ed., *2,000*

Foundational Characters for Daily Use (Richang yingyong jichu erqian zi) [日常應用基礎二千字] (n.p., 1938 [November]), Rare Book and Manuscript Library, Columbia University, Papers of the International Institute of Rural Reconstruction, MS COLL/IIRR.

66. Kennedy, "A Minimum Vocabulary in Modern Chinese," 591.

67. Zhou Houkun, "Patent Document for Common Usage Typewriter Tray (*Tongsu dazipan shangqueshu*) [通俗打字盤商榷書]," *Educational Review (Jiaoyu zazhi)* [教育雜誌] 9, no. 3 (March 1917): 12–14. See also *Biographical Dictionary of Chinese Christianity*, http://www.bdcconline.net/en/stories/d/dong-jingan.php

68. Zhou Houkun, "Diagram Explaining Production of Chinese Typewriter (*Chuangzhi Zhongguo daziji tushuo*) [創制中國打字機圖說]," 28, 31. Praising his Christian missionary predecessor by name, Zhou made clear his opinion that "the honor of being the 'inventor of the first Chinese typewriter' should go to [Devello Sheffield]." At the same time, however, Zhou claimed autonomy on behalf of his own invention, declaring his unawareness of the American's machine during his development process.

69. "Diagram Explaining Production of Chinese Typewriter (*Chuangzhi Zhongguo daizji tushuo*) [創制中國打字機圖說]," *Zhonghua gongchengshi xuehui huikan* 2, no. 10 (1915): 15–29; "Chinaman Invents Chinese Typewriter Using 4,000 Characters," *New York Times* (July 23, 1916), SM15.

70. Ibid.

71. Thomas Sammons, "Chinese Typewriter of Unique Design," *Department of Commerce Bureau of Foreign and Domestic Commerce, Commerce Reports* 3, nos. 154–230 (May 24, 1916): 20.

72. Ibid.

73. "It Takes Four Thousand Characters to Typewrite in Chinese," *Popular Science Monthly* 90, no. 4 (April 1917): 599.

74. See Abigail Markwyn, "Economic Partner and Exotic Other: China and Japan at San Francisco's Panama-Pacific International Exposition," *Western Historical Quarterly* 39, no. 4 (2008): 444, 454–459.

75. *Temporary Catalogue of the Department of Fine Arts Panama-Pacific International Exposition: Official Catalogue of Exhibitors*, rev. ed. (San Francisco: The Wahlgreen Co., 1915), 32.

76. Key information comes from Qi Xuan's exclusion act case file. A Chinese Exclusion Act Case File was a dossier created when a person of Chinese ancestry (both citizens and noncitizens) entered the United States, and then expanded upon during any subsequent reentry. Such records were maintained between the years 1884 and 1943, initiated as part of federal anti-Chinese immigration restrictions enacted during the Arthur administration, and finally repealed in 1943. Exclusion Act dossiers typically included photographs, certificates containing basic demographic information, and interrogation transcripts.

77. Elsewhere, Qi Xuan is reported to be a student in Shanxi Province. See "A Chinese Typewriter," *Peking Gazette* (November 1, 1915), 3; and "A Chinese Typewriter," *Shanghai Times* (November 19, 1915), 1.

78. In his article on Qi Xuan's machine, C.C. Chang refers to a Professor "Brayns," but this is almost certainly a typo. Employment and faculty records from New York University at the time make reference, however, to a "William Remington Bryans," professor of engineering at NYU at this time. See C.C. Chang, "Heun Chi Invents a Chinese Typewriter," *Chinese Students' Monthly* 10, no. 7 (April 1, 1915): 459.

79. The second cylinder was surfaced with paper, upon which was written a key to locate these glyphs, organized into 110 groups.

80. Heun Chi [Qi Xuan], "Apparatus for Writing Chinese," United States Patent no. 1260753 (filed April 17, 1915; patented March 26, 1918).

81. Medhurst, *China: Its State and Prospects*, 558.

82. Robert S. Brumbaugh, "Chinese Typewriter," United States Patent no. 2526633 (filed September 25, 1946; patented October 24, 1950).

83. Wang Kuoyee, "Chinese Typewriter," United States Patent no. 2534330 (filed March 26, 1948; patented December 19, 1950).

84. For a terrific and vividly illustrated introduction to early twentieth-century graphic design in China, see Scott Minick and Jiao Ping, *Chinese Graphic Design in the Twentieth Century* (London: Thames and Hudson, 1990).

85. Johanna Drucker, *The Visible Word: Experimental Typography and Modern Art, 1909–1923* (Chicago: University of Chicago Press, 1997).

86. Chang, "Heun Chi Invents a Chinese Typewriter," 459.

87. "4,200 Characters on New Typewriter; Chinese Machine Has Only Three Keys, but There Are 50,000 Combinations; 100 Words in TWO HOURS; Heuen Chi, New York University Student, Patents Device Called the First of Its Kind," *New York Times* (March 23, 1915), 6.

88. *Official Congressional Directory* (Washington, DC: United States Congress, 1916 [December]), 377.

89. "4,200 Characters on New Typewriter," 6.

90. Ibid.

91. Ibid.

92. "The Newest Inventions," *Washington Post* (March 21, 1917), 6.

93. See, for example, "An Explanation of the Chinese Typewriter and the Twentieth-Century War Effort (*Zhongguo daziji zhi shuoming yu ershi shiji zhi zhanzheng liqi*) [中國打字機之說明與二十世紀之戰爭力氣]," *Huanqiu* [環球 *The World's Chinese Students' Journal*] 1, no. 3 (September 15, 1916): 1–2.

94. Chang, "Heun Chi Invents a Chinese Typewriter," 459.

95. Zhang Yuanji, diary entry, May 16, 1919, in *Zhang Yuanji quanji* [張元濟], vol. 6 (*juan* 6), 56 ["山西留學紐約紀君製有打字機,雖未見其儀器,而所打之字則甚明晰,似此周

厚坤所製為優"]. "Mr. Ji, a Shanxi native studying overseas in New York, has built a typewriter. Although I haven't seen the machine, the characters it types are quite clear. It appears this machine might surpass that of Zhou Houkun." "Shanxi liuxue Niuyue Ji jun zhiyou daziji, sui wei jian qi yiqi, er suo da zhi zi ze shen mingxi, si ci Zhou Houkun suo zhi wei you [山西留學紐約紀君製有打字機,雖未見其儀器,而所打之字則甚明晰,似此周厚坤所製為優]."

96. H.K. Chow [Zhou Houkun], "The Problem of a Typewriter for the Chinese Language," *Chinese Students' Monthly* (April 1, 1915), 435–443.

97. Zhou Houkun, *"Chuangzhi Zhongguo daziji tushuo,"* 31.

98. Ibid. All usage is as it appears in the original.

99. Qi Xuan [Heuen Chi], "The Principle of My Chinese Typewriter," *Chinese Students' Monthly* 10, no. 8 (May 1, 1915): 513–514.

100. Ibid.

101. Ibid.

102. Ibid.

103. "New Invention of a Chinese Typing Machine (*Zhongguo dazi jiqi zhi xin faming*) [中國打字機器新發明]," *Tongwen bao* [通問報] 656 (1915): 8.

104. Zhou would receive 160 yuan a month, and thereafter be retained as a consultant following his move to Nanjing. As Commercial Press continued to develop the machine, Zhou would work for the press three months out of the year, receiving a total of 600 yuan for his services. Zhang Yuanji, diary entry, March 1, 1916 (Wednesday), in *Zhang Yuanji quanji* [張元濟], vol. 6 (*juan* 6), 19–20. This is the earliest entry referring to Zhou Houkun.

CHAPTER 4: WHAT DO YOU CALL A TYPEWRITER WITH NO KEYS?

1. The surface of each character slug measured approximately one-half centimeter squared, fitted upon a base that measured some one-and-a-half centimeters tall. The base of the machine, constructed in wood and used to store various tools and implements for the machine—a cleaning brush, a pair of tweezers, a small wrench—measured 41 centimeters deep, 47.3 centimeters wide, and 6 centimeters high. Atop this sat the machine. Attached to the front of the machine was the character tray guide, a thin, metal-rimmed glass frame in which a paper guide to the characters was housed. The glass frame was to scale with the tray bed, but with the edge measuring some 45 centimeters in width and 25.5 centimeters in height. The platen of the machine measured 36.5 centimeters wide.

2. *User's Manual for Chinese Typewriter Manufactured by Shanghai Commercial Press* (*Shanghai yinshuguan zhizao Huawen daziji shuomingshu*) [上海印書館製造華文打字機說明書] (Shanghai: Commercial Press, 1917 [October]).

3. Christopher A. Reed, *Gutenberg in Shanghai: Chinese Print Capitalism, 1876–1937* (Honolulu: University of Hawai'i Press, 2004).

4. "Summer Vacation at the China Railway School (*Zhonghua tielu xuexiao shujia*) [中華鐵路學校暑假]," *Shenbao* (July 5, 1916), 10.

5. "Zhou and Wang are Quintessential Scholars (*Zhou Wang liang jun juexue*) [周王兩君絕學]," *Shenbao* (July 24, 1916), 10. While Zhou's audience was doubtless intrigued by this young man's invention, part of his appeal was also his status as a returned student who enjoyed firsthand experience of the United States. Whether by invitation or his own initiative, part of Zhou's presentation on his Chinese typewriter was given over to a brief lecture on "Black Schools" in the United States, accompanied by a set of lantern slides recently purchased by the Provincial Education Committee on the subject of "American Blacks" (*Meiguo Heiren*).

6. Zhang Yuanji, diary entry, January 9, 1917, in *Zhang Yuanji quanji*, vol. 6 (*juan 6*), 141.

7. Zhang Yuanji, diary entry, May 26, 1919 (Tuesday), in *Zhang Yuanji quanji*, vol. 7 (*juan 7*), 71.

8. "News from the Nanyang University Engineering Association (*Nanyang daxue gongchenghui jinxun*) [南洋大學工程會近訊]," *Shenbao* (November 10, 1922), 14.

9. See "Jiangsu Department of Industry Hires Zhou Houkun as Advisor (*Su shiyeting pin Zhou Houkun wei guwen*) [蘇實業廳聘周厚坤為顧問]," *Shenbao* (October 21, 1923), 15. In a 1922 article, Zhou Houkun is referred to by his courtesy name Pengxi [朋西]. "Wuchang Inspection Office Examines Hanyeping Remittances (*Wuchang jianting zi cha Hanyeping jiekuan*) [武昌檢廳咨查漢冶萍解欵]," *Shenbao* (August 26, 1922), 15.

10. Zhang Yuanji, diary entry, May 26, 1919 (Tuesday), in *Zhang Yuanji quanji*, vol. 7 (*juan 7*), 71.

11. "Commercial Press Establishes Chinese Typewriting Class (*Shangwu yinshuguan she Huawen daziji lianxi ke*) [商務印書館設華文打字機練習課]," *Education and Vocation* (*Jiaoyu yu zhiye*) [教育與職業] 10 (1918): 8.

12. Shu Changyu [舒昌鈺] (aka Shu Zhendong [舒震東]), "Thoughts While Researching a Typewriter for China (*Yanjiu Zhongguo daziji shi zhi ganxiang*) [研究中國打字機時之感想]," *Tongji* 2 (1918): 156; Zhang Yuanji, diary entry, February 24, 1919 (Monday), in *Zhang Yuanji quanji*, vol. 7 (*juan 7*), 30. Zhang Yuanji also contemplated a multinational manufacturing plan, in which Shu Zhendong might have precision parts manufactured in the United States, and presumably then have them assembled in China. Zhang Yuanji, diary entry, April 16, 1920 (Friday), in *Zhang Yuanji quanji*, vol. 7 (*juan 7*), 205. Zhang Yuanji told Bao Xianchang to have precision parts created in the United States by Shu Zhendong, then returned to Shanghai for assembly. Apparently Bao asked Shu Zhendong, but Shu thought this would be a hassle, and that it was unnecessary to accelerate the process. See Zhang Yuanji, diary entry, April 19, 1920 (Monday), in *Zhang Yuanji quanji*, vol. 7 (*juan 7*), 205.

13. Ping Zhou, "A Record of Viewing the Chinese Typewriter Manufactured by Commercial Press (*Canguan shangwu yinshuguan zhizao Huawen daziji ji*) [參觀商務印書館制造華文打字機記]," *Shangye zazhi* [商業雜誌] 2, no. 12 (1927): 1–4.

14. Dissatisfied with this first model, Shu subsequently set out on a tour of Europe and the United States, with the express purpose of visiting factories and developing novel approaches to the problem of Chinese typewriting. Upon his return to Shanghai, he set about developing an improved model of Chinese typewriter, a process that went through five iterations, or "styles" (*shi*). A timeline of models outlined in a later manual cites 1919 as the year in which the "third style" of Shu Zhendong machine was released.

15. John A. Lent and Ying Xu, "Chinese Animation Film: From Experimentation to Digitalization," in Ying Zhu and Stanley Rosen, *Art, Politics, and Commerce in Chinese Cinema* (Hong Kong: Hong Kong University Press, 2010), 112.

16. The pair later went on to produce a series of animated shorts, as well as *The Camel's Dance* (*Luotuo xianwu*) (1935) and the country's first full-length animated feature, *Princess Iron Fan* (*Tieshan gongzhu*) (1941).

17. Ping Zhou, "A Record of Viewing the Chinese Typewriter," 2.

18. Gan Chunquan and Xu Yizhi, eds., *Essential Knowledge for Secretaries: Requirements of a Chinese Typist* (*Shuji fuwu bibei: yiming Huawen dazi wenshu yaojue*) [書記服務必備一名華文打字文書要決] (Shanghai: Renwen yinshuguan, 1935), 25–30.

19. Ping Zhou, "A Record of Viewing the Chinese Typewriter," 2. In addition to the more than seven thousand characters on the common use tray and the secondary and tertiary use boxes, moreover, the machine was also outfitted with the English alphabet, Arabic numerals, the Chinese phonetic alphabet (*zhuyin*), and Western-style punctuation marks.

20. Zhang Yuanji, diary entry, April 16, 1920 (Friday), in *Zhang Yuanji quanji*, vol. 7 (*juan 7*), 205.

21. "Consulate Purchases Chinese Typewriter (*Lingshiguan zhi Hanwen daziji*) [領事館置漢文打字機]," *Chinese Times* (*Dahan gongbao*) [大漢公報] (May 18, 1925), 3.

22. "Business News (*Shangchang xiaoxi*) [商場消息]," *Shenbao* (October 27, 1926), 19–20.

23. "Report on Machinery on Display at the Department of Forestry and Agriculture (*Chenlie suo jishu nonglinbu yanjiu tan*) [陳列所機術農林部研究談]," *Shenbao* (November 27, 1921), 15.

24. *Shenbao* (May 2, 1924), 21. For information on Song Mingde, see SMA Q235-1-1875 (April 6, 1933), 18–20.

25. Shu, "Thoughts While Researching a Typewriter for China," 156.

26. Ibid.

27. Ibid.

28. For example, see "Preparations for a Chinese Typewriting Class (*Chouban Huawen daziji xunlianban*) [籌辦華文打字機訓練班]," *Henan jiaoyu* (*Henan Education*) [河南教育] 1, no. 6 (1928): 4.

29. As Gail Hershatter notes, we still know relatively little about the world of Republican-era professional women, a notable exception being the 1999 study by Wang

Zheng. See Gail Hershatter, *Women in China's Long Twentieth Century* (Berkeley: University of California Press, 2007); and Wang Zheng, *Women in the Chinese Enlightenment: Oral and Textual Histories* (Berkeley: University of California Press, 1999).

30. Christopher Keep, "The Cultural Work of the Type-Writer Girl," *Victorian Studies* 40, no. 3 (Spring 1997): 405.

31. Communication between Tianjin Municipal Government Department of Education (*Tianjin shi zhengfu jiaoyuju*) [天津市政府教育局] and the International Typing Institute (*Guoji dazi chuanxisuo*) [國際打字傳習所], TMA J110-1-838 (July 6, 1946), 1–15; Communication between Tianjin Municipal Government Department of Education (*Tianjin shi zhengfu jiaoyuju*) [天津市政府教育局] and Junde Chinese Typing Institute (*Junde Huawen dazi zhiye buxi xuexiao*) [峻德華文打字職業補習學校], TMA J110-1-808 (March 5, 1948), 1–12; "Beijing Baoshan and Guangde Chinese Typing Supplementary Schools (*Beijing shi sili Baoshan, Guangde Huawen dazi buxi xuexiao guanyu xuexiao tianban qiyong qianji baosong li'an biao jiaozhiyuan lüli biao he xuesheng mingji chengji biao chengwen ji shi jiaoyuju de zhiling (fu: gaixiao jianzhang, zhaosheng jianzhang he xuesheng chengji biao)* [北京市私立寶善、廣德華文打字補習學校關於學校天辦啟用鈐記報送立案表教職員履歷表和學生名籍成績表呈文及市教育局的指令(附:該校簡章、招生簡章和學生成績表)]," BMA J004-002-00579 (July 1, 1938); "Regarding the Foundation of the Guangdewen Typing Supplementary School in Beiping (*Guanyu chuangban Beiping shi sili Guangdewen dazi buxi xueshe de chengwen ji gai she jianzhang deng yiji shehui ju de piwen*) [關於創辦北平市私立廣德文字打字補習學社的呈文及該社簡章等以及社會局的批文]," BMA J002-003-00754 (May 1, 1938); "Curriculum Vitae of Teaching Staff and Student Roster of the Private Yucai Chinese Typing Vocational Supplementary School in Beiping (*Beiping shi sili Yucai Huawen dazike zhiye buxi xuexiao zhijiaoyuan lüli biao, xuesheng mingji biao*) [北平市私立育才華文打字科職業補習學校職教員履歷表、學生名籍表]," BMA J004-002-00662 (July 31, 1939); "Petition Submitted by the Private Yadong Japanese-Chinese Typing Supplementary School of Beijing to the Beijing Special Municipality Bureau of Education Regarding Inspection of Student Grading List, Curriculum Plan, and Teaching Hours for the Regular-Speed Division of the Sixteenth Term, and the Order Received in Response from the Bureau of Education (*Beijing shi sili Yadong Ri Huawen dazi buxi xuexiao guanyu di'shiliu qi putong sucheng ge zu xuesheng chengjibiao, kecheng yuji ji shouke shi shu qing jianhe gei Beijing tebieshi jiaoyuju de cheng yiji jiaoyuju de zhiling*) [北京市私立亞東日華文打字補習學校關於第十六期普通速成各組學生成績表、課程預計及授課時數請鑒核給北京特別市教育局的呈以及教育局的指令]," BMA J004-002-01022 (January 31, 1943); Private Yadong Japanese-Chinese Typing Supplementary School of Beijing Student Roster (*Beijing shi sili Yadong Ri Hua wen dazi buxi xuexiao xuesheng mingji biao*) [北京市私立亞東日華文打字補習學校學生名籍表], BMA J004-002-01022 (November 7, 1942); Private Shucheng Typing Vocational Supplementary School of Beijing Student Roster (*Beijing shi sili shucheng dazike zhiye buxi xuexiao xuesheng mingji biao*) [北京市私立樹成打字科職業補習學校學生名籍表], BMA J004-002-01091 (March 23, 1942); Private Yanjing Chinese Typing Supplementary School of Beijing Student Roster (*Beijing sili Yanjing Huawen dazi buxi xuexiao xuesheng mingji biao*) [北京私立燕京華文打字補習學校學生名籍表], BMA J004-001-00805 (November 1,

1946); SMA R48-1-287; Tianjin Municipality Eighth Educational District Mass Education Office Eighth Term Chinese Typing Accelerated Class Graduation Name List (*Tianjin shili di'ba shejiaoqu minzhong jiaoyuguan di'ba qi Huawen dazi suchengban biye xuesheng mingce*) [天津市立第八社教區民眾教育館第八期華文打字速成班畢業學生名冊], TMA J110-3-740 (November 25, 1948), 1–2; "Regulations of the Commercial Typing and Shorthand Training Institute (*Shangye dazi suji chuanxisuo jianzhang*) [商業打字速記傳習所簡章]," SMA Q235-1-1844 (June 1932), 49–56; SMA Q235-1-1871.

32. "Mr. Hui's Chinese-English Typing Institute (*Hui shi Hua Ying wen dazi zhuanxiao*) [惠氏華英打字專校]," SMA Q235-1-1847 (1932), 26–49.

33. "Victory and Success Typing Academy (*Jiecheng dazi chuanxisuo*) [捷成打字傳習所]," SMA Q235-1-1848 (1933), 50–70.

34. The principal of the institute was Zhu Hongjuan, a Shanghai resident born around 1908, under whose direction worked a staff of five instructors—two men and three women. Zhu Fei, a young woman of nineteen years, born around 1925, and potentially the younger sister of Zhu Hongjuan, served as an instructor at the main campus. Zhang Guoliang, also at the main campus, was born around 1900. By 1944, Huanqiu boasted an enrollment of over 300 students, prompting the school to expand later to a second location at 652 Taishan Road. See memo from Zhu Hongjuan [朱鴻雋], principal of the Huanqiu Typing Academy of Shanghai to a Mr. Lin [林], Shanghai Special Municipality education bureau chief. SMA R48-1-287 (October 27, 1944), 1–11.

35. "Contacts for School Graduates Where Currently Employed (*Benxiao lijie biyesheng fuwu tongxun lu*) [本校歷屆畢業生服務通訊錄]," SMA Q235-3-503 (n.d.), 6–8.

36. "Preparations for a Chinese Typewriting Class (*Chouban Huawen daziji xunlianban*) [籌辦華文打字機訓練班]," *Henan Education* (*Henan jiaoyu*) [河南教育] 1, no. 6 (1928): 4.

37. "Contacts for School Graduates Where Currently Employed," 6–8.

38. "Chinese Typing Institute Founded in Shenyang by Sun Qishan (*Zai Shen chuangli Huawen dazi lianxisuo zhi Sun Qishan jun*) [在瀋創立華文打字練習所之孫岐山君]," *Great Asia Pictorial* (*Daya huabao*) [大亞畫報] 244 (August 10, 1930): 2; "Chinese-English Typing Institute Girls' Chinese Typing Class Beginning of the Year Ceremony Photo (*Hua-Ying dazi chuanxisuo nüzi Huawen daziban shiyeshi sheying*) [華英打字傳習所女子華文打字班始業式攝影]," *Chenbao Weekly Pictorial* (*Chenbao xingqi huabao*) [晨報星期畫報] 2, no. 95 (1927): 2; "Female Student in the Beiping Chinese-English Typewriting Institute Chinese Accelerated Class (*Beiping Hua-Ying daziji chuanxisuo Huawen suchengban nüsheng*) [北平華英打字傳習所華文速成班女生]," *Eastern Times Photo Supplement* (*Tuhua shibao*) [圖畫時報] 517 (December 2, 1928), front page; Photograph of Female Students of the Beijing Hua-Ying Typing Institute (*Beijing Hua-Ying dazi chuanxisuo*), *Chenbao Weekly Pictorial* (*Chenbao xingqi huabao*) [晨報星期畫報] 2, no. 100 (1927): 2; "Ms. Zhang Rongxiao of the Chinese-English Typing Institute (*Hua-Ying dazi xuexiao Zhang Rongxiao nüshi*) [華英打字學校張蓉孝女士]," *The Angel Pictorial, The Most Beautiful and Cheerful Pictorial in North China* (*Anqi'er*) [安琪儿] 3, no. 1 (1929): 1.

39. "Photograph of Chinese Vocational School Student Ye Shuyi Practicing on the Chinese Typewriter (*Ye Shuyi nüshi Zhonghua zhiye xuexiao xuesheng lianxi Huawen daziji shi zhi ying*) [葉舒綺女士中華職業學校學生練習華文打字機時之影]," *Shibao* [時報] 734 (1931): 3.

40. Wang Xiaoting [王小亭], "The New Woman: A Glance at the Professional Women of Shanghai (*Xin nüxing: Shanghai zhiye funü yi pie*) [新女性:上海職業婦女一瞥]," *Liangyou* [良友] 120 (1936): 16.

41. "Women Vie to Realize Their Promise (*Funü qun zheng qu guangming*) [婦女羣爭取光明]," *Zhanwang* [展望] 15 (1940): 18.

42. "Graduates of Hwa-yin Type-writing School, Peiping (*Beiping Hua Ying dazi xuexiao di'wu jie biyesheng*) [北平華英打字學校第五屆畢業生]," *Shibao* [時報] 620 (1930): 2. English and Chinese both in original.

43. "Photograph of First Entering Class of Female Students at the Liaoning Chinese Typing Institute (*Liaoning Huawen dazi lianxisuo di'yi qi nüxueyuan jiuxue jinian sheying*) [遼寧華文打字練習所第一期女學員就學之紀念攝影]," *Great Asia Pictorial* (*Daya huabao*) [大亞畫報] 244 (August 10, 1930): 2.

44. Sharon Hartman Strom, *Beyond the Typewriter: Gender, Class, and the Origins of Modern American Office Work, 1900–1930* (Chicago: University of Illinois Press, 1992), 177–179.

45. Margery W. Davies, *Woman's Place Is at the Typewriter: Office Work and Office Workers 1870–1930* (Philadelphia: Temple University Press, 1982), 54.

46. Ibid., 53.

47. Ibid.; Strom, *Beyond the Typewriter*.

48. Roger Chartier, *Forms and Meanings: Texts, Performances, and Audiences from Codex to Computer* (Philadelphia: University of Pennsylvania Press, 1995), 19.

49. For the late imperial period, noteworthy exceptions include Lucille Chia, *Printing for Profit: The Commercial Publishers of Jianyang, Fujian* (Cambridge, MA: Harvard University Asia Center, 2003); Kai-wing Chow, *Publishing, Culture, and Power in Early Modern China* (Stanford: Stanford University Press, 2004); Cynthia Brokaw and Kai-wing Chow, eds., *Printing and Book Culture in Late Imperial China* (Berkeley: University of California Press, 2005); Joseph P. McDermott, *A Social History of the Chinese Book: Books and Literati Culture in Late Imperial China* (Hong Kong: Hong Kong University Press, 2006); and Cynthia Brokaw and Christopher Reed, eds., *From Woodblocks to the Internet: Chinese Publishing and Print Culture in Transition, circa 1800 to 2008* (Leiden: Brill, 2010).

50. *Teaching Materials for the Chinese Typewriter* (*Huawen dazi jiangyi*) [華文打字講義], n.p., n.d. (produced pre-1928, circa 1917); Zhou Yukun, ed., *Chinese Typing Method* (*Huawen dazi fa*) (Nanjing: Bati yinshuasuo, 1934); Gan Chunquan and Xu Yizhi, eds., *Essential Knowledge for Secretaries: Requirements of a Chinese Typist* (*Shuji fuwu bibei: yiming Huawen dazi wenshu yaojue*) [書記服務必備一名華文打字文書要決] (Shanghai: Shanghai zhiye zhidaosuo, 1935); Tianjin Chinese Typewriter Company (*Tianjin Zhonghua daziji gongsi*) [天津中華打字機公司], ed., *Chinese Typewriter Training*

Textbook, vol. 1 (*Zhonghua daziji shixi keben—shangce*) [中華打字機實習課本–上冊] (Tianjin: Donghua qiyin shuaju [東華齊印刷局], 1943); Tianjin Chinese Typewriter Company (*Tianjin Zhonghua daziji gongsi*) [天津中華打字機公司], ed., *Chinese Typewriter Training Textbook,* vol. 2 (*Zhonghua daziji shixi keben—xiace*) [中華打字機實習課本–下冊] (Tianjin: Donghua qiyin shuaju [東華齊印刷局], 1943); People's Welfare Typewriter Manufacturing Company (*Minsheng daziji zhizaochang*), ed., *Practice Textbook* (*Lianxi keben*) (n.p.: n.p., c. 1940s).

51. In this one example, we begin to gain a sense of how a typing drill worked. In this lesson, and through the introduction of roughly one hundred characters, the student began in the periphery of the tray bed, before sweeping into the center and into the periphery opposite. The subsequent characters drew the typist downward, before returning him or her to the core and beginning to trace out a kind of central, lexical "spine." The next sequence of characters sent the typist briefly toward the upper right, before returning once again to the core, to supply further vertebrae. Over the subsequent ten characters, no time was to be spent along the edges— instead, seven further segments would be added to the central column, which by now was beginning to take shape in a robust, contiguous band toward the bottom of the tray. Three subsequent characters then brought the typist into the immediate neighborhood of this spine, followed by another ten that launched back out into the periphery, and across sweeping, looping distances. Secondary spines began to take shape, of course, one in the left periphery and the second immediately to the right of the robust core already formed. By the time the next ten characters had been typed, the peripheral spine had been consolidated further. Great sweeping motions were then mixed with tighter, jagged maneuvers. A new technique also appeared, in which new characters were introduced that themselves fell along the curves already traversed by the typist in earlier sequences. These newly introduced characters had already been "passed by," as it were, falling perfectly along an arc first traced some number of steps previously. Shortly thereafter, another character along the same arc was introduced, and then another. By the conclusion of just the first one hundred or so characters, clear patterns were already evident. The typist had by now fully haled into existence the central spine of the tray bed, as well as warming up peripheral zones to the left and the upper right. Larger swaths of the tray bed remained cold and unvisited, particularly in the upper right and upper mid-left. And yet one could be sure that subsequent lessons, or perhaps on-the-job practice, would bring them into play later on.

52. Zhou Houkun, *Chinese Typing Method.*

53. Jacques Derrida, *Of Grammatology* (Baltimore: Johns Hopkins University Press, 1976); Johanna Drucker, *The Visible Word: Experimental Typography and Modern Art, 1909–1923* (Chicago: University of Chicago Press, 1997).

54. Fan Jiling, *Fan's Wanneng-Style Chinese Typewriter Practice Textbook* (*Fan shi wanneng shi Zhongwen daziji shixi fanben*) [范氏萬能式中文打字機實習範本] (Hankou: Fan Research Institute Publishing House (*Fan shi yanjiusuo yinhang*) [范氏研究所印行], 1949), 10.

55. Zhou Houkun, *Chinese Typing Method*. There was a deeply political and ideological dimension to this regimen as well, most notably in the student's introduction to the "secondary usage character box." In a typewriting instruction manual from the 1930s, at a time when the Chiang Kai-shek regime was launching military and propagandist campaigns against its Communist foes, the first lesson in the course had students reproduce the last will and testament of Sun Yat-sen—the deceased symbolic father of the 1911 revolution. The second guided them through the *Manifesto of the Chinese Nationalist Party at the First National Congress* (*Zhongguo guomindang di'yi ci quanguo daibiao dahui xuanyan*). And the third offered a short essay on revolution and liberation. Subsequent lessons moved students through a carefully crafted history of China itself, touching upon the Opium Wars, colonialism, the Boxer Uprising, the Three People's Principles, the 1911 revolution, Yuan Shikai, the Warlord period, Manchukuo, labor, and poverty. Students were meant to repeat these lessons, learning them over and over and ingraining the particular lexical geometries of key political terms and names into their bodies. The first of these lessons was crafted such that the entire passage could be transcribed without the typist ever once needing to retrieve an "uncommon" character from the "secondary character box"—every character the student needed was on the tray bed. Further into one's training, however, exercises began to introduce characters that forced one to travel, tweezers in hand, to the box of secondary or tertiary usage characters. Some of these were less common characters, while others served subtle political functions. Two of the earliest "secondary usage characters" to be introduced, for example, were none other than "yuan" (袁) and "kai" (凱) of "Yuan Shikai," the great betrayer of the 1911 revolution who served as president of the Republic only to dissolve parliament, declare himself emperor, destabilize the fragile government, and leave a political vacuum after his untimely death that inaugurated China's decade-long "warlord period." His name had no place on the tray bed.

56. "At Last—A Chinese Typewriter—A Remington," *Remington Export Review*, n.d., 7. No date appears on the copy housed in the Hagley Museum collection, although the drawing of a Chinese keyboard diagram included within the article is dated February 10, 1921. Hagley Museum and Library, Accession no. 1825, Remington Rand Corporation, Records of the Advertising and Sales Promotion Department, Series I Typewriter Div., subseries B, Remington Typewriter Company, box 3, vol. 3.

57. Robert McKean Jones, "Arabic—Remington No. 10," Hagley Museum and Library, Accession no. 1825, Remington Rand Corporation, Records of the Advertising and Sales Promotion Department, Series I Typewriter Div., subseries B, Remington Typewriter Company, box 3, vol. 3.

58. Paul T. Gilbert, "Putting Ideographs on Typewriter," *Nation's Business* 17, no. 2 (February 1929): 156.

59. Robert McKean Jones, "Urdu—Keyboard no. 1130—No. 4 Monarch" (March 13, 1918), Hagley Museum and Library, Accession no. 1825, Remington Rand Corporation, Records of the Advertising and Sales Promotion Department, Series I Typewriter Div., subseries B, Remington Typewriter Company, box 3, vol. 2; Robert McKean Jones, "Turkish—Keyboard no. 1132—No. 4 Monarch" (February 27, 1920),

Hagley Museum and Library, Accession no. 1825, Remington Rand Corporation, Records of the Advertising and Sales Promotion Department, Series I Typewriter Div., subseries B, Remington Typewriter Company, box 3, vol. 2; Robert McKean Jones, "Arabic—Remington No. 10" (September 20, 1920), Hagley Museum and Library, Accession no. 1825, Remington Rand Corporation, Records of the Advertising and Sales Promotion Department, Series I Typewriter Div., subseries B, Remington Typewriter Company, box 3, vol. 3.

60. "Chu Yin Tzu-mu Keyboard—Keyboard no. 1400" (February 10, 1921), Hagley Museum and Library, Accession no. 1825, Remington Rand Corporation, Records of the Advertising and Sales Promotion Department, Series I Typewriter Div., subseries B, Remington Typewriter Company, box 3, vol. 3. In 1921, Remington also created a "Chinese Romanized" keyboard based upon the Wade System. On October 20, 1921, the Shanghai-based Mustard and Company reported about the lack of interest in the machine. The home office took note briefly: "had circularized Missionaries Teachers etc etc but found no demand for above." "Chinese Romanized—Keyboard no. 141," Hagley Museum and Library, Accession no. 1825, Remington Rand Corporation, Records of the Advertising and Sales Promotion Department, Series I Typewriter Div., subseries B, Remington Typewriter Company, box 3, vol. 1. See also "Chinese Phonetic on a Typewriter," *Popular Science* 97, no. 2 (August 1920): 116.

61. Robert McKean Jones, "Typewriting Machine," United States Patent no. 1646407 (filed March 12, 1924; patented October 25, 1927).

62. Ibid.

63. Gilbert, "Putting Ideographs on Typewriter," 156.

64. John Cameron Grant and Lucien Alphonse Legros, "A Method and Means for Adapting Certain Chinese Characters, Syllabaries or Alphabets for use in Type-casting or Composing Machines, Typewriters and the Like," Great Britain Patent Application no. 2483 (filed January 30, 1913; patented October 30, 1913).

65. Another early attempt at alphabetic Chinese typewriting (circa 1914) was made by Walter Hillier. See "Memorandum by Sir Walter Hillier upon an alphabetical system for writing Chinese, the application of this system to the typewriter and to the Linotype or other typecasting and composing machines, and its adaptation to the braille system for the blind" (London: William Clowes and Sons, Limited, n.d.). Readers are encouraged to keep an eye out for forthcoming work by Andrew Hillier, who is in the midst of writing a biography of the Hillier family and its connection to the British Empire.

66. John DeFrancis, *Nationalism and Language Reform in China* (Princeton: Princeton University Press, 1950), chapter 4.

67. J. Frank Allard, "Type-Writing Machine," United States patent no. 1188875 (filed January 13, 1913; patented June 27, 1916).

68. "Chinese Alphabet Has Been Reduced," *Telegraph-Herald* (April 11, 1920), 23.

69. "Chinese Phonetic on a Typewriter" (advertisement for the Hammond Multiplex), *Popular Science* 97, no. 2 (August 1920): 116.

70. Special thanks to Michael Hill for alerting me to this volume.

71. "Obituary: Robert McKean Jones. Inventor of Chinese Typewriter Was Able Linguist," *New York Times* (June 21, 1933), 18.

72. Photograph of Fong Sec, *Asia: Journal of the American Asiatic Association* 19, no. 11 (November 1919): front matter, photograph by Methodist Episcopal Centenary Commission. Still other examples abound. The inaugural issue of *Shanghai Puck* featured a short ditty of its own, "An American View of the Chinese Typewriter," reprinted from Kenneth L. Roberts's 1916 piece in *Life* magazine, entitled "A Reason Why the Chinese Business Man May Soon Be Tired." See "An American View of the Chinese Typewriter," *Shanghai Puck* 1, no. 1 (September 1, 1918): 28; and "A Reason Why the Chinese Business Man May Soon Be Tired," *Life* 68 (1916): 272.

73. A photograph of Zhang Xiangling appeared in *Who's Who in China* 3rd ed. (Shanghai: China Weekly Review, 1925), 31.

74. "Doings at the Philadelphia Commercial Museum," *Commercial America* 19 (April 1923): 51. For more on the Philadelphia Commercial Museum, see Steven Conn, "An Epistemology for Empire: The Philadelphia Commercial Museum, 1893–1926," *Diplomatic History* 22, no. 4 (1998): 533–563. The shipment from China is confirmed in *Annual Report of the Philadelphia Museums, Commercial Museum* (Philadelphia: Commercial Museum, 1923), 9. In 1923, Zhang explored such themes further in his preface to Julius Su Tow's *The Real Chinese in America.* "As a whole," Zhang began, "the American people have never been given an opportunity to know the Chinese truly and fully." China was home to "trading corporations, banks and steamship lines," he continued, news of which would surprise the average American man or woman. "Chinese are as intelligent and respectable as any other people in the world and ... are not merely laundrymen!" See Julius Su Tow, *The Real Chinese in America* (New York: Academy Press, 1923), editorial introduction by Ziangling Chang [Zhang Xianglin], xi–xii.

75. Photograph 2429 in *Descriptions of the Commercial Press Exhibit* (Shanghai: Commercial Press, ca. 1926), in City of Philadelphia, Department of Records, record group 232 (Sesquicentennial Exhibition Records), 232-4-8.1, "Department of Foreign Participation," box A-1474, folder 8, series 29, "China, Commercial Press Exhibit"; "China, Commercial Press Exhibit," City of Philadelphia, Department of Records, record group 232 (Sesquicentennial Exhibition Records), 232-4-8.1 "Department of Foreign Participation," box A-1474, folder 8, series folder 29.

76. *Descriptions of the Commercial Press Exhibit,* 56.

77. Born in Shanghai, Zhang studied at St. John's University in Shanghai, and later at Columbia University. He served as associate editor of the *Peking Daily News* between 1913 and 1915, and later as associate-level secretary in the Ministries of Communication, Interior, and Foreign Affairs. "Who's Who in China: Biographies of Chinese Leaders," *China Weekly Review* (Shanghai), 1936: 6–7. See also photograph 2308 in *Descriptions of the Commercial Press Exhibit.* The Chairman of the Chinese Commission was Tinsin C. Chow.

78. *Descriptions of the Commercial Press Exhibit,* 41.

79. Ibid., 56.

80. "Consul General Stationed in the United States Writes to Inform of Philadelphia Competition (*Zhu Mei zonglingshi hangao Feicheng saihui qingxing*) [駐美總領事函告費城賽會情形]," *Shenbao* (January 14, 1927), 9; "List of Awards-General n.d.," City of Philadelphia, Department of Records, record group 232 (Sesquicentennial Exhibition Records), 232-4-6.4 (Jury of Awards-Files), box a-1472, folder 17, series folder 1.

81. Qian Xuantong, "China's Script Problem from Now On (*Zhongguo jinhou zhi wenzi wenti*) [中國今後之文字問題]," *Xin qingnian* 4, no. 4 (1918): 70–77.

82. Qian Xuantong [錢玄同], "Why Must We Advocate a 'Romanized National Language?' (*Weishenme yao tichang 'guoyu luoma zi'?*) [為什麼要提倡國語羅馬字?]," *Xinsheng* 1, no. 2 (December 24, 1926). In *The Writings of Qian Xuantong (Qian Xuantong wenji)* [钱玄同文集], volume 3 (Beijing: Zhongguo renmin daxue chubanshe, 1999), 387.

83. Gilbert Levering, "Chinese Language Typewriter," *Life* 2311 (February 17, 1927), 4.

CHAPTER 5: CONTROLLING THE KANJISPHERE

1. Email communication, James Yee, July 6, 2009.

2. The London visit was precipitated by another email message. "Sorry if this is not of interest," it began. "My mum has a chinese typewriter and is about to scrap it as we have not been able to find a home for it. It is a Superwriter315SR with several trays of characters—still in working order." The family was remodeling the floors of their home, I would later learn, prompting them to inventory and purge some of their belongings. With great reluctance, the mother of the household had agreed to part with her beloved machine. She entrusted her daughter, Maria, to find it a new home. "It seems a waste to throw it out," the email continued. "Mum used to use it all the time and was pretty fast." Email from Maria Tai to the author, May 14, 2010. Names have been changed.

3. Sherry Turkle, ed., *Evocative Objects: Things We Think With* (Cambridge, MA: MIT Press, 2007).

4. "CJK" is sometimes extended into "CJKV," to include Vietnamese. See Ken Lunde, *CJKV Information Processing: Chinese, Japanese, Korean & Vietnamese Computing* (Sebastopol, CA: O'Reilly, 2009).

5. Ryōshin Minami, "Mechanical Power and Printing Technology in Pre–World War II Japan," *Technology and Culture* 23, no. 4 (1982): 609–624; Daqing Yang, "Telecommunication and the Japanese Empire: A Preliminary Analysis of Telegraphic Traffic," *Historical Social Research* 35, no. 1 (2010): 68–69; Miyako Inoue, "Stenography and Ventriloquism in Late Nineteenth Century Japan," *Language and Communication* 31 (2011): 181–190; Patricia L. Maclachlan, *The People's Post Office: The History and Politics of the Japanese Postal System, 1871–2010* (Cambridge, MA: Harvard Asia Center,

2012); Seth Jacobowitz, *Writing Technology in Meiji Japan: A Media History of Modern Japanese Literature and Visual Culture* (Cambridge, MA: Harvard Asia Center, 2015).

6. Ryōshin Minami, "Mechanical Power and Printing Technology in Pre–World War II Japan," 609–624.

7. In Tokyo and Osaka, the five largest newspaper outfits nearly quintupled in annual circulation between 1881 and 1891, from approximately twelve to fifty million subscribers. This figure more than doubled between 1891 and 1901, reaching a combined annual circulation of 119,368,000 for Tokyo *Asahi*, Tokyo *Nichinichi*, *Yomiuri*, *Ōsaka Asahi*, and *Ōsaka Mainichi*. This represents an annual rate of growth of 32.9 percent from 1881 to 1891 and 13.4 percent from 1891 to 1901. Unlike their counterpart publishers in the United Kingdom and the United States, moreover, sectors of the Japanese publishing industry leapfrogged steam power altogether, moving directly from manual labor to the electrification of rotary presses. The Tokyo *Asahi*, for example, moved directly to electrification, taking advantage of a technology that had only recently been introduced in Western contexts. Ryōshin Minami, "Mechanical Power and Printing Technology in Pre–World War II Japan," 617–619.

8. Yang, "Telecommunication and the Japanese Empire," 68–69.

9. "Denshin jigō [電信字号]," "Extension Selskabet—Japansk Telegrafnøgle," 1871. Arkiv nr. 10.619. In "Love og vedtægter med anordninger," GN Store Nord A/S SN China and Japan Extension Telegraf. Rigsarkivet [Danish National Archives], Copenhagen, Denmark.

10. Yang, "Telecommunication and the Japanese Empire," 68–69.

11. Andre Schmid, *Korea between Empires* (New York: Columbia University Press, 2002), 57–59.

12. Ibid., 67–69.

13. Ernest Gellner, *Nations and Nationalism* (Ithaca: Cornell University Press, 1983); Benedict Anderson, *Imagined Communities: Reflections on the Origin and Spread of Nationalism*, rev. ed. (New York: Verso, 1991).

14. In one of the many ironies of the era, *Hwangsong sinmun* in many ways led the charge in the critique of character-based script, and yet did not feature a single editorial written in the Korean vernacular during its thirteen years of publication. See Schmid, *Korea between Empires*, 17.

15. *Proposal for the Abolition of Chinese Characters* (*Kanji gohaishi no gi*) [漢字ご廃止の議]; Seeley, *A History of Writing in Japan*, 138–139.

16. "Hiragana no setsu" [平仮名の説] (On Hiragana); Seeley, *A History of Writing in Japan*, 139.

17. Watabe Hisako [渡部久子], *Japanese Typewriter Textbook* (*Hōbun taipuraitā tokuhon*) [邦文タイプライター讀本] (Tokyo: Sūbundō [崇文堂], 1929), 6–7. The choice of the term *taipuraitā* here is important, with Kurosawa adopting the phoneticized loan word rather than a translation into kanji characters. My thanks to Joshua Fogel for alerting me to the importance of this point.

18. Dave Sheridan, Memo to Sales Staff regarding Remington Japanese Typewriter, Hagley Museum and Library, accession 1825, Remington Rand Corporation, Records of the Advertising and Sales Promotion Department, Series I Typewriter Div. Subseries B Remington Typewriter Company, box 3, folder 6, "Keyboards and Typestyles—Correspondence, 1906."

19. Teishin Ministry Electricity Bureau (*Teishin-shō denmu kyoku*) [遞信省電務局], ed., *Japanese Typewriting* (*Wabun taipuraichingu*) [和文タイプライチング] (Tokyo: Teishin kyōkai, 1941), 25–27, 43.

20. Kamo Masakazu [加茂正一], *Taipuraitā no chishiki to renshū* タイプライターの知識と練習 (Tokyo: Bunyūdō Shoten [文友堂書店], 1923), front matter. For a wonderful compendium of typewriter art in the west, see Barrie Tullet, *Typewriter Art: A Modern Anthology* (London: Laurence King Publishers, 2014).

21. Sheridan, Memo to Sales Staff regarding Remington Japanese Typewriter.

22. Sukeshige Yanagiwara, "Type-writing Machine," United States Patent no. 1206072 (filed February 1, 1915; patented November 28, 1916). Assignor to Underwood Typewriter Company; Burnham Stickney, "Typewriting Machine," United States Patent no. 1549622 (filed February 9, 1923; patented August 11, 1925).

23. 1915 Tsugi Kitahara postcard, author's collection. Similarly, in response to the Underwood patent by Burnham Stickney, Remington tasked its senior keyboard designer Robert McKean Jones to the katakana project—the company's "master typographer," whom we encountered in chapter 4. See Robert McKean Jones, "Typewriting Machine," United States Patent no. 1687939 (filed May 19, 1927; patented October 16, 1928).

24. *The Teaching of Words* (*Moji no oshie*) [文字の教え]; Seeley, *A History of Writing in Japan*, 141.

25. "List of Characters for General Use" (*Jōyō kanji hyō*) [常用漢字表]; Interim Committee on the National Language (*Rinji kokugo chōsakai*) [臨時国語調査会].

26. *Sanzenji jibiki, The Three Thousand Character Dictionary* (*Sanzenji jibiki*) [三千字字引]; Seeley, *A History of Writing in Japan*, 141, 146–147.

27. "Advocating the Restriction of the Number of Kanji" (*Kanji seigen o teishō su*) [漢字制限を提唱す]; Seeley, *A History of Writing in Japan*, 146. Still other early examples include Ōki Takatō [大木喬任] (1831–1899), Japan's first minister of education, who in 1872 commissioned a common usage kanji dictionary—the two-volume *Newly Compiled Dictionary* (*Shinsen jishō* 新選辞書)—containing 3,167 characters. Seeley, *A History of Writing in Japan*, 142.

28. A second approach to the question of Japanese linguistic modernization focused on Romanization, led by figures such as Nanbu Yoshikazu (南部義籌, 1840–1917), Nishi Amane (西周, 1827–1897), and Ōtsuki Fumihiko (大槻文彦, 1847–1928). In 1869, Nanbu published his treatise "Argument for Amending the National Language." In 1885, the Romanization Club and *Rōmaji Magazine* adapted the Romanization method first developed by James Curtis Hepburn (1815–1911). Among the many competing systems for rendering Japanese in the Latin alphabet was the

so-called "Japan Style" (*Nipponshiki*) developed by Tanakadate Aikitsu (田中舘愛橘, 1856–1952). Seeley, *A History of Writing in Japan*, 139–140; Nanette Gottlieb, "The Rōmaji Movement in Japan," *Journal of the Royal Asiatic Society* 20, no. 1 (2010): 75–88; *Shūkokugo ron* [修国語論] (*Argument for Amending the National Language*); *Rōmaji kai* (Romanization Club) and *Rōmaji zasshi*.

29. Watabe Hisako [渡部久子], *Japanese Typewriter Textbook* (*Hōbun taipuraitā tokuhon*) [邦文タイプライター讀本] (Tokyo: Sūbundō [崇文堂], 1929), 6–7.

30. Kyota Sugimoto, "Type-Writer," United States Patent no. 1245633 (filed November 7, 1916; patented November 6, 1917).

31. Toshiba Japanese Typewriter. Manufactured c. 1935. Peter Mitterhofer Schreibmaschinenmuseum/Museo delle Macchine da Scrivere. Partschins (Parcines), Italy, "Macchina da Scrivere Giapponese Toshiba."

32. Nippon Typewriter Company [日本イプライター株式会社], ed., *Character Index for Japanese Typewriter* (Hōbun taipuraitā-yō moji no sakuin) [邦文タイプライター用文字の索引] (Tokyo: n.p., 1917); Hisao Yamada, "A Historical Study of Typewriters and Typing Methods; from the Position of Planning Japanese Parallels," *Journal of Information Processing* 2, no. 4 (February 1980): 175–202; Hisao Yamada and Jiro Tanaka, "A Human Factors Study of Input Keyboard for Japanese Text," *Proceedings of the International Computer Symposium* (Taipei: National Taiwan University, 1977), 47–64.

33. Raja Adal, "The Flower of the Office: The Social Life of the Japanese Typewriter in Its First Decade," presentation at the Association for Asian Studies Annual Meeting, March 31–April 3, 2011.

34. Janet Hunter, "Technology Transfer and the Gendering of Communications Work: Meiji Japan in Comparative Historical Perspective," *Social Science Japan Journal* 14, no. 1 (Winter 2011): 1–20. See also Kae Ishii, "The Gendering of Workplace Culture: An Example from Japanese Telegraph Operators," *Bulletin of Health Science University* 1, no. 1 (2005): 37–48; Brenda Maddox, "Women and the Switchboard," in *The Social History of the Telephone*, ed. Ithiel de Sola Pool (Cambridge, MA: MIT Press, 1977), 262–280; Susan Bachrach, *Dames Employées: The Feminization of Postal Work in Nineteenth-Century France* (London: Routledge, 1984); Michele Martin, *"Hello, Central?": Gender, Technology and Culture in the Formation of Telephone Systems* (Montreal: McGill-Queens University Press, 1991); Ken Lipartito, "When Women Were Switches: Technology, Work, and Gender in the Telephone Industry, 1890–1920," *American Historical Review* 99, no. 4 (1994): 1074–1111; Alisa Freedman, Laura Miller, and Christine R. Yano, eds., *Modern Girls on the Go: Gender, Mobility, and Labor in Japan* (Stanford: Stanford University Press, 2013).

35. *Investigation of Women's Occupations in Tokyo and Osaka* (*Tōkyō Ōsaka ryōshi ni okeru shokugyō fujo chōsa*) [東京大阪両市に於ける職業婦女調査] (Tokyo: n.p., 1927), 4–11.

36. For the journal's 1940s-era generic "women's content," see Nishida Masaaki [西田正秋], "Japanese-Style Female Beauty of Today (*Kyō no nipponteki joseibi*)

[今日の日本的女性美]," *Taipisuto* [タイピスト] 17, no. 7 (July 1942): 2–5. For content more focused on Japanese typewriting and its civilizational importance, see Omi Hironobu [小見博信], "Japanese Culture and the Mission of the Japanese Typewriter (*Nipponbunka to Hōbun taipuraitā no shimei*) [日本文化と邦文タイプライターの使命]," *Taipisuto* [タイピスト] 17, no. 11 (November 1942): 12–13. Collections for this periodical are incomplete, creating challenges for identifying the exact publication period. The year 1925 is calculated by reverse dating based upon the issue, number, and date information for extant issues. The periodical ran for approximately two decades.

37. Kyota Sugimoto, "Type-Writer," United States Patent no. 1245633 (filed November 7, 1916; patented November 6, 1917).

38. Yusaku Shinozawa, "Typewriter," United States Patent no. 1297020 (filed June 19, 1918; patented March 11, 1919). Reports of Japanese-built Chinese machines surfaced as early as 1914. In that year, readers of the Chinese journal *Progress (Jinbu)* learned of a Kanbun, Chinese-character typewriter prototype under development by a Japanese engineer. Born in Fukuoka, Sakai Yasujiro pursued his undergraduate studies in Electrical Engineering at the University of California. Following graduation in 1904, Sakai went on to work for Westinghouse Electric and Manufacturing Company, where he specialized in automation and assigned a host of patents to his patron company. Sakai returned to Japan in 1913, going on to serve as Consulting Engineer at Takata and Company in Tokyo and Chief Engineer at the Yasukawa Electric Works. Upon returning to Japan from Pennsylvania, Sakai's interests soon drifted from questions of machine automation to the automation of the Japanese language—specifically of kanji. In Japan, he commenced work on a kanji typewriter. Wan Zhang [綰章], "Invention of a New Chinese Character Typewriter (*Hanwen daziji zhi xin faming*) [漢文打字機之新發明]," *Progress: A Journal of Modern Civilization (Jinbu)* [進步] 6, no. 1 (1914): 5; "Notice Regarding Department of Electrical Engineering, University of California," *Journal of Electricity* 41, no. 1 (1918): 515; *Bulletin* (Berkeley: University of California, 1910), 65; Frank Conrad and Yasudiro Sakai, "Impedance Device for Use with Current-Rectifiers," United States Patent no. 1075404 (filed January 10, 1912; patented October 14, 1913); Yasudiro Sakai, "Stop Cock," United States Patent no. 1001455 (filed December 10, 1910; patented August 22, 1911); Yasudiro Sakai, "Electrical Terminal," United States Patent no. 1049404 (filed January 7, 1911; patented January 7, 1913); Yasudiro Sakai, "Vapor Electric Device," United States Patent no. 1101665 (filed December 30, 1910; patented June 30, 1914); Yasudiro Sakai, "Vapor Electric Apparatus," United States Patent no. 1148628 (filed June 14, 1912; patented August 3, 1915); Yasudiro Sakai, "Armature Winding," United States Patent no. 1156711 (filed February 3, 1910; patented October 12, 1915). See also "Shunjiro Kurita," *Who's Who in Japan* 13–14 (1930): 8. In 1917 readers of *Shenbao* learned that, on July 12, the Chinese Youth Association in Shanghai would be hosting representatives of the Mitsui Trading Company, there to showcase a typewriter that could handle not only Japanese but also Chinese. See "Testing Chinese Typewriters (*Shiyan Huawen daziji*) [試驗華文打字機]," *Shenbao* (July 12, 1917), 11.

39. The same held true for early inventors of Chinese typewriters, in the opposite direction. In a telling passage from his 1915 patent, Qi Xuan, the inventor we first met in chapter 3, gestured toward his own transnational ambitions. "My method and arrangement," Qi mentioned almost in passing, "possess great advantage, not only in Chinese, but in all languages in which the words are composed of radicals and not of letters, such for instance as Japanese and Korean." See Chi, Heuen [Qi Xuan], "Apparatus for Writing Chinese," United States Patent no. 1260753 (filed April 17, 1915; patented March 26, 1918).

40. Douglas Howland, *Borders of Chinese Civilization: Geography and History at Empire's End* (Durham: Duke University Press, 1996), 44.

41. Ibid., 54.

42. Mary Badger Wilson, "Fleet-Fingered Typist," *New York Times* (December 2, 1923), SM2.

43. Ibid.

44. "Stenographer Has a Tough Job," *Ludington Daily News* (April 8, 1937), 5.

45. "A Line O' Type or Two," *Chicago Daily Tribune* (August 31, 1949), 16.

46. Oriental Typewriter Character Handbook (*Tōyō taipuraitā moji binran: Nigōki-yō*) [東洋タイプライター文字便覧：　弐号機用] (Tokyo: Oriental Typewriter Co. [東洋タイプライター]), 1923; Morita Torao [森田虎雄], *Japanese Typewriter Textbook* (*Hōbun taipuraitā kyokasho*) [邦文タイプライター教科書] (Tokyo: Tokyo Women's Foreign Language School [東京女子外國語學校], 1934), 8–9.

47. Reed, *Gutenberg in Shanghai*, 128.

48. Y. Tak Matsusaka, "Managing Occupied Manchuria," in *Japan's Wartime Empire*, ed. Peter Duus, Ramon H. Myers, and Mark R. Peattie (Princeton: Princeton University Press, 1996), 112–120.

49. Li Xianyan [李獻延], ed., *Newest Stationery Forms* (*Zuixin gongwen chengshi*) [最新公文程式] (Xinjing: Fengtian Typing Institute [奉天打字專門學校], circa 1932). Owing to missing pages, an exact year cannot be assigned. The prevalence of typing samples listed as "Datong Year One" (*Datong yuannian*) strongly suggest a publication date of 1932 or thereabouts.

50. Ibid., 43–48.

51. Ibid., 1. 一國有一國的公文程式,一時代有一時代的公文程式,都是隨著國情和習慣而演進的;那末,述說公文程式的書籍,也要隨著時代而改革的,這是一定的理。打字員是專任謄錄公文的人員,所以打字員更要隨時學習新的公文程式,才能適合時代,才能供職工作。

52. "Accounts of Visiting America (*Lü Mei guancha tan*) [旅美觀察談]," *Shenbao* (April 3, 1919), 14.

53. According to one biography, Yu Binqi did not receive a formal degree from Waseda University. See "Native of Xiaoshan was Pioneer of Chinese Table Tennis and Swimming (*Xiaoshanren huo shi Zhongguo pingpang qiu ji youyong yundong zhuyao kaichuang zhe*) [萧山人或是中国乒乓球及游泳运动主要开创者]," May 2012, http://www.xsnet.cn/news/shms/2012_5/1570558.shtml

54. "Swimming Expert Yu Binqi (*Youyong zhuanjia Yu Binqi nanshi*) [游泳專家俞斌祺男士]," *Nan pengyou* 1, no. 10 (1932): reverse cover.

55. "A Positive Review of Yu-Style Chinese Typewriter (*Yu shi Zhongwen daziji zhi haoping*) [俞式中文打字機之好評]," *Zhongguo shiye* [中國實業] 1, no. 6 (1935): 1158.

56. Yu Shuolin [俞碩霖], "The Birth of the Yu-Style Typewriter (*Yu shi daziji de yansheng*) [俞式打字机的诞生]," *Old Kids Blog* (*Lao xiaohai shequ*) [老小孩社区] (June 3, 2010), http://www.oldkids.cn/blog/blog_con.php?blogid=124277 (accessed June 13, 2011); Yu Shuolin [俞碩霖], "Yu-Style Typewriter Factory (*Yu shi daziji zhizaochang*) [俞式打字机制造厂]," *Old Kids Blog* (*Lao xiaohai shequ*) [老小孩社区] (June 6, 2010), http://www.oldkids.cn/blog/blog_con.php?blogid=130431 (accessed June 13, 2011).

57. "Contacts for School Graduates Where Currently Employed," 6–8.

58. "A Positive Review of Yu-Style Chinese Typewriter," 1158.

59. Jin Shuqing joined the school in 1934. The outbreak of Japanese aggressions in 1931 also seems to have taken a toll on Yu Binqi's family and personal affairs, as related by Yu Binqi's son. At the time, Yu maintained relations with both his wife and a mistress surnamed Wu. In the wake of the 1932 incident, however, Yu Binqi's son, mother, and grandmother took refuge in the family's native place of Sushan in Zhejiang. Yu Binqi himself stayed behind in Shanghai, presumably to continue overseeing his business. When the family was reunited in the fall of 1932, Yu Binqi separated from his wife, who took up residence elsewhere with Yu Binqi's son. Somewhere around this time, Yu Binqi placed an ad for a secretary, to which a young woman named Jin Shuqing (金淑清) responded. Gradually, Jin displaced Wu as Yu Binqi's favorite, until finally she became his companion and mistress. Jin also pushed for marriage, ultimately prompting a legal divorce between Yu and his estranged wife. See Yu Shuolin [俞碩霖], "Yu-Style Chinese Typewriter Unlimited Company (*Yu shi daziji wuxian gongsi*) [俞式打字机无限公司]," *Old Kids Blog* (*Lao xiaohai shequ*) [老小孩社区] (June 7, 2010), http://www.oldkids.cn/blog/blog_con.php?blogid=130576 (accessed June 13, 2011).

60. In 1934, Yu Binqi teamed up with Chinese Stenography inventor Yang Bingxun [杨炳勋], expanding his school to offer instruction in both areas. The school was located on Shanchang Alley, Kayi Road [卡億路善昌里]. "Chinese Typewriter Inventor and Chinese Stenography Inventor Jointly Run Special School (*Zhongwen daziji suji famingren heban zhuanxiao*) [中文打字機速記發明人合辦專校]," *Shenbao* (September 5, 1934), 16. Huang Juede was a graduate of Yu's school, joining the staff in 1935 as an administrator and as assistant instructor of typing.

61. Yu's school also helped give rise to additional Chinese typing schools, staffed with Yu's graduates and centered around the Yu-style machine. In 1935, for example, one of Yu's graduates—a young man by the name of Sheng Jiping—was hired to oversee a newly formed Chinese typewriting class at a middle school in Zhejiang province. Important for our purposes is the fact that the class used a "Yu-style Chinese typewriter" (*Yu shi Zhongwen daziji*). "Business Class Adds Course in Chinese Typing (*Shangke tianshe Zhongwen daziji kecheng*) [商科添設中文打字機課程]," *Zhejiang shengli Hangzhou gaoji zhongxue xiaokan* [浙江省立杭州高级中学校刊] 119 (1935): 841.

62. See Yu Shuolin, "Two Types of Chinese Typewriter (*Liang zhong Zhongwen daziji*) [两种中文打字机]," *Old Kids Blog (Lao xiaohai shequ)* [老小孩社区] (February 8, 2010), www.oldkids.cn/blog/blog_con.php?blogid=116181 (accessed May 12, 2013).

63. "Resist Japan and Save the Nation Movement (*Kang Ri jiuguo yundong*) [抗日救國運動]," *Shenbao* (November 8, 1931), 13–14.

64. "Yu Binqi Defends Himself to the Resist Japan Association (*Yu Binqi xiang kang Ri hui* shenban) [俞斌祺向抗日會伸辨]," *Shenbao* (November 9, 1931), 11.

65. "Resist Japan Association Standing Committee 17th Meeting Memo (*Kang Ri hui changwuhuiyi ji di'shiqi ci*) [抗日會常務會議記第十七次]," *Shenbao* (November 12, 1931), 13. For more on patriotic consumption campaigns of the era, see Jeffrey N. Wasserstrom, *Student Protests in Twentieth-Century China: The View from Shanghai* (Stanford: Stanford University Press, 1997), 176–178, 190–191; Karl Gerth, *China Made: Consumer Culture and the Creation of the Nation* (Cambridge, MA: Harvard Asia Center, 2003); and Mark W. Frazier, *The Making of the Chinese Industrial Workplace: State, Revolution, and Labor Management* (Cambridge: Cambridge University Press, 2006), 47.

66. "The Chinese Typewriter: Resisting Japan and Supporting China (*Kang Ri sheng zhong zhi Zhongwen daziji*) [抗日聲中之中文打字機]," *Shenbao* (January 26, 1932), 12.

67. "Steel Types Invented and Used in Chinese Typewriters (*Daziji yong gangzhi suozi faming*) [打字機用鋼質活字發明]," *Shenbao* (September 3, 1932), 16; "Chinese Typewriter Steel Types Invented (*Faming Zhongwen daziji gangzhi huozi*) [發明中文打字機鋼質活字]," *Shenbao* (September 10, 1932), 16; "Yu Binqi Invents Steel Type (*Yu Binqi faming gangzhi zhuzi*) [俞斌祺發明鋼質鑄字]," *China Industry (Zhongguo shiye)* [中國實業] 1, no. 5 (1935): 939.

68. "Chinese Typewriter Company Donates Profits to Northeast Refugees (*Zhongwen daziji choukuan juanzhu dongbei nanmin*) [中文打字機抽欵捐助東北難民]," *Shenbao* (December 18, 1932), 14.

69. "Donation to Northeast Refugees: Buy One Chinese Typewriter, Donate Thirty Yuan (*Choukuan juanzhu dongbei nanmin gou Zhongwen daziji yi jia ke choujuan sanshi yuan banfa*) [抽欵捐助東北難民購中文打字機一架可抽捐三十圓辦法]," *Shenbao* (December 23, 1932), 12.

70. "Yu's Chinese Typewriter: Profits Donated to Flooded Districts (*Yu shi Zhongwen daziji ticheng chong shuizai yici*) [俞氏中文打字機提成充水災義賑]," *Shenbao* (December 22, 1935), 12.

71. "New Chinese Typewriter: Seeking Cooperation with Capitalists (*Xin faming Zhongwen daziji mi zibenjia hezuo*) [新發明中文打字機覓資本家合作]," *Shenbao* (August 25, 1933), 14. These included Xieda [協大], Jingda [精大], Jiangchang [降昌], Daming [大明], and Luguiji [卢桂記]. At the time, Yu-style machines were already being manufactured by the Haishang Domestic Product Factory [海上國貨工廠]. This factory cited its inability to keep pace with production requirements and consumer demand, thereafter entering into a shared production arrangement with the above-mentioned consortium. "Chinese Typewriter Patent Approved: Five Major Factories Enthusiastically Join in Production (*Zhongwen daziji zhuanli hezhun wuda chang jiji zhizao*) [中文打字機專利核准五大工廠積極製造]," *Shenbao* (March 17, 1934), 13.

72. "Yu's Chinese Typewriter's Revolution (*Yu shi Zhongwen daziji zhi da gexin*) [俞氏中文打字機之大革新]," *Shenbao* (May 9, 1934), 12.

73. "Hongye Company Agency: Yu's Chinese Typewriters Sell Well (*Hongye gongsi jingli Yu shi Zhongwen daziji changxiao*) [宏業公私經理俞氏中文打字機暢銷]," *Shenbao* (December 8, 1934), 14.

74. "Yu's Typewriter: Stencil Paper Printing Success (*Xin faming Yu shi daziji lazhi youyin chenggong*) [新發明俞氏打字機蠟紙油印成功]," *Shenbao* (January 28, 1935), 12.

75. "Yu's Chinese Typewriter Is Five Times More Convenient than Writing and Copying by Hand (*Yu shi Zhongwen daziji wubei yu shanxie*) [俞氏中文打字機五倍于繕寫]," *Shenbao* (July 7, 1936), 15.

76. Yu Binqi had collaborated with fellow inventor Lin Zeren to establish the Chinese Inventors Association, seeking others in Shanghai and beyond to join. "Yu Binqi and Others Will Establish China Inventors Association (*Yu Binqi deng zuzhi Zhongguo famingren xiehui*) [俞斌祺等組織中國發明人協會]," *Book Prospects (Tushu zhanwang)* [圖書展望] 1, no. 8 (April 28, 1936): 83; "Chinese Inventors Association Preparatory Conference Held Yesterday (*Zhongguo famingren xiehui zuori kai choubeihui*) [中國發明人協會昨日開籌備會]," *Shenbao* (February 1, 1937), 20.

77. "Ping-Pong Match Supporting Suiyuan, Souvenir Medals Granted (*Suiyuan pingpang zeng jinianzhang*) [綏援乒乓贈紀念章]," *Shenbao* (January 9, 1937), 10.

78. Chinese Second Historical Archives (*Zhongguo di'er lishi dang'anguan*), ed., *Historical Materials on the Old Chinese Maritime Customs, 1859–1948*, vol. 112 (1932) (*Zhongguo jiu haiguan shiliao*) [中國舊海關史料] (Beijing: Jinghua Press [京華出版社], 2001); Chinese Second Historical Archives (*Zhongguo di'er lishi dang'anguan*), ed., *Historical Materials on the Old Chinese Maritime Customs, 1859–1948*, vol. 114 (1933) (*Zhongguo jiu haiguan shiliao*) [中國舊海關史料] (Beijing: Jinghua Press [京華出版社], 2001); Chinese Second Historical Archives (*Zhongguo di'er lishi dang'anguan*), ed., *Historical Materials on the Old Chinese Maritime Customs, 1859–1948*, vol. 118 (1935) (*Zhongguo jiu haiguan shiliao*) [中國舊海關史料] (Beijing: Jinghua Press [京華出版社], 2001); Chinese Second Historical Archives (*Zhongguo di'er lishi dang'anguan*), ed., *Historical Materials on the Old Chinese Maritime Customs, 1859–1948*, vol. 122 (1936) (*Zhongguo jiu haiguan shiliao*) [中國舊海關史料] (Beijing: Jinghua Press [京華出版社], 2001); Chinese Second Historical Archives (*Zhongguo di'er lishi dang'anguan*), ed., *Historical Materials on the Old Chinese Maritime Customs, 1859–1948*, vol. 126 (1937) (*Zhongguo jiu haiguan shiliao*) [中國舊海關史料] (Beijing: Jinghua Press [京華出版社], 2001); Chinese Second Historical Archives (*Zhongguo di'er lishi dang'anguan*), ed., *Historical Materials on the Old Chinese Maritime Customs, 1859–1948*, vol. 130 (1938) (*Zhongguo jiu haiguan shiliao*) [中國舊海關史料] (Beijing: Jinghua Press [京華出版社], 2001); Chinese Second Historical Archives (*Zhongguo di'er lishi dang'anguan*), ed., *Historical Materials on the Old Chinese Maritime Customs, 1859–1948*, vol. 134 (1939) (*Zhongguo jiu haiguan shiliao*) [中國舊海關史料] (Beijing: Jinghua Press [京華出版社], 2001); Chinese Second Historical Archives (*Zhongguo di'er lishi dang'anguan*), ed., *Historical Materials on the Old Chinese Maritime Customs, 1859–1948*, vol. 138 (1940) (*Zhongguo jiu haiguan shiliao*) [中國舊海關史料] (Beijing: Jinghua Press [京華出版社],

2001); Chinese Second Historical Archives (*Zhongguo di'er lishi dang'anguan*), ed., *Historical Materials on the Old Chinese Maritime Customs, 1859–1948,* volume 142 (1941) (*Zhongguo jiu haiguan shiliao*) [中國舊海關史料] (Beijing: Jinghua Press [京華出版社], 2001); Chinese Second Historical Archives (*Zhongguo di'er lishi dang'anguan*), ed., *Historical Materials on the Old Chinese Maritime Customs, 1859–1948,* vol. 144 (1942) (*Zhongguo jiu haiguan shiliao*) [中國舊海關史料] (Beijing: Jinghua Press [京華出版社], 2001).

79. Daqing Yang, "Telecommunication and the Japanese Empire: A Preliminary Analysis of Telegraphic Traffic," *Historical Social Research* 35, no. 1 (2010): 66–89.

80. Ibid., 70–71.

81. Parks Coble, *Chinese Capitalists in Japan's New Order: The Occupied Lower Yangzi, 1937–1945* (Berkeley: University of California Press, 2003).

82. "The Six Patriotic Women Arrive in Tianjin (*Tenshin e tsuita aikoku roku josei*) [天津へ着いた愛国六女性]," *Asahi Shinbun* [朝日新聞] (January 4, 1938), 10; "The Six Patriotic Women of Tianjin (*Tenshin no aikoku roku josei*) [天津の愛国六女性]," *Asahi Shinbun* [朝日新聞] (February 5, 1938), 10; "Heading to the South Seas as a Typist (*Taipisuto toshite nanyō e*) [タイピストとして南洋へ]," *Asahi Shinbun* [朝日新聞] (August 24, 1939), E6.

83. "The Key Lies in Mengjiang: Seeing Ōba Sachiko Off," *Taipisuto* [タイピスト] 16, no. 10 (1941): 10–11. For further patriotic content about the typewriter, see "A Sailor's Inspection of the Typewriter (*Suihei-san no taipuraitā kengaku*) [水兵さんのタイプライター見学]," *Taipisuto* [タイピスト] 16, no. 7 (July 1940): 16.

84. Sakurada was the recipient of the 1940 Akutagawa Prize for his novel *Hiraga Gennai.*

85. Sakurada Tsunehisa [櫻田常久], *The Army Typist (Jūgun taipisuto)* [従軍タイピスト] (Tokyo: Akamon shobō [赤門書房], 1941).

86. Makimasa [牧正], "Diary of a Garrison in South China (*Nanshi chūgun ki*) [南支駐軍記]," *Taipisuto* [タイピスト] 17, no. 1 (February 1942): 16–25.

87. "Typist Fever in Taiwan, Extremely Successful (*Taiwan no taipisuto netsu: kiwamete seikyō*) [台湾のタイピスト熱:極めて盛況]," *Taipisuto* [タイピスト] 16, no. 8 (1941): 11.

88. Nippon Typewriter Company Divisions, *Taipisuto* [タイピスト] 17, no. 10 (October 1942—Showa 16): 54.

89. "Japanese, Manchu, Chinese, Mongolian All-Script Typewriter (*Ri Man Hua Meng wen gezhong daziji*) [日滿華蒙文各種打字機]," *Far East Trade Monthly (Yuandong maoyi yuebao)* [遠東貿易月報] 7 (1940): reverse cover.

90. "The First Manchuria-Wide Typist Competition Held in Xinjing," *Taipisuto* [タイピスト] 16, no. 10 (1941): 2–7. The event was held on May 12, 1941. For later competitions, see "Mantetsu Type Competition Results (*Zen Mantetsu jousho kyōgi taikai no seiseki*) [全満鉄淨書競技大會の成績]," *Taipisuto* [タイピスト] 17, no. 10 (October 1942—Showa 16): 6–11; and Yuji Riichi [湯地利市]; "On the Mantetsu Type Competition (*Mantetsu no taipu kyōgi ni tsuite*) [満鉄のタイプ競技に就て]," *Taipisuto* [タイピスト] 18, no. 10 (October 1943—Showa 17): 2–3.

91. *Improved Shu Zhendong Style Chinese Typewriter Manual* (*Gailiang Shu shi Huawen daziji shuomingshu*) [改良舒式華文打字機說明書] (Shanghai: Shangwu yinshu guan, 1938), University of Pennsylvania Archives—W. Norman Brown Papers (UPT 50 B879), box 10, folder 5. There were also efforts to electrify the Chinese typewriter, yet with minimum effect. Chinese typewriters remained mechanical well into the 1980s. See "Inventing an Electric Chinese Typewriter (*Faming dianli Zhongwen daziji*) [發明電力中文打字機]," *Capital Electro-Optical Monthly* (*Shoudu dianguang yuekan*) [首都電光月刊] 61 (1936): 9; and "Electric Chinese Typewriter a Success (*Dianqi Zhongwen daziji chenggong*) [電氣中文打字機成功]," *Shoudu dianguang yuekan* [首度電光月刊] 74 (April 1, 1937): 10.

92. Personal communication from Shu Chonghui to the author, following Shu Chonghui's February 6, 2010, interview with Tao Minzhi; letter from Tao Minzhi to author, February 11, 2010.

93. "Shanghai Individual Ping-Pong Competition: Ouyang Wei Wins Championship, Youth Yang Hanhong Wins Second Place (*Quan Hu geren pingpang sai Ouyang Wei guanjun xiaojiang Yang Hanhong de Yajun*) [全滬個人乒乓賽歐陽維冠軍小將楊漢宏得亞軍]," *Shenbao* (June 7, 1943), 4; "Ping-Pong Association Established Yesterday (*Pingpang lianhehui zuo zhengshi chengli*) [乒乓聯合會昨正式成立]," *Shenbao* (December 6, 1943), 2; "Japanese Navy Festival Celebration, Sports Meeting Today: Track and Field, Football, Basketball, and Volleyball (*Qingzhu Ri haijun jie jinri yundong dahui you tianjing zulan paiqiu deng jiemu*) [慶祝日海軍節今日運動大會有田徑足籃排球等節目]," *Shenbao* (May 27, 1944), 3. The family's home life may have degraded as well. In the wake of his parents' divorce, Yu Binqi's son Shuolin left home, setting out for Suzhou with his wife to start a family. Shuolin brought two Yu-style machines with him, with the hopes of perhaps starting a typing school in his new city. As for the Yu Binqi outfit in Shanghai, this was left to the managerial care of Tao Minzhi, a young woman with experience at the Suzhou Great China Typewriter Company (*Da Zhonghua dazishe*). Tao Minzhi would manage the store on Beijing Road through 1949 and the Chinese Communist victory. See Yu Shuolin [俞碩霖], "The Last Yu-Style Typewriter (*Zui hou de Yu shi daziji*) [最后的俞式打字机]," *Old Kids Blog* (*Lao xiaohai shequ*) [老小孩社区] (June 9, 2010), http://www.oldkids.cn/blog/blog_con.php?blogid=130259 (accessed June 13, 2011); and letter from Tao Minzhi to author, February 11, 2010.

94. Coble, *Chinese Capitalists in Japan's New Order: The Occupied Lower Yangzi*, 113. See also Poshek Fu, *Passivity, Resistance, and Collaboration: Intellectual Choices in Occupied Shanghai, 1937–1945* (Stanford: Stanford University Press, 1993).

95. Timothy Brook, *Collaboration: Japanese Agents and Local Elites in Wartime China* (Cambridge, MA: Harvard University Press, 2007); Margherita Zanasi, "Globalizing Hanjian: The Suzhou Trials and the Post–WWII Discourse on Collaboration," *American Historical Review* 113, no. 2 (June 2008): 731–751.

96. C.Y. Chao Typewriting Maintenance Department (*Zhao Zhangyun Daziji xiuli bu*) [趙章云打字機修理部]. See Memo from Shanghai Municipal Council Secretary to "All Departments and Emergency Offices," signed by Takagi, entitled "Cleaning of Typewriters, Calculators, etc.—1943," SMA U1-4-3586 (April 2, 1943), 35. Receipt

from C.Y. Chao for Cleaning Services Sent to Secretariat Office, SMA U1-4-3582 (October 12, 1943), 5.

97. Huanqiu Chinese Typewriter Manufacturing Company (*Huanqiu Huawen daziji zhizaochang*) [環球華文打字機製造廠] located at 169 Yuanmingyuan Road; Chang Yah Kee Typewriter Company (*Zhang Xieji daziji gongsi*) [張協記打字機公司] located at 187 Qipu Road (七浦路); Ming Kee Typewriter Company (*Mingji daziji hang*) [銘記打字機行] at 412A Jiangsu Road; price quotations from typewriter companies to the General Office, First District of the Government of Shanghai, SMA R22-2-776 (December 21, 1943), 1–28.

98. "China Standard Typewriter Mfg. Co," SMA U1-4-3582 (August 7, 1943), 11–13.

99. Memo from Treasurer to Secretary General, entitled "Public Works Department—Chinese Typewriters," SMA U1-4-3582 (August 12, 1943), 9.

100. Memo from the Secretary's Office, Municipal Council to the Director entitled "Chinese Typewriters," SMA U1-4-3582 (July 13, 1943), 6. Indeed, even following the end of the war, the retrofitting of Japanese machines for Chinese usage continued. According to one price estimate submitted to the Shanghai Municipal Police Administration (*Shanghai shi jingchaju*), Yu's Chinese Typewriter Manufacturing Company on 279 Peking Road offered a combined price of 32,000 yuan for a box of 2,500 character slugs and "changing one Japanese typewriter" (*gaizao Riwen daziji yi jia*). Letter from Yu's Chinese Typewriter Mfg. Co. to the Shanghai Municipal Police Administration (*Shanghai shi jingchaju*) [上海市警察局], SMA Q131-7-1368 (December 13, 1945), 4.

101. This school trained students for careers as preschool instructors. See http://blog.sina.com.cn/s/blog_4945b4f80101rtfb.html

102. "Regarding the Foundation of the Guangdewen Typing Supplementary School in Beiping (*Guanyu chuangban Beiping shi sili Guangdewen dazi buxi xueshe de chengwen ji gai she jianzhang deng yiji shehui ju de piwen*) [關於創辦北平市私立廣德文打字補習學社的呈文及該社簡章等以及社會局的批文]," BMA J002-003-00754 (May 1, 1938). Each day, students trained for two hours, at a time to be arranged in accordance with their own schedules and the opening hours of the institute. Students enrolled in the accelerated program continued their studies for one and a half months, paying 15 yuan per month. Those in the normal class completed their work in two months, paying 8 yuan per month.

103. "Beijing Jiyang Typing School Temporarily Ceases Operation (*Beijing shi sili Jiyang Huawen dazi buxi xuexiao zanxing tingban*) [北京市私立暨陽華文打字補習學校暫行停辦]," BMA J002-003-00636 (January 1, 1939).

104. "Regarding the Foundation of the Guangdewen Typing Supplementary School in Beiping (*Guanyu chuangban Beiping shi sili Guangdewen dazi buxi xueshe de chengwen ji gai she jianzhang deng yiji shehui ju de piwen*) [關於創辦北平市私立廣德文打字補習學社的呈文及該社簡章等以及社會局的批文]," BMA J002-003-00754 (May 1, 1938).

105. See "北京市私立東亞華文打字科職業補習學校常年經費預算表" in "北京東亞華文打字職業學校關於創辦學校請立案的呈及市教育局的指令," BMA J004-002-00559 (September 30, 1939).

106. "Regarding the Foundation of the Guangdewen Typing Supplementary School in Beiping (*Guanyu chuangban Beiping shi sili Guangdewen dazi buxi xueshe de chengwen ji gai she jianzhang deng yiji shehui ju de piwen*) [關於創辦北平市私立廣德文打字補習學社的呈文及該社簡章等以及社會局的批文]," BMA J002-003-00754 (May 1, 1938).

107. Zhou's associate at the school, twenty-three-year-old Shaoxing native Zhang Menglin, was also a graduate of the East Asia Japanese-Chinese Typing School. Zhang Menglin [張孟鄰] is referred to elsewhere as Zhang Yuyan [張玉鄢]. See "Curriculum Vitae of Teaching Staff of the Private Yucai Chinese Typing Vocational Supplementary School in Beiping (*Beiping shi sili Yucai Huawen dazike zhiye buxi xuexiao zhi jiaoyuan lüli biao*) [北平市私立育才華文打字科職業補習學校教員履歷表]" and "Student Roster of the Private Guangde Supplementary School in Beiping (*Beiping shi sili Guangde buxi xuexiao xuesheng mingji biao*) [北平市私立廣德補習學校學生名籍表]," in "Curriculum Vitae of Teaching Staff and Student Roster of the Private Yucai Chinese Typing Vocational Supplementary School in Beiping (*Beiping shi sili Yucai Huawen dazike zhiye buxi xuexiao zhijiaoyuan lüli biao, xuesheng mingji biao*) [北平市私立育才華文打字科職業補習學校職員履歷表、學生名籍表]," BMA J004-002-00662 (July 31, 1939).

108. The name or location of the Japanese language school is not provided in source materials. The associate Li brought with him experience as the former principal of his own typing school, the Tianjin-based Chenguang Typing School. See "Beijing Baoshan and Guangde Chinese Typing Supplementary Schools (*Beijing shi sili Baoshan, Guangde Huawen dazi buxi xuexiao*) [北京市私立寶善、廣德華文打字補習學校]," BMA J004-002-00579 (July 1, 1938).

109. At the East Asia Japanese-Chinese Typewriting Supplementary School, Chinese-language typing instruction was overseen by Shen Lianzhen, a thirty-eight-year-old woman, also hailing from Lianyang county in Fengtian, and also a graduate of the Fengtian Japanese-Chinese Typing Institute. See "Petition Submitted by the Private Yadong Japanese-Chinese Typing Supplementary School of Beijing to the Beijing Special Municipality Bureau of Education Regarding Inspection of Student Grading List, Curriculum Plan, and Teaching Hours for the Regular-Speed Division of the Sixteenth Term, and the Order Received in Response from the Bureau of Education (*Beijing shi sili Yadong Ri Huawen dazi buxi xuexiao guanyu di'shiliu qi putong sucheng ge zu xuesheng chengjibiao, kecheng yuji ji shouke shi shu qing jianhe gei Beijing tebieshi jiaoyuju de cheng yiji jiaoyuju de zhiling*) [北京市私立亞東日華文打字補習學校關於第十六期普通速成各組學生成績表、課程預計及授課時數請鑒核給北京特別市教育局的呈以及教育局的指令]," BMA J004-002-01022 (January 31, 1943).

110. Lori Watt, *When Empire Comes Home: Repatriates and Reintegration in Postwar Japan* (Cambridge, MA: Harvard University Asia Center, 2009).

111. People's Welfare Typewriter Manufacturing Company (*Minsheng daziji zhizaochang*) [民生打字機製造廠], ed., *Practice Textbook* (*Lianxi keben*) [練習課本], n.p.: c. 1940s, cover.

112. Fan Jiling [范繼舲], *Fan's Wanneng-Style Chinese Typewriter Practice Textbook* (*Fan shi wanneng shi Zhongwen daziji shixi fanben*) [范氏萬能式中文打字機實習範本]

(Hankou: Fan Research Institute Publishing House (*Fan shi yanjiusuo yinhang*) [范氏研究所印行], 1949), i.

113. "Red Star Chinese Typewriter Plant 1952 Construction Plan (*Hongxing daziji chang yi jiu wu er nian ji jian jihua*) [紅星打字機廠一九五二年基建計劃]," TMA X77-1-415 (1952), 13–17, 16. See also Tianjin People's Government Local-State Jointly Run Hongxing Factory (*Tianjin shi renmin zhengfu difang guoying huyeju Hongxing huchang*) [天津市人民政府地方國營互業局紅星互廠], "Report on the Improvement of the Chinese Typewriter Character Chart (*Huawen daziji zibiao gaijin baogao*) [華文打字機字表改進報告]," TMA J104-2-1639 (October 1953), 29–39.

114. "Shanghai Chinese Typewriter Manufacturing Joint Venture Marketing Plan (*Shanghai Zhongwen daziji zhizaochang lianyingsuo chanxiao jihua*) [上海中文打字机製造廠聯營所產銷計劃]," SMA S289-4-37 (December 1951), 65. The domestic industry was further challenged by pricing challenges and market instability in the wake of the Second World War, the Civil War, and the revolution of 1949. See "Regulating the Pricing Problem of Domestically Produced Chinese Typewriters (*Wei tiaozheng guochan Zhongwen daziji shoujia wenti*) [為調整國產中文打字機售價問題]," report sent from the Chinese Typewriter Manufacturers Business Association (*Zhongwen daziji zhizao chang shanglian*) to the Shanghai Cultural and Educational Supplies Trade Association (*Shanghai shi wenjiao yongpin tongye gonghui*), SMA B99-4-124 (January 15, 1953), 52–90.

115. The domestic Chinese typewriter manufacturing and retail industry was fractured among many companies, including the Yu-style Typewriter Manufacturing Plant, the Jingyi Chinese Typewriter Manufacturing Plant, the Wanneng Typewriter Company, the China Typewriter Manufacturing Company, and the Minsheng Chinese Typewriter Manufacturing Plant, among many others. At least five others were founded between 1949 and 1951, moreover. These included a new Yu-style Typewriter Manufacturing Plant, founded in April 1949, which employed 55 people to manufacture the Wanneng-style typewriter; the Ziqiu Industrial Plant, founded on January 4, 1950, which employed 47 people and manufactured both Chinese typewriters and alloy characters; the Wenhua Chinese Typewriter Manufacturing Plant, founded in September 1950, which employed a total of 42 workers primarily for the manufacture of Chinese typewriters; the Jingyi Typewriter Manufacturing Plant, founded October 1, 1951, which employed 16 people and manufactured Chinese typewriters and office supplies; and the Wanling Scientific Apparatus Manufacturing Plant, founded in September 1951, which employed 12 people and focused primarily on typewriters. "Draft Charter of the Shanghai Chinese Typewriter Manufacturing Plant Joint Management Organization (*Shanghai Zhongwen daziji zhizaochang lianyingsuo zuzhi zhangcheng cao'an*) [上海中文打字機製造廠聯營所組織章程草案]," SMA S289-4-37 (December 1951).

116. "Draft Charter of the Shanghai Chinese Typewriter Manufacturing Plant Joint Management Organization (*Shanghai Zhongwen daziji zhizaochang lianyingsuo zuzhi zhangcheng cao'an*) [上海中文打字機製造廠聯營所組織章程草案]," in "Shanghai Chinese Typewriter Manufacturing Joint Venture Marketing Plan (*Shanghai Zhongwen daziji zhizaochang lianyingsuo chanxiao jihua*) [上海中文打字機製造廠聯營所產銷計劃],"

SMA S289-4-37 (December 1951); "A Large-Scale Public-Private Typewriter Store Newly Opened in Shanghai (*Shanghai xin kai yi jia guimo juda de gongsi heying daziji dian*) [上海新开一家規模巨大的公私打字机店]," *Xinhuashe xinwen gao* [新華社新聞稿] 2295 (1956).

117. "Draft Charter of the Shanghai Chinese Typewriter Manufacturing Plant Joint Management Organization (*Shanghai Zhongwen daziji zhizaochang lianyingsuo zuzhi zhangcheng cao'an*) [上海中文打字機製造廠聯營所組織章程草案]," SMA S289-4-37 (December 1951).

118. Shanghai Calculator and Typewriter Factory (*Shanghai jisuanji daziji chang*) [上海計算機打字機廠], ed., "Evaluation Report on the Double Pigeon Brand DHY Model Chinese Typewriter (*Shuangge pai DHY xing Zhongwen daziji jianding baogao*) [双鸽牌DHY型中文打字机鉴定报告]," SMA B155-2-284 (April 24, 1964), 4; Shanghai Calculator and Typewriter Factory (*Shanghai jisuanji daziji chang*) [上海計算機打字機廠], ed., "Double Pigeon Brand Chinese Typewriter Improvement and Trial Production Summary Report (*Shuangge pai Zhongwen daziji gaijin shizhi jishu zongjie*) [双鸽牌中文打字机改进试制技术总结]," SMA B155-2-282 (March 22, 1964), 11–14; Shanghai Calculator and Typewriter Factory (*Shanghai jisuanji daziji chang*) [上海計算機打字機廠], ed., "Double Pigeon Brand Chinese Typewriter Internal Evaluation Report (*Shuangge pai Zhongwen daziji chang nei jianding baogao*) [双鸽牌中文打字机厂内鉴定报告]," SMA B155-2-282 (March 22, 1964), 9–10.

119. Shanghai Calculator and Typewriter Factory (*Shanghai jisuanji daziji chang*) [上海計算機打字機廠], ed., "Evaluation Report on the Double Pigeon Brand DHY Model Chinese Typewriter (*Shuangge pai DHY xing Zhongwen daziji jianding baogao*) [双鸽牌DHY型中文打字机鉴定报告]," SMA B155-2-284 (April 24, 1964), 4.

120. Ibid.

121. Ibid., 1.

CHAPTER 6: QWERTY IS DEAD! LONG LIVE QWERTY!

1. Matthew Fuller, *Behind the Blip: Essays on the Culture of Software* (Sagebrush Education Resources, 2003).

2. What is more, users can customize the input experience in a variety of ways, effectively multiplying the number of potential input schemes. This simple fact becomes even more intriguing when we consider that, in English typing, there is but one way to enter the phrase in question: *t-y-p-e-w-r-i-t-e-r*.

3. "Front Views and Profiles: Miss Yin at the Console," *Chicago Daily Tribune* (October 10, 1945), 16.

4. Ibid.

5. By comparison, the Olivetti MS25 measured approximately 14.2 inches wide and 14.2 inches deep.

6. "Chinese Project: The Lin Yutang Chinese Typewriter," Smithsonian, n.d., multiple dates encompassed (1950), 1.

7. Ann Blair, *Too Much to Know: Managing Scholarly Information before the Modern Age* (New Haven: Yale University Press, 2010); Mary Elizabeth Berry, *Japan in Print: Information and Nation in the Early Modern Period* (Berkeley: University of California Press, 2006).

8. Chen Youxun [陳有勳], "Report on a Comparison of Four-Corner Look-Up and Radical Look-Up (*Sijiao haoma jianzifa yu bushou jianzifa de bijiao shiyan baogao*) [四角號碼檢字法與部首檢字法的比較實驗報告]," *Educational Weekly (Jiaoyu zhoukan)* [教育周刊] 177 (1933): 14–20.

9. "Among the difficulties of the Chinese language," H.L. Wang argued in 1916, "the lack of an Index System is probably one of the most serious." "A large collection of books will lose much of their usefulness unless they can be arranged in such wise that their contents and their locations be readily found." John Wang (H.L. Huang), "Technical Education in China," *Chinese Students' Monthly* 11, no. 3 (January 1, 1916): 209–214.

10. Were we to recreate this on film, a fitting technique might be *vertigo zoom*, the camera dollying away from the Chinese subject all while maintaining our focus on his visage. The impression produced would be one of a stationary subject around whom space itself warps and tunnels, conveying a moment of newly realized alienation.

11. J.J.L. Duyvendak, "Wong's System for Arranging Chinese Characters. The Revised Four-Corner Numeral System," *T'oung Pao* 28, no. 1/2 (1931): 71–74.

12. Cai Yuanpei [蔡元培], "An Introduction to Point/Vertical Stroke/Horizontal Stroke/Slanting Stroke Character Retrieval System (*Jieshao dian zhi heng xie jianzifa*) [介紹點直橫斜檢字法]," *Modern Student (Xiandai xuesheng)* [現代學生] (April 1, 1931), 1–8. See also "Methods Regarding the Character Retrieval Problem (*Duiyu jianzifa wenti de banfa*) [對於檢字法問題的辦法]," *Eastern Miscellany (Dongfang zazhi)* [東方雜誌] 20, no. 23 (1923): 97–100.

13. Jiang Yiqian [蔣一前], *History of the Development of Character Retrieval Systems in China and a Table of Seventy-Seven New Character Retrieval Systems (Zhongguo jianzifa yange shilüe ji qishiqi zhong xin jianzifa biao)* [中國檢字法沿革史略及七十七種新檢字法表] (n.p.: Zhongguo suoyin she, 1933). Among these, at least twenty-six systems were based upon strokes.

14. For more on Gao Mengdan, see Reed, *Gutenberg in Shanghai*.

15. Lin Yutang, "Explanation of Chinese Character Index (*Hanzi suoyin zhi shuoming*) [漢字索引制說明]," *Xin Qingnian* 4, no. 25 (1918): 128–135; Lin Yutang, "Last-Stroke Character Retrieval Method (*Mobi jianzifa*)," in *Collected Essays on Linguistics (Yuyanxue luncong)* [語言學論叢] (Taibei: Wenxing shudian gufen youxian gongsi, 1967), 284.

16. Lin Yutang, "Explanation of Chinese Character Index," 128–135; Lin Yutang, "Last-Stroke Character Retrieval Method (*Mobi jianzifa*)," 284.

17. Jiang Yiqian, *History of the Development of Character Retrieval Systems in China*, 1.

18. For a fascinating study of evolutionism and its place in modern Chinese thought, see Andrew F. Jones, *Developmental Fairy Tales: Evolutionary Thinking and Modern Chinese Culture* (Cambridge, MA: Harvard University Press, 2011).

19. Jiang Yiqian, *History of the Development of Character Retrieval Systems in China*, 1.

20. Chen Lifu, *Storm Clouds Over China: The Memoirs of Ch'en Li-fu, 1900–1993*, ed. Sidney Chang and Ramon Myers (Stanford: Hoover Institute Press, 1994), 42.

21. Ibid., 70.

22. Ibid., 19–23, 71.

23. As his exemplar to demonstrate the proliferation of specialized products, Chen Lifu cited the example of rose-scented soap.

24. Chen Lifu, *Storm Clouds Over China*, 32.

25. *The Principle and Practice of Five-Stroke Indexing System for Chinese Characters* (*Wubi jianzifa de yuanli ji yingyong*) [五筆檢字法的原理及應用] (Shanghai: Zhonghua shuju, 1928), and *How to Locate Clan Surnames* (*Xingshi su jianzifa*) [姓氏速檢字法]. Following an emotional farewell dinner, and his transfer of responsibilities over the confidential section to Chen Bulei (1890–1948), Chen Lifu dedicated himself to, among other things, the refinement and development of his Five-Stroke system. He published *The Principle and Practice of Five-Stroke Indexing System for Chinese Characters* and *How to Locate Clan Surnames*—an application of his method to 500 Chinese surnames.

26. Du Dingyou, "On the Psychology of How the Masses Search for Characters (*Minzhong jianzi xinli lunlüe*) [民眾檢字心理論略]," 340–350, in *Selected Library Science Writings of Du Dingyou* (*Du Dingyou tushuguanxue lunwen xuanji*) [杜定友圖書館學論文選集], ed. Qian Yaxin [钱亚新] and Bai Guoying [白国应] (Beijing: Shumu wenxian chubanshe, 1988). Originally in *Education and the Masses* (*Jiaoyu yu minzhong*) [教育與民眾] 6, no. 9 (1925).

27. Here, Du is referring to the code 0033, by which the character *lian* is categorized within the Four-Corner system.

28. Du Dingyou, "On the Psychology of How the Masses Search for Characters," 340–350.

29. Incidentally, as an inveterate entrepreneur, Du Dingyou laid claim to having invented the penultimate character of this title: the single character 圕 he used to represent the polysyllabic term "tushuguan" (or "library"), conventionally written with three Chinese characters, 圖書館.

30. Lin Yutang, "Features of the Invention," Archives of John Day Co., Princeton University, box/folder 14416, call no. CO123 (c. October 14, 1931), 3. "Fact and Fiction: A Chinese Typewriter (*Shishi feifei: Hanwen daziji*) [是是非非:漢文打字機]," *Nanhua wenyi* [南華文藝] 1, no. 7/8 (1932): 103.

31. Lin Yutang, "Features of the Invention," 3.

32. Lin Yutang received a special grant of $3,500 for his typewriter project from the China Foundation for the Promotion of Education and Culture. See Edward Hunter, "Increasing Program of China Foundation," *China Weekly Review* (August 8, 1931), 379.

33. Lin Yutang, "Features of the Invention," 3.

34. Ibid., 5.

35. Ibid.

36. Archives of the John Day Company, Princeton University, box 144, folder 6, no. C0123. The interior of Lin Yutang's mechanism, he at first proposed, would be based upon the Hammond typewriter, which featured curved type sets. Lin proposed to outfit these mechanical type sets with his radicals, phonetics, and full-body characters, and then to build them up on the page. He would arrange these components onto 30 type bars, of 32 columns each, each column having 4 positions. "This bar and column indicates the group to which a character belongs." In the case of left-right characters, the carriage would not advance. This he termed "The Mechanical Design." Ibid., 4–5.

37. Ibid., 2. Indeed, Lin Yutang's relationship to McKean Jones on this particular account might be more than incidental. When we turn the pages of Lin's 1931 letter over, we discover five ever-so-faint letters: *RMcKJ*, the signature of Robert McKean Jones, appearing next to a stamp that reads "Remington Typewriter Company." Here, then, was the "master typographer" from Remington we met in chapter 4, who years earlier had patented his failed Chinese typewriter system premised upon the Chinese phonetic alphabet.

38. Ibid.," 4. A further collection of three letters from Lin Yutang to Richard Walsh, Jr. in the winters of 1937 and 1938 brings to light the continuation of this hidden history of Lin's early experiments with Chinese typewriting, and the model he developed before MingKwai. In the first of these three letters, dated December 16, 1937, Lin recounted that John Marshall of the Rockefeller Foundation expressed his belief that the realization of such a project "would take a professor of engineering with a keen inventive mind," and recommended that "probably MIT is the place— with some other institutions (International Business Machines Co.—a brilliant man over there who invents radio-teletype)." Lin confessed to Selskar Gunn that he had "no idea how much money would be needed for building a second model," and that he would not be able to return to focusing on the project until completion of his present book manuscript, most likely that of *Mulan*. Letter from Lin Yutang to Richard Walsh and Pearl S. Buck dated December 16, 1937. Archives of John Day Co., box 144, folder 6, call no. C0123. In a pair of letters dated almost precisely one year hence, Lin wrote again to Walsh, broaching once more the question of the typewriter. Writing from his home on 59 rue Nicolo, Lin outlined his ongoing conversations with Gunn and Marshall, with whom Lin planned to meet the following day to discuss what he referred to as "my scheme." The second letter, dated December 21, 1938, recounted details of the meeting with Gunn, the letter making reference to an enclosure that, unfortunately, does not survive within the archival

record: two photographs of what Lin refers to as "the first model." "As the secret lies in the keyboard and the linguistic work, you can show all this material to proper interested parties. An application for patent was filed through a New York lawyer, brother of Horace Mann of St. John's University, Shanghai, in the Winter of 1931." Letter from Lin Yutang to Richard Walsh and Pearl S. Buck, December 13, 1938, sent from Paris. Letter from Lin Yutang to Richard Walsh and Pearl S. Buck, December 13, 1938, sent from Paris. Archives of John Day Co., Princeton University, box 144, folder 6, call no. C0123.

39. "Chinese Project: The Lin Yutang Chinese Typewriter," Mergenthaler Linotype Company Records, 1905–1993, Archives Center, National Museum of American History, Smithsonian Institution, box 3628, multiple dates in 1950 listed, 5.

40. Lin Taiyi, *Biography of Lin Yutang* (*Lin Yutang zhuan*) (Taibei: Lianjing chubanshe, 1989), 235.

41. Ibid.

42. Ibid.

43. Ibid.

44. Other coverage included Yang Mingshi [楊名時], "The Principles of Lin Yutang's Chinese Typewriter (*Lin Yutang shi Huawen daziji de yuanli*) [林語堂式華文打字機的原理]," *Guowen guoji* [國文國際] 1, no. 3 (1948): 3.

45. Lin Taiyi, *Biography of Lin Yutang*, 236.

46. Ibid.

47. Perhaps the only executives who would have been familiar with the idea of male type operators were the executives of Mergenthaler Linotype, insofar as Linotype operation remained a male domain throughout its history. And yet Lin Yutang's MingKwai machine was not a Linotype—it was a clerical typewriter, and as such, they too would undoubtedly have equated it with the machines that adorned the desks of their managerial offices, not the printing trenches wherein "linos" set type at blazing speed for global newspaper audiences.

48. Lin Taiyi, *Biography of Lin Yutang*, 236–237.

49. Ibid.

50. The exact cause of the problem was not specified in any source I have encountered.

51. Lin Taiyi, *Biography of Lin Yutang*, 237–238.

52. Quoted in clipping from *Chinese Journal* (May 26, 1947), included in Archives of John Day Company, Princeton University, box 236, folder 14, number CO123.

53. "Lin Yutang Invents Chinese Typewriter: Will Do in an Hour What Now Takes a Day," *New York Herald Tribune* (August 22, 1947), 13.

54. After this celebratory news, the Italian engineer who helped Lin Yutang design MingKwai began to make claims that MingKwai was properly his invention, not Lin's. Unnamed in Lin Taiyi's biography, the Italian engineer threatened to issue a lawsuit against Lin Yutang. As Lin Taiyi remarked, both she and her father found it

amazing that an Italian man who spoke no Chinese could claim such a thing. Lin Yutang sought out a lawyer, but it appears that the threat of lawsuit passed without event. There is no further mention of the Italian engineer or his claim in Lin Taiyi's biography, nor in other sources I have encountered. Lin Taiyi, *Biography of Lin Yutang*, 238.

55. "Chinese Put on Typewriter by Lin Yutang," *Los Angeles Times* (August 22, 1947), 2.

56. "Just How Smart Are We," *Daily News New York* (September 2, 1947), clipping included in Archives of John Day Company, Princeton University, box 236, folder 14, number CO123; Harry Hansen, "How Can Lin Yutang Make His New Typewriter Sing?," *Chicago Daily Tribune* (August 24, 1947), C4; "Lin Yutang Invents Chinese Typewriter: Will Do in an Hour What Now Takes a Day," *New York Herald Tribune* (August 22, 1947), 13; "New Typewriter Will Aid Chinese: Invention of Dr. Lin Yutang Can Do a Secretary's Day's Work in an Hour," *New York Times* (August 22, 1947), 17.

57. Letter from Pearl S. Buck to Lin Yutang, May 4, 1947, Pearl S. Buck International Archive.

58. Martin W. Reed, "Lin Yutang Typewriter," Mergenthaler Linotype Company Records, 1905–1993, Archives Center, National Museum of American History, Smithsonian Institution.

59. Ibid.

60. Ibid.

61. Ibid.

62. "Chinese Project: The Lin Yutang Chinese Typewriter," Mergenthaler Linotype Company Records, 1905–1993, Archives Center, National Museum of American History, Smithsonian Institution, box 3628, multiple dates in 1950 listed, 4.

63. Ibid., 4–5.

64. Ibid., 5.

65. "Psychological Warfare, EUSAK Compound, Seoul, Korea (1952)," National Archives and Records Group, ARC identifier 25967, local identifier 111-LC-31798.

CHAPTER 7: THE TYPING REBELLION

1. Thomas S. Mullaney, *Coming to Terms with the Nation: Ethnic Classification in Modern China* (Berkeley: University of California Press, 2010).

2. *Ha'erbin shi renmin zhengfu gong'anju zhengzhi baoweichu* (哈尔滨市人民政府公安局政治保卫处). Special thanks to Michael Schoenhals for allowing me to view select archival records.

3. As shown by Antonia Finnane; *Beijing shi fu shipin shangyeju dang zu.*

4. For example, the August 30, 1955 typewritten report entitled "Qing shi guo-qingjie taixiang shunxu wenti (請示國慶節抬像順序問題)," produced by the Zhong

Gong Hebei shengwei jiyaochu [中共河北省委機要處],1955. As evidenced in archival collections compiled by Daniel Leese.

5. As found in the archival collections of Jacob Eyferth. Chinese Communist Party Baodi County Committee (*Zhongguo gongchandang Baodi xian weiyuanhui*), "Survey Report on Living Conditions and Political Attitudes in Baodi County (*Zhonggong Baodi xianwei guanyu nongmin sixiang qingkuang de diaocha baogao*) [中共宝鸡县委关于农民思想情况的调查报告]," August 17, 1957.

6. As shown in Benno Weiner's work.

7. "Whatever Work Aims to Complete and Not to Fail the Five Year Plan, All That Work Is Glorious! (*Renhe laodong, dou shi wancheng wunian jihua buke queshaode laodong, dou shi guangrongde laodong!*) [任何勞動,都是完成五年计划不可缺少的劳动,都是光荣的劳动!]," designed by Zhou Daowu [周道悟], March 1956, PC-1956-013, private collection, chineseposters.net.

8. *The Poems of Chairman Mao* (*Mao Zhuxi shici*) [毛主席詩詞], 1968. Type-and-mimeograph edition by self-identified Red Guard, released on or around International Workers' Day, author's personal collection; *Long Live Chairman Mao Thought. Selections from 1957 and 1958* (*Mao Zhuxi sixiang wansui 1957 nian–1958 nian wenji*) [毛主席思想万岁 1957年–1958年文集], c. 1958. Type-and-mimeograph edition by members of the Yunnan University Mao Zedong-ism Artillery Regiment Foreign Language Division Propaganda Group (*Yunnan Daxue Mao Zedong zhuyi pao bing tuan waiyu fentuan xuanchuan zu*) [云南大学毛泽东主义炮兵团外语分团宣传组], author's collection.

9. Liansu Meng, "The Inferno Tango: Gender Politics and Modern Chinese Poetry, 1917–1980," PhD dissertation, University of Michigan, 2010, 1, 233–234.

10. "A New Typing Record (*Dazi xin jilu*) [打字新記錄]," *People's Daily* (*Renmin ribao*) (November 23, 1956), 2.

11. Ibid.

12. Ingrid Richardson, "Mobile Technosoma: Some Phenomenological Reflections on Itinerant Media Devices," *fiberculture* 6 (December 10, 2005); Ingrid Richardson, "Faces, Interfaces, Screens: Relational Ontologies of Framing, Attention and Distraction," *Transformations* 18 (2010).

13. Pierre Bourdieu, *The Logic of Practice* (Stanford: Stanford University Press, 1992), 57.

14. "Kaifeng Typesetter Zhang Jiying Diligently Improves Typesetting Method, Establishes New Record of 3,000-plus Characters per Hour (*Kaifeng paizi gongren Zhang Jiying nuli gaijin paizifa chuang mei xiaoshi san qian yu zi xin jilu*) [開封排字工人張繼英努力改進排字法創每小時三千余字新紀錄]," *People's Daily* (*Renmin ribao*) (December 16, 1951). Reprinted from article in *Henan Daily* (*Henan Ribao*).

15. Zhang Jiying first worked at the Zhengzhou Xinhua Publishing Company (*Zhengzhou Xinhua yinshuachang*).

16. Zhang Jiying, "How Did I Raise My Work Efficiency? (*Wo de gongzuo xiaolü shi zenme tigao de?*) [我的工作效率是怎麼提高的]," in Zhongnan People's Press (*Zhongnan*

renmin chubanshe) [中南人民出版社], ed., *The Zhang Jiying Typesetting Method* (*Zhang Jiying jianzifa*) [張繼英揀字法] (Hankou: Zhongnan renmin chubanshe, 1952), 20.

17. Li Zhongyuan and Liu Zhaolan, "Kaifeng Typesetter Zhang Jiying's Advanced Work Method (*Kaifeng paizi gongren Zhang Jiying de xianjin gongzuofa*) [開封排字工人張繼英的先進工作法]," *People's Daily* (*Renmin ribao*) (March 10, 1952), 2–4.

18. My thanks to Kamran Naim for alerting me to the etymology of the term.

19. Franz Schurmann, *Ideology and Organization in Communist China* (Berkeley: University of California Press, 1966), 59–68; Alan P.L. Liu, *Communications and National Integration in Communist China* (Berkeley: University of California Press, 1971), 139.

20. Li Zhongyuan and Liu Zhaolan, "Kaifeng Typesetter Zhang," 2–3.

21. Zhang Jiying, "How Did I Raise My Work Efficiency?"

22. Sigrid Schmalzer, *The People's Peking Man: Popular Science and Human Identity in Twentieth-Century China* (Chicago: University of Chicago Press, 2008), 126–128.

23. Zhang Jiying, "How Did I Raise My Work Efficiency?," 21.

24. "Model Workers from Across China Attend International Workers' Day Ceremony (*Ge di lai Jing canjia 'wuyi' jie guanli de laodong mofan*) [各地來京參加五一節觀禮的勞動模範]," *People's Daily* (*Renmin ribao*) (May 7, 1952), 3; Zhang Jiying. "How Did I Raise My Work Efficiency?," 21.

25. He Jiceng [何繼曾], *Elements of Type Setting* (*Paizi qianshuo*) [排字淺說] (Shanghai: Commercial Press, 1959), 41.

26. Despite certain successes in Jinggangshan, hometown favorite Zhang Jiying still reigned supreme, however. Exploding his 1952 record of 3,820 characters per hour, Zhang reportedly reached 4,890, 5,538, and then 6,252 characters per hour during the years of the Great Leap. See Wang Shigeng [王世庚], "Zhang Jiying Sets Another Typesetting Record (*Zhang Jiying zai chuang jianzi xin jilu*) [张继英再创拣字新纪录]," *Henan ribao* [河南日报], (March 30, 1959), 1.

27. Journalism Research Institute of the People's University of China (*Zhongguo renmin daxue xinwenxue yanjiusuo*) [中国人民大学新闻学研究所], ed., *Typesetting for Newspapers* (*Baozhi de paizi he pinpan*) [报纸的排字和拼盘] (Shanghai: Commercial Press, 1958).

28. When combined, these four characters compose the phrase *jiejue wenti*, or "to resolve a problem."

29. When combined, these characters form the phrases *jianshe zuguo* ("develop the motherland") and *tigao changliang* ("increase production").

30. When combined, these characters form the phrases *renmin gongshe* ("people's commune") and *shehui zhuyi* ("socialism"), via the shared character *she*.

31. Journalism Research Institute of the People's University of China, ed., *Typesetting for Newspapers*, 29. Note that "dingzhen xuxian" is better known as "dingzhen xuma" and is a Chinese word game in which a group of players must produce four-character idiomatic expressions, each person required to think of one that begins

with the character that the previous player's expression ended with. The editors drew parallels between this method and "thimble linking" (*dingzhen xuma*), a poetry game dating back to at least the *Shijing*, and popular in the Song and Yuan dynasties. In thimble linking, which bears certain commonalities to the technique of *coblas capfinidas* in medieval Provençal troubadour poetry, the same character that appears at the close of one verse is repeated at the beginning of the next. The game likely evolved out of oral poetry traditions, as a form of memory aid. Nicholas Morrow Williams, "A Conversation in Poems: Xie Lingyun, Xie Huilian, and Jiang Yan," *Journal of the American Oriental Society* 127, no. 4 (2007): 491–506.

32. Journalism Research Institute of the People's University of China, ed., *Typesetting for Newspapers*, 29.

33. Ibid., 30.

34. *Huawen daziji wenzi pailie biao*. Accompanies the *Huawen dazi jiangyi* (n.p., n.d.; produced pre-1928, circa 1917).

35. The same system of organization governs the arrangement of the character *nan* (南). Directly above this character sits the character *hu* (湖), forming "Hunan" (湖南). To the immediate left of *nan* sits *yun* (雲), forming "Yunnan" (雲南). And directly beneath *nan* sits *he* (河), forming "Henan" (河南). Perhaps the most elaborately interlinked and positioned character is that of *xi* (西), meaning "west." Adjacent to this character are those of *zang* (藏), *shaan* (陝), *guang* (廣), and *shan* (山), forming the names Xizang (Tibet), Shaanxi, Guangxi, and Shanxi.

36. Chen Guangyao [陳光垚], *Essays on the Simplification of Chinese Characters (Jianzi lunji)* [簡字論集] (Shanghai: Commercial Press, 1931), 91–92.

37. Shen Yunfen [沈蕴芬], "I Love the Work Allocated Me by the Party with All My Heart (*Wo re'ai dang fenpei gei wo de gongzuo*) [我熱愛黨分配給我的工作]," *People's Daily (Renmin ribao)* (November 30, 1953), 3.

38. "Introduction to the 'New Typing Method' ('*Xin dazi caozuo fa' jieshao*) ['新打字操作法'介紹]," *People's Daily (Renmin ribao)* [人民日报] (November 30, 1953), 3.

39. Ibid.

40. *Quanguo qingnian shehui zhuyi jianshe jiji fenzi dahui* (全国青年社会主义建设积极分子大会). Shen is cited by name in a speech by Hu Yaobang commemorating the 35th anniversary of the May Fourth Movement—the speech was reprinted in the *People's Daily* on May 4, 1954. See Hu Yaobang, "Determine to Be Active Builders and Protectors of Socialism (*Lizhi zuo shehui zhuyi de jiji jianshezhe he baowei zhe*) [立志作社会主义的积极建设者和保卫者]," *People's Daily (Renmin ribao)* (May 4, 1954), 2. She is also cited by name in Cheng Yangzhi [程養之], ed., *Chinese Typing Practice Textbook (Zhongwen dazi lianxi keben)* [中文打字練習課本] (Shanghai: Commercial Press, 1956); and in Deng Zhixiu [邓智秀], "The Achievements of Beijing Model Worker Shen Yunfen (*Beijing laomo Shen Yunfen de shiji*) [北京劳模沈蕴芬的事迹]" (Beijing: Dianli laomowang [March 6, 2006]), www.sjlmw.com/html/beijing/20060306/2193.html (accessed January 23, 2010).

41. Fa-ti Fan, "Redrawing the Map: Science in Twentieth-Century China," *Isis* 98 (2007): 524–538; Schmalzer, *The People's Peking Man*, 8; Joel Andreas, *Rise of the Red Engineers: The Cultural Revolution and the Origins of China's New Class* (Stanford: Stanford University Press, 2009).

42. Fan, "Redrawing the Map"; Schmalzer, *The People's Peking Man*.

43. Schmalzer, *The People's Peking Man*, 126.

44. Franz Schurmann, *Ideology and Organization in Communist China* (Berkeley: University of California Press, 1966), 59; Michael Schoenhals, *Doing Things with Words in Chinese Politics* (Berkeley: Institute of East Asian Studies, University of California, 1992).

45. Schurmann, *Ideology and Organization in Communist China*, 58.

46. Cheng Yangzhi [程養之], ed., *Chinese Typing Manual for the Wanneng-Style Typewriter* (*Wanneng shi daziji shiyong Zhongwen dazi shouce*) [万能式打字机适用中文打字手册] (Shanghai: Commercial Press, 1956), 1–2.

47. Ibid.

48. Ibid., 2.

49. Tianjin People's Government Local-State Jointly Run Hongxing Factory (*Tianjin shi renmin zhengfu difang guoying huyeju Hongxing huchang*) [天津市人民政府地方國營互業局紅星互廠], "Report on the Improvement of the Chinese Typewriter Character Chart (*Huawen daziji zibiao gaijin baogao*) [華文打字機字表改進報告]," TMA J104-2-1639 (October 1953), 10.

50. Tianjin People's Government Local-State Jointly Run Hongxing Factory, "Report on the Improvement of the Chinese Typewriter Character Chart," 31. The example given: if two characters within the "hand-radical" category tended to appear together in actual usage to form a common two-character compound—such as *da* (打) and *quan* (拳), which together form the compound *daquan* meaning "to fight" or "to box"—tray bed designers should place them adjacently, even if this meant disrupting stroke-count organization of other characters within the hand-radical class.

51. *Wanneng Style Chinese Typewriter Basic Character Table* (*Wanneng shi Zhongwen daziji jiben zipan biao*) [万能式中文打字机基本字盘表]. Included as appendix in Cheng Yangzhi, *Chinese Typing Manual for the Wanneng Style Typewriter*.

52. Tianjin People's Government Local-State Jointly Run Hongxing Factory, "Report on the Improvement of the Chinese Typewriter Character Chart," 10–11.

53. "Reformed" (*Gexin*) Chinese Typewriter Tray Bed of 1956.

54. Ronald Kline and Trevor Pinch, "Users as Agents of Technological Change: The Social Construction of the Automobile in the Rural United States," *Technology and Culture* 37 (1996): 763–795.

55. This move quite likely afforded a measure of job security as well. While state authorities and manufacturers were eager to promote standardization—and with it *interchangeability*—an operator could not be easily replaced after having reorganized the tray bed into a deeply personal, individualized configuration.

56. Wang Guihua [王桂華] and Lin Gensheng [林根生], eds., *Chinese Typing Technology (Zhongwen dazi jishu)* [中文打字技術] (Nanjing: Jiangsu renmin chubanshe, 1960).

57. Ibid., 11.

58. It bears emphasizing that, circa 1960, the development of predictive text tray beds had clearly reached a sufficient degree of commonality and sophistication that Wang and Lin were able to present not merely one technique, but five.

59. Wang Guihua and Lin Gensheng, *Chinese Typing Technology*, 8.

60. Chinese typewriter formerly employed at the United Nations (Geneva), Double Pigeon style, manufactured in 1972 by the Shanghai Calculator and Typewriter Factory *(Shanghai jisuanji daziji chang)*, housed at the Musée de la Machine à Écrire, Lausanne, Switzerland; Chinese typewriter formerly employed at UNESCO (Paris), Double Pigeon style, manufactured in 1971 by the Shanghai Calculator and Typewriter Factory *(Shanghai jisuanji daziji chang)*, housed at the Musée de la Machine à Écrire, Lausanne, Switzerland.

61. On "tinkering," see Adele E. Clarke and Joan Fujimura, "What Tools? Which Jobs? Why Right?" in *The Right Tools for the Job: At Work in Twentieth Century Life Sciences*, ed. Adele E. Clarke and Joan Fujimira (Princeton: Princeton University Press, 1992), 7. For "reconfigurations," see Lucy Suchman, *Human-Machine Reconfigurations: Plans and Situated Actions* (Cambridge: Cambridge University Press, 2006).

62. Chinese typists in the post-1949 period pushed their experiments in vernacular taxonomy to their very extremes, going so far as to rearrange punctuation marks and numerals into predictive clusters. In the case of punctuation, rather than grouping the comma, period, question mark, semicolon, etc., together, as on Chinese typewriter tray beds in the first half of the twentieth century, both the UNESCO and UN typists located them adjacent to those characters with which they either frequently or always appeared. The case of the question mark (?) is the most illustrative. Insofar as Chinese expresses interrogatives through the use of a limited set of particles, which themselves are normally followed by a question mark, the UNESCO typist decided to place these particles and the question mark together. Specifically, the question mark on the machine is flanked by *ma* (吗), the most common interrogative particle, used to transform a statement into a question without changing word order; *ba* (吧), used at the end of an utterance to indicate a "How about *x*?" suggestion; and *ne* (呢), used at the end of a sentence to indicate a rhetorical question, a suggestion, and certain other contextually specific meanings. As with the example of "Mao Zedong" earlier, moreover, the differences between the UNESCO and UN machines are equally revealing. Even as these typists shared certain taxonomic instincts, these instincts manifested in significantly different tray bed configurations. Song Mei Lee-Wong, "Coherence, Focus and Structure: The Role of Discourse Particle *ne*," *Pragmatics* 11, no. 2 (2001): 139–153.

63. Geoffrey Bowker and Susan Leigh Star, *Sorting Things Out: Classification and Its Consequences* (Cambridge, MA: MIT Press, 1999). Note that a heat map is a data visualization wherein data values are represented as color values within a matrix.

64. Geoffrey C. Bowker, *Memory Practices in the Sciences* (Cambridge, MA: MIT Press, 2005).

65. *Chinese Typewriter Tray Bed Comprehensive Character Arrangement Reference Table* (*Zhongwen daziji zipan zi zonghe pailie cankaobiao*) [中文打字机字盘字综合排列参考表], appendix of Zhu Shirong [朱世荣], ed., *Manual for Chinese Typists* (*Zhongwen dazi-yuan shouce*) [中文打字员手册] (Chongqing: Chongqing chubanshe, 1988).

66. Natural-language tray beds became so popular at the time, in fact, that the method traveled all the way to the offices of Cambridge University. During my visit in July 2013, Charles Aylmer, head of the Chinese Department at the Cambridge University Library, kindly showed me the natural-language Chinese typewriter tray bed he himself had designed decades earlier.

CONCLUSION: TOWARD A HISTORY OF CHINESE COMPUTING AND THE AGE OF INPUT

1. Brian Rotman, *Becoming Beside Ourselves: The Alphabet, Ghosts, and Distributed Human Being* (Durham: Duke University Press, 2008).

2. Eric Singer, "Sonic Banana: A Novel Bend-Sensor-Based MIDI Controller," *Proceedings of the 2003 Conference on New Interfaces for Musical Expression,* Montreal, Canada, 2003.

3. It is worth noting that, as these keystrokes are sent back and forth across fiber optic cables, in theory they become susceptible to the very kinds of surveillance we have become ever more cognizant of in the wake of recent revelations—none as stirring and unsettling perhaps as the Edward Snowden leak. Despite the public's increased wariness of surveillance over private communications, however, we have collectively overlooked the ways in which the shifting ground of technology has now made possible new and potentially more invasive forms of surveillance: the ability to spy on a person as he uses Microsoft Word or TextEdit, perhaps as easily as when he sends SMS messages or emails. See Thomas S. Mullaney, "How to Spy on 600 Million People: The Hidden Vulnerabilities in Chinese Information Technology," *Foreign Affairs* (June 5, 2016), https://www.foreignaffairs.com/articles/china/2016-06-05/how-spy-600-million-people

4. David Bonavia, "Coming to Grips with a Chinese Typewriter," *Times* (London) (May 8, 1973), 8.

5. Philip Howard, "When Chinese Is a String of Two-Letter Words," *Times* (London) (January 16, 1978), 12.

6. The description reads: "A special exhibition celebrating the eccentric and bizarre in the museum's collection. A clockwork bird-scarer, a box of dust, a Chinese typewriter and a battered old suitcase are just some of the objects with stories to tell." "Eccentricity: Unexpected Objects and Irregular Behavior," special exhibition, Museum of the History of Science, Oxford, England, 2011.

7. Chinese typewriter formerly employed by Wang Shou-ling in Uppsala, Double Pigeon style, manufactured c. 1970, housed at the Tekniska Museet, Stockholm, cataloged as "Skrivmaskin/Chinese Typewriter," inventory no. TM44032 Klass 1417.

8. The original reads: "Hieroglyfer är bildtecken, picktogram ... Dagens trafikskyltar är en sorts piktogram. Även kinesisk skrift är piktografisk och har tiotusentals tecken ... Till skillnad från den piktografiska skriften har de alfabetiska ganska få tecken, 'bokstäver.' Det gor att det är mycket lättare att utveckla tryckteknik om man använder alfabetisk skrift." Visit by author, summer 2010.

9. *The Simpsons,* season 13, episode 1304, "A Hunka Hunka Burns in Love," December 2, 2001.

10. Exchange between Nora (Knoxville, Tennessee) and Cecil on straightdope.com, 1995.

BIBLIOGRAPHY OF SOURCES

MACHINES

Adji Saka Pallava Typewriter. Manufactured circa 1911 (Germany/Meteor Co.). Peter Mitterhofer Schreibmaschinenmuseum/Museo delle Macchine da Scrivere, Partschins (Parcines), Italy.

Blickensderfer Oriental Hebrew Typewriter. Manufactured circa 1900 (United States). Peter Mitterhofer Schreibmaschinenmuseum/Museo delle Macchine da Scrivere, Partschins (Parcines), Italy.

Caligraph Typewriter. Manufactured 1892 (United States). Peter Mitterhofer Schreibmaschinenmuseum/Museo delle Macchine da Scrivere, Partschins (Parcines), Italy.

Chinese Typewriter Formerly Employed at Chinese Church in London. Superwriter style, manufactured in 1971 by the Chinese Typewriter Company Limited Stock Corporation (*Zhongguo daziji zhizaochang*) [中 國 打 字 機 製 造 倉], a subsidiary of Japanese Business Machines Limited (*Nippon keieiki kabushikigaisha*) [日 本 經 營 機 株 式 會 社]. Author's personal collection.

Chinese Typewriter Formerly Employed at the First Chinese Baptist Church in San Francisco, California. Double Pigeon style, manufactured in 1971. Author's personal collection.

Chinese Typewriter Formerly Employed at the United Nations (Geneva). Double Pigeon style, manufactured in 1972 by the Shanghai Calculator and Typewriter Factory (*Shanghai jisuanji daziji chang*), housed at the Musée de la Machine à Écrire, Lausanne, Switzerland.

Chinese Typewriter Formerly Employed at UNESCO (Paris). Double Pigeon style, manufactured in 1971 by the Shanghai Calculator and Typewriter Factory (*Shanghai jisuanji daziji chang*), housed at the Musée de la Machine à Écrire, Lausanne, Switzerland.

Chinese Typewriter Formerly Employed by Translator Philippe Kantor at l'École des Mines and Other Institutions in Paris. Double Pigeon style, manufactured in 1972 by the Shanghai Calculator and Typewriter Factory (*Shanghai jisuanji daziji chang*), housed at the Musée des Arts et Métiers, Paris, cataloged as "Machine à écrire asiatique," inventory no. 43582-0000, location code ZAB35TRE E02.

Chinese Typewriter Formerly Employed by Wang Shou-ling in Uppsala. Double Pigeon style, manufactured circa 1970, housed at the Tekniska Museet, Stockholm, cataloged as "Skrivmaskin/Chinese Typewriter," inventory no. TM44032 Klass 1417.

Chinese Typewriter Formerly Owned by Georges Charpak. Double Pigeon style, manufactured in 1992 by the Shanghai Calculator and Typewriter Factory (*Shanghai jisuanji daziji chang*), housed at the Musée des Arts et Métiers, Paris, cataloged as "Machine à écrire à caractères chinois," inventory no. 44566-0001, location code ZAB35TRE E02.

Hall Typewriter. Manufactured circa 1880 (United States). Peter Mitterhofer Schreibmaschinenmuseum/Museo delle Macchine da Scrivere, Partschins (Parcines), Italy.

Heady Index Typewriter. Manufactured 1924. Peter Mitterhofer Schreibmaschinenmuseum/Museo delle Macchine da Scrivere, Partschins (Parcines), Italy.

Ideal Polyglott German-Russian Typewriter. Manufactured circa 1904 (Germany). Peter Mitterhofer Schreibmaschinenmuseum/Museo delle Macchine da Scrivere, Partschins (Parcines), Italy.

Improved Shu Zhendong-Style Chinese Typewriter (*Gailiang Shu shi Huawen daziji*) [改良舒式華文打字機]. Manufactured circa 1935, housed at the Huntington Library, San Marino, California.

Japanese Typewriter Formerly Owned by H.S. Watanabe. Manufactured in the 1930s by the Nippon Typewriter Company (Tokyo). Author's personal collection.

Malling Hansen Typewriter. Manufactured circa 1867 (Denmark?/ Hans Rasmus Johan Malling-Hansen). Peter Mitterhofer Schreibmaschinenmuseum/Museo delle Macchine da Scrivere, Partschins (Parcines), Italy.

Mignon Schreibmaschine Model 2. Manufactured 1905 (Germany). Peter Mitterhofer Schreibmaschinenmuseum/Museo delle Macchine da Scrivere, Partschins (Parcines), Italy.

Olivetti M1. Manufactured circa 1911 (Italy). Peter Mitterhofer Schreibmaschinenmuseum/Museo delle Macchine da Scrivere, Partschins (Parcines), Italy.

Orga Privat Greek Typewriter. Manufactured circa 1923 (Germany). Peter Mitter-hofer Schreibmaschinenmuseum/Museo delle Macchine da Scrivere, Partschins (Parcines), Italy.

Remington 2. Manufactured circa 1878 (United States). Peter Mitterhofer Schreib-maschinenmuseum/Museo delle Macchine da Scrivere, Partschins (Parcines), Italy.

Remington 10 Hebrew Typewriter. Manufactured circa 1910 (United States). Peter Mitterhofer Schreibmaschinenmuseum/Museo delle Macchine da Scrivere, Par-tschins (Parcines), Italy.

Remington Siamese Typewriter. Manufactured circa 1925 (United States). Peter Mit-terhofer Schreibmaschinenmuseum/Museo delle Macchine da Scrivere, Partschins (Parcines), Italy.

Simplex Typewriter. Manufactured circa 1901 (United States). Peter Mitterhofer Schreibmaschinenmuseum/Museo delle Macchine da Scrivere, Partschins (Parcines), Italy.

Tastaturen-Verzeichnis für die Mignon-Schreibmaschine. AEG-Deutsche Werke, Berlin, Germany. Peter Mitterhofer Schreibmaschinenmuseum/Museo delle Mac-chine da Scrivere, Partschins (Parcines), Italy.

Toshiba Japanese Typewriter. Manufactured circa 1935. Peter Mitterhofer Schreib-maschinenmuseum/Museo delle Macchine da Scrivere, Partschins (Parcines), Italy.

Underwood 5 Russian Typewriter. Manufactured circa 1900 (United States). Peter Mitterhofer Schreibmaschinenmuseum/Museo delle Macchine da Scrivere, Par-tschins (Parcines), Italy.

Yost 2. Manufactured circa 1882 (United States/George Washington Newton Yost). Peter Mitterhofer Schreibmaschinenmuseum/Museo delle Macchine da Scrivere, Partschins (Parcines), Italy.

Yost 10. Manufactured circa 1887 (United States). Musée de la Machine à Écrire, Lausanne, Switzerland.

CHINESE-LANGUAGE SOURCES

"Agriculture and Commercial Bureau Grants Five-Year Patent Rights to Typewriter Invented by Qi Xuan (*Nongshang bu zi Qi Xuan suo faming daziji pin zhunyu zhuanli wunian*) [農商部咨祁暄所發明打字機品准予專利五年]." *Zhonghua quanguo shanghui lianhehui huibao* [中華全國商會聯合報會報] 3, no. 1 (1916): 8.

Baoji County Committee of the Chinese Communist Party (*Zhongguo gongchandang Baoji xian weiyuanhui*) [中国共产党宝鸡县委员会]. "Survey Report on Living Conditions and Political Attitudes in Baoji County (*Zhonggong Baoji xianwei guanyu nongmin sixiang qingkuang de diaocha baogao*) [中共宝鸡县县委关于农民思想情况的调查报告]." August 17, 1957.

"Beijing Baoshan and Guangde Chinese Typing Supplementary Schools (*Beijing shi sili Baoshan, Guangde Huawen dazi buxi xuexiao*) [北京市私立寶善、廣德華文打字補習學校]." BMA J004-002-00579 (July 1, 1938).

"Beijing Jiyang Typing School Temporarily Ceases Operation (*Beijing shi sili Jiyang Huawen dazi buxi xuexiao zanxing tingban*) [北京市私立暨陽華文打字補習學校暫行停辦]." BMA J002-003-00636 (January 1, 1939).

Biaozhun zhengkai mingmi dianma xinbian [標準正楷明密電碼新編]. Shanghai: Da bei dianbao gongsi—Da dong dianbao gongsi, 1937. Rigsarkivet [Danish National Archives]. Copenhagen, Denmark. 10619 GN Store Nord A/S. 1870–1969 Kode- og telegrafbøger. Kodebøger 1924–1969.

Biography of Yu Binqi. *Shanghai Notables* (1943).

"Business Class Adds Course in Chinese Typing (*Shangke tianshe Zhongwen daziji kecheng*) [商科添設中文打字機課程]." *Zhejiang shengli Hangzhou gaoji zhongxue xiaokan* [浙江省立杭州高级中学校刊] 119 (1935): 841.

"Business News (*Shangchang xiaoxi*) [商場消息]." *Shenbao* [申報] (October 27, 1926), 19–20.

Chen Guangyao [陳光垚]. *Essays on the Simplification of Chinese Characters (Jianzi lunji)* [簡字論集]. Shanghai: Commercial Press, 1931.

Chen Wangdao [陈望道]. *Collected Articles of Chen Wangdao on Language and Writing (Chen Wangdao yuwen lunji)* [陈望道语文论集]. Shanghai: Shanghai jiaoyu chubanshe, 1980.

Chen Youxun [陳有勳]. "Report on a Comparison of Four-Corner Look-Up and Radical Look-Up (*Sijiao haoma jianzifa yu bushou jianzifa de bijiao shiyan baogao*) [四角號碼檢字法與部首檢字法的比較實驗報告]." *Educational Weekly (Jiaoyu zhoukan)* [教育周刊] 177 (1933): 14–20.

Cheng Yangzhi [程養之], ed. *Chinese Typing Manual for the Wanneng-Style Typewriter (Wanneng shi daziji shiyong Zhongwen dazi shouce)* [万能式打字机适用中文打字手册]. Shanghai: Commercial Press, 1956. Appendix includes *Wanneng-Style Chinese Typewriter Basic Character Table (Wanneng shi Zhongwen daziji jiben zipan biao)* [万能式中文打字机基本字盘表].

"China Standard Typewriter Mfg. Co." SMA U1-4-3582 (August 7, 1943), 11–13.

"Chinese Inventors Association Preparatory Conference Held Yesterday (*Zhongguo famingren xiehui zuori kai choubeihui*) [中國發明人協會昨日開籌備會]." *Shenbao* [申報] (February 1, 1937), 21.

"A Chinese Typewriter (*Huawen daziqi*) [華文打字器]." *Half Weekly (Ban xinqi bao)* [半星期報] 18 (July 3, 1908): 39.

"Chinese Typewriter: A New Invention (*Zhongguo dazi jiqi zhi xin faming*) [中國打字機器之新發明]." *Tongwen bao* [通間報] 656 (1915): 8.

Chinese Typewriter Character Arrangement Table (*Huawen daziji wenzi pailie biao*) [華 文 打 字 機 文 字 排 列 表]. Character table included with *Teaching Materials for the Chinese Typewriter* (*Huawen dazi jiangyi*) [華 文 打 字 講 義], n.p., n.d. (produced pre-1928, circa 1917).

"Chinese Typewriter Continues to Be Exempted from Export Tariff (*Huawen daziji jixu mian chukou shui*) [華 文 打 字 機 繼 續 免 出 口 稅]." *Gongshang banyuekan* [工 商 半 月 刊] 2, no. 17 (1930): 12.

"Chinese Typewriter: Lin Yutang's Improvement Successful, in Time Will Match English (*Zhongwen daziji: Lin Yutang gailiang chenggong, xu shi yu Yingwen xiangdeng*) [中 文 打 字 機: 林 語 堂 改 良 成 功, 需 時 與 英 文 相 等]." *Dazhong wenhua* 3 (1946): 21.

"Chinese Typewriter Promoted in South China Seas (*Huawen daziji tuixiao nanyang*) [華 文 打 字 機 推 銷 南 洋]." *Shenbao* [申 報] (May 2, 1924), 21.

"The Chinese Typewriter: Resisting Japan and Supporting China (*Kang ri sheng Zhong zhi Zhongwen daziji* [抗 日 聲 中 之 中 文 打 字 機]." *Shenbao* [申 報] (January 26, 1932), 12.

Chinese Typewriter Tray Bed Comprehensive Character Arrangement Reference Table (*Zhongwen daziji zipan zi zonghe pailie cankaobiao*) [中 文 打 字 机 字 盘 字 综 合 排 列 参 考 表]. Appendix of Zhu Shirong [朱 世 榮], ed. *Manual for Chinese Typists* (*Zhongwen daziyuan shouce*) [中 文 打 字 員 手 册. Chongqing: Chongqing chubanshe, 1988.

Chow, H.K. [Zhou Houkun]. "Specifications for Asphalt Cement." *Gongcheng* (*Zhongguo gongcheng xuehui huikan*) 5, no. 4 (1930): 517–521.

Commercial Press [商 务 印 书 馆], ed. *Ninety-Five Years of the Commercial Press, 1897–1992/The Commercial Press and Me* (*Shangwu yinshuguan jiushiwu nian, 1897– 1992/Wo he Shangwu yinshuguan*) [商 务 印 书 馆 九 十 五 年/我 和 商 务 印 书 馆]. Beijing: Commercial Press, 1992.

"Commercial Press Establishes Chinese Typewriting Class (*Shangwu yinshuguan she Huawen daziji lianxi ke*) [商 務 印 書 館 設 華 文 打 字 機 練 習 課]." *Education and Vocation* (*Jiaoyu yu zhiye*) [教 育 與 職 業] 10 (1918): 8.

Communication between Tianjin Municipal Government Department of Education (*Tianjin shi zhengfu jiaoyuju*) [天 津 市 政 府 教 育 局] and Junde Chinese Typing Institute (*Junde Huawen dazi zhiye buxi xuexiao*) [峻 德 華 文 打 字 職 業 補 習 學 校]. TMA J110-1-808 (March 5, 1948), 1–12.

Communication between Tianjin Municipal Government Department of Education (*Tianjin shi zhengfu jiaoyuju*) [天 津 市 政 府 教 育 局] and the International Typing Institute (*Guoji dazi chuanxisuo*) [國 際 打 字 傳 習 所]. TMA J110-1-838 (July 6, 1946), 1–15.

"Consulate Purchases Chinese Typewriter (Lingshiguan zhi Hanwen daziji) [領事館置漢文打字機]." Chinese Times (Tai Hon Kong Bo) [大漢公報] (May 18, 1925), 3. [Published in Vancouver, Canada.]

"Curriculum Vitae of Teaching Staff and Student Roster of the Private Yucai Chinese Typing Vocational Supplementary School in Beiping (Beiping shi sili Yucai Huawen dazike zhiye buxi xuexiao zhijiaoyuan lüli biao, xuesheng mingji biao) [北平市私立育才華文打字科職業補習學校職教員履歷表、學生名籍表]." BMA J004-002-00662 (July 31, 1939).

Dan Jinrong [單錦蓉]. "Suggested Improvements to the Method by which Character Tray Beds are Organized on Chinese Typewriters (Jianyi gaijin Zhongwen daziji zipan pailie fangfa) [建議改進中文打字機字盤排列方法]." People's Daily (Renmin ribao) [人民日報] (July 23, 1952).

Deng Zhixiu [邓智秀]. "The Achievements of Beijing Model Worker Shen Yunfen (Beijing laomo Shen Yunfen de shiji) [北京劳模沈蕴芬的事迹]." Beijing: Dianli laomowang [电力劳模网] (March 6, 2006), http://www.sjlmw.com/html/beijing/20060306/2193.html (accessed January 23, 2010).

"Diagram Explaining Production of Chinese Typewriter (Chuangzhi Zhongwen daziji tushuo) [創制中文打字機圖說]." Guohuo yuebao [國貨月報] 2 (1915): 1–12.

"Diagram Explaining Production of Chinese Typewriter (Chuangzhi Zhongguo daziji tushuo) [創制中國打字機圖說]." Zhonghua gongchengshi xuehui huikan [中華工程師學會會刊] 2, no. 10 (1915): 15–29.

"Discounted Prices Will No Longer Be Available for Typewriters (Daziji jianjia jijiang zaizhi) [打字機廉價即將截止]." Shenbao [申報] (April 24, 1927), 15.

"Draft Proposal Regarding Students Taking Exam to Study Abroad in the United States (Kaoshi liu Mei xuesheng cao'an) [考試留美學生草案]." Shenbao [申報] (August 8, 1910), 5–6.

Du Dingyou [杜定友]. "On the Psychology of How the Masses Search for Characters (Minzhong jianzi xinli lunlüe) [民眾檢字心理論略]." In Selected Library Science Writings of Du Dingyou (Du Dingyou tushuguanxue lunwen xuanji) [杜定友图书馆学论文选集], edited by Qian Yaxin [钱亚新] and Bai Guoying [白国应]. Beijing: Shumu wenxian chubanshe, 1988, 340–350. Originally in Education and the Masses (Jiaoyu yu minzhong) [教育與民眾] 6, no. 9 (1925).

Editorial Board of Zhonghua Books (Zhonghua shuju bianjibu) [中华书局编辑部], eds. Zhonghua Books and Me (Wo yu Zhonghua shuju) [我与中华书局]. Beijing: Zhonghua shuju, 2002.

"Electric Chinese Typewriter a Success (Dianqi Zhongwen daziji chenggong) [電氣中文打字機成功]." Shoudu dianguang yuekan [首度電光月刊] 74 (April 1, 1937), 10.

"Exhibition of New Chinese Inventions (Guoren xin faming shiwu zhanlanhui zhi jingguo) [國人新發明事物展覽會之經過]." Zhejiang minzhong jiaoyu [浙江民眾教育] 4, no. 1 (1936): 15–16.

"An Explanation of the Chinese Typewriter and the Twentieth-Century War Effort (*Zhongguo daziji zhi shuoming yu ershi shiji zhi zhanzheng liqi*) [中國打字機之說明與二十世紀之戰爭力氣]." *The World's Chinese Students' Journal* (*Huanqiu*) [環球] 1, no. 3 (September 15, 1916): 1–2.

"Fact and Fiction: A Chinese Typewriter (*Shishi feifei: Hanwen daziji*) [是是非非:漢文打字機]." *Nanhua wenyi* [南華文藝] 1, no. 7/8 (1932): 103.

Fan Jiling [范繼舲]. *Fan's Wanneng-Style Chinese Typewriter Practice Textbook* (*Fan shi wanneng shi Zhongwen daziji shixi fanben*) [范氏萬能式中文打字機實習範本]. Hankou: Fan Research Institute Publishing House (*Fan shi yanjiusuo yinhang*) [范氏研究所印行], 1949.

"A Few Methods for the Creation of Indexes for Chinese Books (*Bianzhi Zhongwen shuji mulu de ji ge fangfa*) [編織中文書記目錄的幾個方法]." *Eastern Miscellany* (*Dongfang zazhi*) [東方雜誌] 20, no. 23 (1923): 86–103.

Fong, M.Q. *Xinbian yi dianma* [新編易電碼]. N.p., 1912.

Gan Chunquan [甘纯权] and Xu Yizhi [徐怡芝], eds. *Essential Knowledge for Secretaries: Requirements of a Chinese Typist* (*Shuji fuwu bibei: yiming Huawen dazi wenshu yaojue*) [書記服務必備一名華文打字文書要決]. Shanghai: Shanghai zhiye zhidaosuo [上海職業指導所], 1935.

Ge Yifu [戈一夫]. "Setting More than 5,000 Characters per Hour (*Mei xiaoshi jianzi wuqian duo*) [每小時揀字五千多]." *People's Daily* (*Renmin ribao*) [人民日报] (June 27, 1958), 4.

He Cun [禾村]. "How Did the Party Branch of the Northern China Military Region Printing Office Oversee the Spread of Advanced Typesetting Methods? (*Huabei junqu zhengzhibu yinshuachang dangzhibu shi zenyang lingdao tuiguang xianjin paizifa de*) [華北軍區政治部印刷廠黨支部是怎樣領導推廣先進排字法的]." *People's Daily* (*Renmin ribao*) [人民日报] (July 18, 1952), 3.

He Gonggan [何公敢]. "*Danti jianzifa* [單體檢字法]." *Eastern Miscellany* (*Dongfang zazhi*) [東方雜誌] 25, no. 4 (1928): 59–72.

He Gongxing [何躬行]. "Chinese Type Founders (*Woguo zhi zhuzi gongren*) [我國之鑄字工人]." *Shizhao yuebao* [時兆月報] 29, no. 4 (1934): 16–17.

He Jiceng [何繼曾]. *Elements of Type Setting* (*Paizi qianshuo*) [排字淺說]. Shanghai: Commercial Press, 1959.

Hong Bingyuan [洪秉淵]. "Research on Appropriate Typesetting Methods (*Heli de paizi fangfa zhi yanjiu*) [合理的排字方法之研究]." *Zhejiang sheng jianshe yuekan* [浙江省建設月刊] 4, no. 8/9 (1931): 22–45.

"Hongye Company Agency: Yu's Chinese Typewriters Sell Well (*Hongye gongsi jingli Yu shi Zhongwen daziji changxiao*) [宏業公司經理俞氏中文打字機暢銷]." *Shenbao* [申報] (December 8, 1934), 14.

Hu Shi [胡 適]. "*Cang huishi zhaji (Xu qianhao)* [藏 晖 室 札 记 (续 前 号)]." *New Youth (Xin qingnian)* [新 青 年] 3, no. 1 (March 1917): 1–5.

Hu Shi [胡 適]. "The Chinese Typewriter—A Record of My Visit to Boston (*Zhongwen daziji—Boshidun ji*) [中 文 打 字 機 – 波 士 頓 記]." In Hu Shi. *Hu Shi xueshu wenji— yuyan wenzi yanjiu* [胡 适 学 术 文 集 – 语 言 文 字 研 究]. Beijing: Zhonghua shuju, 1993.

Hu Yaobang [胡 耀 邦]. "Determine to Be Active Builders and Protectors of Socialism (*Lizhi zuo shehui zhuyi de jiji jianshezhe he baowei zhe*) [立 志 作 社 会 主 义 的 积 极 建 设 者 和 保 卫 者]." *People's Daily (Renmin ribao)* [人 民 日 报] (May 4, 1954), 2.

Hua Sheng [華 生]. "Touring the National Domestic Products Exhibition Held by the Shanghai Chamber of Commerce (*Shishanghui zhuban de guohuo zhanlanhui xunli*) [市 商 會 主 辦 的 國 貨 展 覽 會 巡 禮]." *Shenbao* [申 報] (October 5, 1936), 27785.

Improved Shu Zhendong-Style Chinese Typewriter Manual (Gailiang Shu shi Huawen daziji shuomingshu) [改 良 舒 式 華 文 打 字 機 說 明 書]. Shanghai: Commercial Press, 1938. University of Pennsylvania Archives—W. Norman Brown Papers (UPT 50 B879), box 10, folder 5.

"Introduction to the Achievements of Female Typists in the Shanghai Electric Power Planning Institute of the Water Conservancy and Electric Power Ministry (*Shuili dianli bu Shanghai dianli sheji yuan dazi nügong xiaozu shiji jieshao*) [水 利 電 力 部 上 海 電 力 設 計 院 打 字 女 工 小 組 事 跡 介 紹]." SMA A55-2-326, n.d.

"Introduction to the 'New Typing Method' ('*Xin dazi caozuo fa' jieshao*) ['新 打 字 操 作 法'介 紹]." *People's Daily (Renmin ribao)* [人 民 日 报] (November 30, 1953), 3.

"Inventing an Electric Chinese Typewriter (*Faming dianli Zhongwen daziji*) [發 明 電 力 中 文 打 字 機]." *Capital Electro-Optical Monthly (Shoudu dianguang yuekan)* [首 都 電 光 月 刊] 61 (1936): 9.

"Invention of a Chinese Typewriter (*Zhongwen daziji de faming*) [中 文 打 字 機 的 發 明]." *Xiao pengyou* [小 朋 友] 859 (1947): 22.

"Investigation: The Bitter Story of a Typesetter (*Diaocha: Yige paizi gongren de kuhua*) [調 查 一 個 排 字 工 人 的 苦 話]." *Shanghai Companion (Shanghai huoyou)* [上 海 伙 友] 3 (October 10, 1920): 2–3.

Jiang Yiqian [蔣 一 前]. *History of the Development of Character Retrieval Systems in China and a Table of Seventy-Seven New Character Retrieval Systems (Zhongguo jianzifa yange shilüe ji qishiqi zhong xin jianzifa biao)* [中 國 檢 字 法 沿 革 史 略 及 七 十 七 種 新 檢 字 法 表]. N.p.: Zhongguo suoyin she [中 國 索 引 社 會], 1933.

"Jiangsu Department of Industry Hires Zhou Houkun as Advisor (*Su shiyeting pin Zhou Houkun wei guwen*) [蘇 實 業 廳 聘 周 厚 坤 為 顧 問]." *Shenbao* [申 報] (October 21, 1923), 15.

Journalism Research Institute of the People's University of China (*Zhongguo renmin daxue xinwenxue yanjiusuo*) [中 国 人 民 大 学 新 闻 学 研 究 所], ed. *Typesetting for*

Newspapers (Baozhi de paizi he pinpan) [报 纸 的 排 字 和 拼 盘]. Shanghai: Commercial Press, 1958.

"Joyous News of High Production *(Gaochan xibao)* [高 產 喜 報]." *People's Daily (Renmin ribao)* [人 民 日 报] (August 11, 1958), 3.

"Kaifeng Typesetter Zhang Jiying Diligently Improves Typesetting Method, Establishes New Record of 3,000-plus Characters per Hour *(Kaifeng paizi gongren Zhang Jiying nuli gaijin paizifa chuang mei xiaoshi san qian yu zi xin jilu)* [開 封 排 字 工 人 張 繼 英 努 力 改 進 排 字 法 創 每 小 時 三 千 余 字 新 紀 錄]." *People's Daily (Renmin ribao)* [人 民 日 报] (December 16, 1951). [Reprinted from article in *Henan Daily (Henan Ribao)*].

"A Large-Scale Public-Private Typewriter Store Newly Opened in Shanghai *(Shanghai xin kai yi jia guimo juda de gongsi heying daziji dian)* [上 海 新 開 一 家 規 模 巨 大 的 公 私 合 營 打 字 機 店]." *Xinhuashe xinwen gao* [新 華 社 新 聞 稿] no. 2295 (1956).

Li Gui [李 圭]. *New Records of My Travels Around the World (Huanyou diqiu xinlu)* [環 游 地 球 新 錄]. Translated by Charles Desnoyers as *A Journey to the East: Li Gui's A New Account of a Trip Around the Globe*. Ann Arbor: University of Michigan Press, 2004.

Li Xianyan [李 獻 延], ed. *Newest Stationery Forms (Zuixin gongwen chengshi)* [最 新 公 文 程 式]. Xinjing: Fengtian Typing Institute [奉 天 打 字 專 門 學 校], circa 1932.

Li Zhongyuan [李 中 原] and Liu Zhaolan [劉 兆 蘭]. "Kaifeng Typesetter Zhang Jiying's Advanced Work Method *(Kaifeng paizi gongren Zhang Jiying de xianjin gongzuofa)* [開 封 排 字 工 人 張 繼 英 的 先 進 工 作 法]." *People's Daily (Renmin ribao)* [人 民 日 报] (March 10, 1952), 2.

"Libraries and the Foundation for the Promotion of Education and Culture *(Jiaoyu wenhua jijinhui yu tushuguan shiye)* [教 育 文 化 基 金 會 與 圖 書 館 實 業]." *Journal of the Chinese Library Association (Zhonghua tushuguan xiehui huibao)* [中 華 圖 書 館 協 會 會 報] 7 (August 1931): 10.

Lin Taiyi [林 太 乙]. *The Biography of Lin Yutang (Lin Yutang zhuan)* [林 語 堂 傳]. Taibei: Lianjing chubanshe, 1989.

Lin Yutang [林 語 堂]. "Chinese Typewriter *(Zhongwen daziji)* [中 文 打 字 機]." Translated by Lin Taiyi. *Xifeng* [西 風] 85 (1946): 36–39.

Lin Yutang [林 語 堂]. "Explanation of Chinese Character Index *(Hanzi suoyin zhi shuoming)* [漢 字 索 引 制 說 明]." *New Youth (Xin qingnian)* [新 青 年] 2 (1918): 128–135.

Lin Yutang [林 語 堂] and Chen Houshi [陳 厚 士]. "Invention of a Chinese Typewriter *(Zhongwen daziji zhi faming)* [中 文 打 字 機 之 發 明]." *The World and China (Shijie yu Zhongguo)* [世 界 與 中 國] 1, no. 6 (1946): 32–34.

"Lin Yutang Invents a New Chinese Typewriter *(Lin Yutang faming xin Zhongwen daziji)* [林 語 堂 發 明 新 中 文 打 字 機]." *Wenli xuebao* [文 理 学 报] 1, no. 1 (1946): 102.

"Lin Yutang Invents a New Chinese Typewriter (*Lin Yutang xin faming Zhongwen daziji*) [林語堂新發明中文打字機]." *Shenbao* [申報] (August 23, 1947), 3.

"Lin Yutang Invents Chinese Typewriter (*Lin Yutang faming Zhongwen daziji*) [林語堂發明中文打字機]." *Shenbao* [申報] (January 10, 1947), 8.

"Lin Yutang Invents Chinese Typewriter—Gao Zhongqin Invents Electric Typewriter (*Lin Yutang faming Zhongwen daziji—Gao Zhongqin faming diandong daziji*) [林語堂發明中文打字機—高中芹發明電動打字機]." *Shenbao* [申報] (September 19, 1947), 6.

"Lin Yutang's New Character Retrieval System (*Lin Yutang shi de jianzi xinfa*) [林語堂氏的檢字新法]." *Beixin* [北新] 8 (1926): 3–9.

Lu-Ho Rural Service Bureau (*Luhe xiangcun fuwubu*) [潞河鄉村服務部], ed. *2,000 Foundational Characters for Daily Use (Richang yingyong jichu erqian zi)* [日常應用基礎二千字]. N.p., 1938 (November). Rare Book and Manuscript Library, Columbia University. Papers of the International Institute of Rural Reconstruction. MS COLL/ IIRR.

MacGowan, D. J. *Philosophical Almanac (Bowu tongshu)* [博物通書]. Ningbo: Zhen shan tang, 1851.

"Manual for Improved Shu-Style Chinese Typewriter (*Gailiang Shu shi Huawen daziji shuomingshu*) [改良舒式華文打字機說明書]." Shanghai: Commercial Press, 1938. University of Pennsylvania Archives—W. Norman Brown Papers (UPT 50 B879), box 10, folder 5.

Memo from Zhu Hongjuan [朱鴻雋], principal of the Huanqiu Typing Academy of Shanghai to a Mr. Lin [林], Shanghai Special Municipality education bureau chief. SMA R48-1-287 (October 27, 1944), 1–11.

"Methods Regarding the Character Retrieval Problem (*Duiyu jianzifa wenti de banfa*) [對於檢字法問題的辦法]." *Eastern Miscellany (Dongfang zazhi)* [東方雜誌] 20, no. 23 (1923): 97–100.

Ministry of Communications (*Jiaotong bu*) [交通部]. Ministry of Communications Regulations Regarding National Phonetic Alphabet Telegraphy (*Jiaotongbu guiding guoyin dianbao fashi*) [交通部規定國音電報法式]. N.p., 1928.

Ministry of Communications (*Jiaotong bu*) [交通部]. *Plaintext and Secret Telegraph Code—New Edition (Mingmi dianma xinbian)* [明密電碼新編]. Nanjing: Jinghua yinshu guan [南京印書館], 1933. Rigsarkivet [Danish National Archives]. Copenhagen, Denmark. 10619 GN Store Nord A/S. 1870–1969 Kode- og telegrafbøger. Kodebøger 1924–1969.

Ministry of Communications (*Jiaotong bu*) [交通部]. *Plaintext and Secret Telegraph Code—New Edition (Mingmi dianma xinbian)* [明密電碼新編]. N.p.: Jiaotongbu kanxing [交通部刊行], 1946.

Ministry of Communications (*Jiaotong bu*) [交通部]. *Plaintext and Secret Telegraph Code—New Edition* (*Mingmi dianma xinbian*) [明密碼電報新編]. Shanghai: n.p., 1916.

"MIT China (*Meiguo Masheng ligong daxue Zhongguo*) [美國麻省理工大學中國]." *Shenbao* [申報] (July 15, 1914), 6.

"MIT China (*Meiguo Masheng ligong xuexiao Zhongguo*) [美國麻省理工學校中國]." *Shenbao* [申報] (July 19, 1915), 6.

"Model Workers from Across China Attend International Workers' Day Ceremony (*Ge di lai Jing canjia 'wuyi' jie guanli de laodong mofan*) [各地來京參加五一節觀禮的勞動模範]." *People's Daily* (*Renmin ribao*) [人民日報] (May 7, 1952), 3.

"Mr. Hui's Chinese-English Typing Institute (*Hui shi Hua Ying wen dazi zhuanxiao*) [惠氏華英打字專校]." SMA Q235-1-1847 (1932), 26–49.

"Name List of Students Selected to Travel to America (*Qu ding you Mei xuesheng mingdan*) [取定遊美學生名單]." *Shenbao* [申報] (August 9, 1910), 5.

National Association for the Advancement of Mass Education (*Zhonghua pingmin jiaoyu cujinhui*) [中華平民教育促進會], ed. *Farmer's One Thousand Character Textbook* (*Nongmin qianzi ke*) [農民千字課]. N.p., 1933. Rare Book and Manuscript Library, Columbia University. Papers of the International Institute of Rural Reconstruction. MS COLL/IIRR.

Newest Chinese Typing Glossary (*Zui xin Zhongwen dazi zihui*) [最新中文打字字彙]. Beijing and Tianjin: Nanjing daziji hang [南京打字機行], n.d.

"New Invention (*Xin faming*) [新發明]." *Guofang yuekan* [國防月刊] 2, no. 1 (1947): 2.

"New Invention: A Chinese Typewriter (*Zhongguo daziji zhi xin faming*) [中國打字機之新發明]." *Shenbao* [申報] (August 16, 1915), 10.

"New Invention of a Chinese Typing Machine (*Zhongguo dazi jiqi zhi xin faming*) [中國打字機器新發明]." *Tongwen bao* [通問報] 656 (1915): 8.

"News about China Product Trade Company (*Zhongguo wuchan maoyi gongsi xiaoxi*) [中國物產貿易公司消息]." *Shenbao* [申報] (August 29, 1931), 20.

"A New Typing Record (*Dazi xin jilu*) [打字新記錄]." *People's Daily* (*Renmin ribao*) [人民日報] (November 23, 1956), 2.

"Patent for Yu-Style Chinese typewriter (*Yu shi Zhongwen daziji zhuanli*) [俞氏中文打字機專利]." *Shenbao* [申報] (August 1, 1934), 16.

Peng Yuwen [彭玉雯]. *Shisan jing jizi moben* [十三經集字摹本]. N.p., 1849. Copy housed at the National Library of Australia.

People's Welfare Typewriter Manufacturing Company (*Minsheng daziji zhizaochang*) [民生打字機製造廠], ed. *Practice Textbook* (*Lianxi keben*) [練習課本]. N.p.: circa 1940s.

"Petition Submitted by the Private Yadong Japanese-Chinese Typing Supplementary School of Beijing to the Beijing Special Municipality Bureau of Education Regarding Inspection of Student Grading List, Curriculum Plan, and Teaching Hours for the Regular-Speed Division of the Sixteenth Term, and the Order Received in Response from the Bureau of Education (*Beijing shi sili Yadong Ri Huawen dazi buxi xuexiao guanyu di'shiliu qi putong sucheng ge zu xuesheng chengjibiao, kecheng yuji ji shouke shi shu qing jianhe gei Beijing tebieshi jiaoyuju de cheng yiji jiaoyuju de zhiling*) [北京市私立亞東日華文打字補習學校關於第十六期普通速成各組學生成績表、課程預計及授課時數請鑒核給北京特別市教育局的呈以及教育局的指令]." BMA J004-002-01022 (January 31, 1943).

Ping Zhou [屏周]. "A Record of Viewing the Chinese Typewriter Manufactured by Commercial Press (*Canguan shangwu yinshuguan zhizao Huawen daziji ji*) [參觀商務印書館製造華文打字機記]." *Shangye zazhi* [商業雜誌] 2, no. 12 (1927): 1–4.

"A Positive Review of the Yu-Style Chinese Typewriter (*Yu shi Zhongwen daziji zhi haoping*) [俞式中文打字機之好評]." *China Industry* (*Zhongguo shiye*) [中國實業] 1, no. 6 (1935): 1158.

"Preparations for a Chinese Typewriting Class (*Chouban Huawen daziji xunlianban*) [籌辦華文打字機訓練班]." *Henan jiaoyu* (*Henan Education*) [河南教育] 1, no. 6 (1928): 4.

The Principle and Practice of Five-Stroke Indexing System for Chinese Characters (*Wubi jianzifa de yuanli ji yingyong*) [五筆檢字法的原理及應用]. Shanghai: Zhonghua shuju, 1928.

Private Shucheng Typing Vocational Supplementary School of Beijing Student Roster (*Beijing shi sili shucheng dazike zhiye buxi xuexiao xuesheng mingji biao*) [北京市私立樹成打字科職業補習學校學生名籍表]. BMA J004-002-01091 (March 23, 1942).

Private Yadong Japanese-Chinese Typing Supplementary School of Beijing Student Roster (*Beijing shi sili Yadong Ri Hua wen dazi buxi xuexiao xuesheng mingji biao*) [北京市私立亞東日華文打字補習學校學生名籍表]. BMA J004-002-01022 (November 7, 1942).

Private Yanjing Chinese Typing Supplementary School of Beijing Student Roster (*Beijing sili Yanjing Huawen dazi buxi xuexiao xuesheng mingji biao*) [北京私立燕京華文打字補習學校學生名籍表]. BMA J4-1-805 (November 1, 1946).

Qian Xuantong [錢玄同]. "China's Script Problem from Now On (*Zhongguo jinhou de wenzi wenti*) [中國今後的文字問題]." *New Youth* (*Xin qingnian*) [新青年] 4, no. 4 (1918).

Qian Xuantong [錢 玄 同]. "Why Must We Advocate a 'Romanized National Language'? (*Weishenme yao tichang 'guoyu luoma zi'?*) [為 什 麼 要 提 倡 國 語 羅 馬 字?]." *Xinsheng* 1, no. 2 (December 24, 1926). In *The Writings of Qian Xuantong* (*Qian Xuantong wenji*) [钱 玄 同 文 集]. Vol. 3. Beijing: Zhongguo renmin daxue chubanshe, 1999: 385–391.

"Recent News on the Engineering Society at Nanyang University (*Nanyang daxue gongchenghui jinxun*) [南 洋 大 學 工 程 會 近 訊]." *Shenbao* [申 報] (November 10, 1922), 14.

"Regarding the Foundation of the Guangdewen Typing Supplementary School in Beiping (*Guanyu chuangban Beiping shi sili Guangdewen dazi buxi xueshe de chengwen ji gai she jianzhang deng yiji shehui ju de piwen*) [關 於 創 辦 北 平 市 私 立 廣 德 文 打 字 補 習 學 社 的 呈 文 及 該 社 簡 章 等 以 及 社 會 局 的 批 文]." BMA J002-003-00754 (May 1, 1938).

"Regarding Type-and-Copy Shops (*Guanyu tengxie dazishe*) [关 于 誊 写 打 字 社]." SMA Q235-3-503, n.d.

"Regulating the Pricing Problem of Domestically Produced Chinese Typewriters (*Wei tiaozheng guochan Zhongwen daziji shoujia wenti*) [為 調 整 國 產 中 文 打 字 機 售 價 問 題]." Report sent from the Chinese Typewriter Manufacturers Business Association (*Zhongwen daziji zhizao chang shanglian*) to the Shanghai Cultural and Educational Supplies Trade Association (*Shanghai shi wenjiao yongpin tongye gonghui*). SMA B99-4-124 (January 15, 1953), 52–90.

"Regulations of the Commercial Typing and Shorthand Training Institute (*Shangye dazi suji chuanxisuo jianzhang*) [商 業 打 字 速 記 傳 習 所 簡 章]." SMA Q235-1-1844 (June 1932), 49–56.

"Research on Typesetting Methods in China (*Zhongguo paizi fa zhi yanjiu*) [中 國 排 字 法 之 研 究]." *Dongbei wenhua yuebao* [東 北 文 化 月 報 "The Light of Northeast"] 6, no. 2 (1927): 40–49.

"Resist Japan and Save the Nation Movement (*Kang Ri jiuguo yundong*) [抗 日 救 國 運 動]." *Shenbao* [申 報] (November 8, 1931), 13–14.

"Resist Japan Association Standing Committee 17th Meeting Memo (*Kang Ri hui changwuhuiyi ji di'shiqi ci*) [抗 日 會 常 務 會 議 記 第 十 七 次]." *Shenbao* [申 報] (November 12, 1931), 13.

"Responses to Criticisms and Suggestions Submitted by Readers of the *People's Daily* (*Dui Renmin ribao duzhe piping jianyi de fanying*) [對 人 民 日 報 讀 者 批 評 建 議 的 反 應]." *People's Daily* (*Renmin ribao*) [人 民 日 报] (August 27, 1952), 6.

"Second Tianjin Japanese Typewriter Competition (*Di'er hui Tianjin Bangwen liancheng jingji dahui*) [第 二 回 天 津 邦 文 鍊 成 競 技 大 會]." TMA J128-3-9615 (October 9, 1943).

Shanghai Calculator and Typewriter Factory (*Shanghai jisuanji daziji chang*) [上海計算機打字機廠], ed. *Chinese Typewriter Tray Bed Character Table (Zhongwen daziji zipan zibiao)* [中文打字機字盤字表]. [Accompanies UN Geneva Chinese Typewriter, housed at the Musée de la Machine à Écrire, Lausanne, Switzerland]

Shanghai Calculator and Typewriter Factory (*Shanghai jisuanji daziji chang*) [上海計算機打字機廠], ed. "Double Pigeon Brand Chinese Typewriter Improvement and Trial Production Summary Report (*Shuangge pai Zhongwen daziji gaijin shizhi jishu zongjie*) [双鸽牌中文打字机改进试制技术总结]." SMA B155-2-282 (March 22, 1964), 11–14.

Shanghai Calculator and Typewriter Factory (*Shanghai jisuanji daziji chang*) [上海計算機打字機廠], ed. "Double Pigeon Brand Chinese Typewriter Internal Evaluation Report (*Shuangge pai Zhongwen daziji chang nei jianding baogao*) [双鸽牌中文打字机厂内鉴定报告]." SMA B155-2-282 (March 22, 1964), 9–10.

Shanghai Calculator and Typewriter Factory (*Shanghai jisuanji daziji chang*) [上海計算機打字機廠], ed. "Evaluation Report on the Double Pigeon Brand DHY Model Chinese Typewriter (*Shuangge pai DHY xing Zhongwen daziji jianding baogao*) [双鸽牌DHY型中文打字机鉴定报告]." SMA B155-2-284 (April 24, 1964), 4–8.

"Shanghai Chinese Typewriter Manufacturing Joint Venture Marketing Plan (*Shanghai Zhongwen daziji zhizaochang lianyingsuo chanxiao jihua*) [上海中文打字機製造廠聯營所產銷計劃]." SMA S289-4-37 (December 1951), 65.

Shanghai Private Property Revaluation Jury (*Shanghai shi siying qiye caichan chonggu pingshen weiyuanhui*) [上海市私營企業財產重估評審委員會]. "Revaluation of Property Report: Fusheng Chinese Typewriter Factory (*Chonggu caichan baobiao: Fusheng Huawen daziji zhizaochang*) [重估財產報表: 富勝華文打字機製造廠]." SMA S1-4-436 (December 31, 1950), 22–33.

Sheffield, Devello Z. [謝衛樓]. "Diba zhang taizi duanding liangmin de baoying [第八章太子斷定良民的報應]." *Xiao hai yuebao* [小孩月報] 5, no. 2 (1879): 2–3.

Sheffield, Devello Z. [謝衛樓]. "Di'er zhang minshou youhuo weibei huangdi [第二章民受誘惑違背皇帝]." *Xiao hai yuebao* [小孩月報] 4, no. 3 (1878): 5.

Sheffield, Devello Z. [謝衛樓]. *Important Doctrines of Theology (Shendao yaolun)* [神道要論]. Tongzhou: Tongzhou wenkui qikan yin [通州文魁齊刊印], 1894.

Sheffield, Devello Z. [謝衛樓]. "Shangfa yuyan diliu zhang liangmin quanren fangzhan huigai [賞罰喻言第六章良民勸人放瞻悔改]." *Xiao hai yuebao* [小孩月報] 4, no. 10 (1879): 5.

Sheffield, Devello Z. [謝衛樓]. "Shangfa yuyan disan zhang minshou youhuo fanzui geng shen [賞罰喻言第三章島民受誘惑犯罪更深]." *Xiao hai yuebao* [小孩月報] 4, no. 4 (1878): 6–7.

Sheffield, Devello Z. [謝衛樓]. "Shangfa yuyan diyi zhang daomin shou bawang xiahai [賞罰喻言第一章島民受霸王轄害]." *Xiao hai yuebao* [小孩月報] 4, no. 2 (1878): 3.

Shen Yunfen [沈蘊芬]. "I Love the Work Allocated Me by the Party with All My Heart (*Wo re'ai dang fenpei gei wo de gongzuo*) [我熱愛黨分配給我的工作]." *People's Daily* (*Renmin ribao*) [人民日报] (November 30, 1953), 3.

Shen Yuzhong [沈禹鐘]. "Compositors (*Paizi ren*) [排字人]." *Hong zazhi* [紅雜誌] 2, no. 16 (1923): 1–11.

Shu Changyu [舒昌鈺] (aka Shu Zhendong [舒震東]). "Chinese Typewriter (*Zhongguo daziji*) [中國打字機]." *Tongji* [同濟] 2 (1918): 73–82.

Shu Changyu [舒昌鈺] (aka Shu Zhendong [舒震東]). "Thoughts While Researching a Typewriter for China (*Yanjiu Zhongguo daziji shi zhi ganxiang*) [研究中國打字機時之感想]." *Tongji* [同濟] 2 (1918): 153–156.

Social Service Group (*Shehui fuwuzu*) [社會服務組]. "Those with Typing Experience Can Register at the Social Service Office (*Ji you dazi jishu ke dao shehui fuwuchu dengji*) [既有打字技術可到社會服務處登記]." *People's Daily* (*Renmin ribao*) [人民日报] (March 30, 1949), 4.

Song Mingde [宋明得]. "Chinese Typewriter (*Huawen daziji*) [華文打字機]." *Tongzhou* [同舟] 3, no. 1 (1934): 11–12.

"Summer Vacation at the China Railway School (*Zhonghua tielu xuexiao shujia*) [中華鐵路學校暑假]." *Shenbao* [申報] (July 5, 1916), 10.

Sun Ligan [孫礼干]. *Repair Methods for the Wanneng-Style Chinese Typewriter (Wanneng shi Zhongwen daziji xiulifa*) [萬能式中文打字機修理法]. Shanghai: Shanghai kexue jishu chubanshe [上海科學技術出版社], 1954.

"Swimming Expert Yu Binqi (*Youyong zhuanjia Yu Binqi nanshi*) [游泳專家俞斌祺男士]." *Nan pengyou* [男朋友] 1, no. 10 (1932): reverse cover.

Teaching Materials for the Chinese Typewriter (Huawen dazi jiangyi) [華文打字講義]. N.p., n.d. (produced pre-1928, circa 1917).

T'een-piao-hsin-shu (Nouveau Code du télégraphie Chinoise) (Dianbao xinshu) [電報新書]. April 1872. Housed in personal collection of Henning Høeg Hansen, Copenhagen, Denmark.

Tianjin Chinese Typewriter Company (*Tianjin Zhonghua daziji gongsi*) [天津中華打字機公司], ed. *Chinese Typewriter Training Textbook*. Vol. 1 (*Zhonghua daziji shixi keben—shangce*) [中華打字機實習課本－上冊]. Tianjin: Donghua qiyin shuaju [東華齊印刷局], 1943.

Tianjin Chinese Typewriter Company (*Tianjin Zhonghua daziji gongsi*) [天津中華打字機公司], ed. *Chinese Typewriter Training Textbook*. Vol. 2 (*Zhonghua daziji shixi* keben—xiace) [中華打字機實習課本－下冊]. Tianjin: Donghua qiyin shuaju [東華齊印刷局], 1943.

Tianjin Municipality Eighth Educational District Mass Education Office Eighth Term Chinese Typing Accelerated Class Graduation Name List (*Tianjin shili di'ba shejiaoqu minzhong jiaoyuguan di'ba qi Huawen dazi suchengban biye xuesheng mingce*) [天 津 市 立 第 八 社 教 區 民 眾 教 育 館 第 八 期 華 文 打 字 速 成 班 畢 業 學 生 名 冊]. TMA J110-3-740 (November 25, 1948), 1–2.

Tianjin People's Government Local-State Jointly Run Hongxing Factory (*Tianjin shi renmin zhengfu difang guoying huyeju Hongxing huchang*) [天 津 市 人 民 政 府 地 方 國 營 互 業 局 紅 星 互 廠]. "Report on the Improvement of the Chinese Typewriter Character Chart (*Huawen daziji zibiao gaijin baogao*) [華 文 打 字 機 字 表 改 進 報 告]." TMA J104-2-1639 (October 1953), 29–39.

"Two New Chinese Typewriters Invented (*Zhongwen daziji liang qi xin faming*) [中 文 打 字 機 兩 起 新 發 明]." *Kexue yuekan* [科 學 月 刊] 15 (1947): 23–24.

"Typing Class Begins (*Dazi ban kaike*) [打 字 班 開 課]." *Jinlingguang* [金 陵 光] 6, no. 2 (April 1914): 33.

Universal Word List for the Average Citizen (Ping guomin tongyong cibiao) [平 國 民 通 用 詞 表]. Rare Book and Manuscript Library, Columbia University. Papers of the International Institute of Rural Reconstruction. MS COLL/IIRR, n.d.

User's Manual for Chinese Typewriter Manufactured by Shanghai Commercial Press (Shanghai yinshuguan zhizao Huawen daziji shuomingshu) [上 海 印 書 館 製 造 華 文 打 字 機 說 明 書]. Shanghai: Commercial Press, 1917 (October).

"Victory and Success Typing Specialist School (*Jiecheng dazi zhuanxiao*) [捷 成 打 字 專 校]." SMA Q235-1-1848 (1933), 50–70.

Viguier, Septime Auguste. *Dianbao shuji* [電 報 書 籍]. Shanghai: n.p., 1871. Housed in personal collection of Henning Høeg Hansen, Copenhagen, Denmark.

Viguier, Septime Auguste [威 基 謁] (*Weijiye*). *Dianbao xinshu* [電 報 新 書]. In "Extension Selskabet—Kinesisk Telegrafordbog." 1871. Arkiv nr. 10.619. In "Love og vedtægter med anordninger." GN Store Nord A/S SN China and Japan Extension Telegraf. Rigsarkivet [Danish National Archives], Copenhagen, Denmark.

Viguier, Septime Auguste. *Dianxin huizi* [電 信 彙 字]. Shanghai: Dianji xinju [電 機 信 局], 1870. Rigsarkivet [Danish National Archives], Copenhagen, Denmark. 10619 GN Store Nord A/S. 1870–1969 Kode- og telegrafbøger.

Viguier, Septime Auguste. *T'een-piao-shu-tsieh* (Code de télégraphie Chinoise). Shanghai: n.p., 1871. Housed in personal collection of Henning Høeg Hansen, Copenhagen, Denmark.

Wan Guoding [萬 國 鼎]. "Chinese Character Fundamental Stroke Organization System (*Hanzi mubi pailiefa*) [漢 字 母 笔 排 列 法]." *Eastern Miscellany (Dongfang zazhi)* [東 方 雜 誌] 2 (1926): 75–90.

Wan Zhang [绾 章]. "Invention of a New Chinese Character Typewriter (*Hanwen daziji zhi xin faming*) [漢 文 打 字 機 之 新 發 明]." *Progress: A Journal of Modern Civilization (Jinbu)* [進 步] 6, no. 1 (1914): 1–10.

Wang Guihua [王 桂 華] and Lin Gensheng [林 根 生], eds. *Chinese Typing Technology (Zhongwen dazi jishu)* [中 文 打 字 技 术]. Nanjing: Jiangsu renmin chubanshe, 1960.

Wang Yi [王 怡]. "The Cultural Value of Shorthand (*Suji zai wenhua shang de jiazhi*) [速 記 在 文 化 上 的 價 值]." *Dushu jikan* [讀 書 季 刊] 2, no. 1 (1936): 87–92.

Wang Yi [王 怡]. *New Style Chinese Shorthand Method (Zhongguo xinshi sujishu)* [中 國 新 式 速 記 術]. Shanghai: New Style Shorthand Institute [新 式 速 記 傳 習 所], 1919 [1906].

Wanneng-Style Chinese Typewriter Basic Character Table (Wanneng shi Zhongwen daziji jiben zipan biao) [万 能 式 中 文 打 字 机 基 本 字 盘 表]. Included as appendix in Cheng Yangzhi [程 養 之], ed., *Chinese Typing Manual for the Wanneng-Style Typewriter (Wanneng shi daziji shiyong Zhongwen dazi shouce)* [万 能 式 打 字 机 适 用 中 文 打 字 手 册]. Shanghai: Commercial Press, 1956.

Wu Qizhong [吳 启 中] and Ceng Zhaolun [曾 昭 抡]. "A Countryman's New Invention (*Guoren xin faming*) [國 人 新 發 明]." *Shishi yuebao* [時 事 月 報 告] 11, no. 3 (1934): 18.

Wu Yue [吳 躍] and Guan Honglin [管 洪 林], eds. *Chinese Typing (Zhongwen dazi)* [中 文 打 字]. Beijing: Gaodeng jiaoyu chubanshe, 1989. [Appendix includes "Zhongwen daziji gexin zipanbiao [中 文 打 字 机 革 新 字 盘 表]."]

"Wuchang Inspection Office Examines Hanyeping Remittances (*Wuchang jianting zi cha Hanyeping jiekuan*) [武 昌 檢 廳 咨 查 漢 冶 萍 解 欵]." *Shenbao* [申 報] (August 26, 1922), 15.

Xu Bing [徐 冰]. "From 'Book from the Sky' to 'Book from the Ground' (*Cong Tianshu dao Dishu*) [从 天 书 到 地 书]." Manuscript provided by Xu Bing to author via email, May 15, 2013.

Xue Zhendong [薛 振 东], ed. *Nanhui County Gazetteer (Nanhui xianzhi)* [南 汇 县 志]. Shanghai: Shanghai shi Nanhui xian xianzhi bianzuan weiyuanhui, 1992.

Yang Mingshi [楊 名 時]. "Principles of Lin Yutang's Chinese Typewriter (*Lin Yutang shi Huawen daziji de yuanli*) [林 語 堂 氏 華 文 打 字 機 的 原 理]." *Guowen guoji* [國 文 國 際] 1, no. 3 (1948): 3.

"Yu Binqi and Others Will Establish China Inventors Association (*Yu Binqi deng zuzhi Zhongguo famingren xiehui*) [俞 斌 祺 等 組 織 中 國 發 明 人 協 會]." *Book Prospects (Tushu zhanwang)* [圖 書 展 望] 1, no. 8 (April 28, 1936): 83.

Yu Binqi Chinese Typewriter Character Table (Yu Binqi Zhongwen daziji zibiao) [俞 斌 祺 中 文 打 字 機 字 表]. Shanghai: n.p., circa 1930s (post-1928). Author's personal collection.

"Yu Binqi Defends Himself to the Resist Japan Association (*Yu Binqi xiang kang Ri hui shenban*) [俞斌祺向抗日會伸辨]." *Shenbao* [申報] (November 9, 1931), 11.

"Yu Binqi Invents Steel Type (*Yu Binqi faming gangzhi zhuzi*) [俞斌祺發明鋼質鑄字]." *China Industry* (*Zhongguo shiye*) [中國實業] 1, no. 5 (1935): 939.

Yu Jun [英君]. "China's Typewriter: The Creation of Zhou Houkun (*Zhongguo zhi daziji: Zhou Houkun chuangzao*) [中國之打字機:周厚坤創造]." *Qingsheng zhoukan* [青聲週刊] 4 (1917): 2–3.

Yu Shuolin [俞碩霖]. "The Birth of the Yu-Style Typewriter (*Yu shi daziji de yansheng*) [俞式打字机的诞生]." *Old Kids Blog* (*Lao xiaohai shequ*) [老小孩社区] (June 3, 2010), http://www.oldkids.cn/blog/blog_con.php?blogid=124277 (accessed June 13, 2011).

Yu Shuolin [俞碩霖]. "The Last Yu-Style Typewriter (*Zui hou de Yu shi daziji*) [最后的俞式打字机]." *Old Kids Blog* (*Lao xiaohai shequ*) [老小孩社区] (June 9, 2010), http://www.oldkids.cn/blog/blog_con.php?blogid=130259 (accessed June 13, 2011).

Yu Shuolin [俞碩霖]. "Two Types of Chinese Typewriter (*Liang zhong Zhongwen daizji*) [两种中文打字机]." *Old Kids Blog* (*Lao xiaohai shequ*) [老小孩社区] (February 8, 2010), http://www.oldkids.cn/blog/blog_con.php?blogid=116181 (accessed May 12, 2013).

Yu Shuolin [俞碩霖]. "Yu-Style Chinese Typewriter Patent (*Yu shi Zhongwen daziji de zhuanli*) [俞式中文打字机的专利]." *Old Kids Blog* (*Lao xiaohai shequ*) [老小孩社区] (June 3, 2010), http://www.oldkids.cn/blog/blog_con.php?blogid=104202 (accessed June 13, 2011).

Yu Shuolin [俞碩霖]. "Yu-Style Chinese Typewriter Unlimited Company (*Yu shi daziji wuxian gongsi*) [俞式打字机无限公司]." *Old Kids Blog* (*Lao xiaohai shequ*) [老小孩社区] (June 7, 2010), http://www.oldkids.cn/blog/blog_con.php?blogid=130576 (accessed June 13, 2011).

Yu Shuolin [俞碩霖]. "Yu-Style Typewriter Factory (*Yu shi daziji zhizaochang*) [俞式打字机制造厂]." *Old Kids Blog* (*Lao xiaohai shequ*) [老小孩社区] (June 6, 2010), http://www.oldkids.cn/blog/blog_con.php?blogid=130431 (accessed June 13, 2011).

"Yu's Chinese Typewriter: Profits Donated to Flooded Districts (*Yu shi Zhongwen daziji ticheng chong shuizai yizhen*) [俞氏中文打字機提成充水災義賑]." *Shenbao* [申報] (December 22, 1935), 12.

Yu-Style Typing Glossary (*Yu shi dazi zihu*) [俞式打字字彙]. Shanghai: *Yu shi Zhongwen daziji faxingsuo* [俞式中文打字機發行所], 1951.

Zhang Feng [張風]. "Stroke-Order Character Retrieval Method (*Bishun jianzifa*) [筆順檢字法]." *Yiban* [一般] 1, no. 4 (1927).

Zhang Jiying [張繼英]. "How Did I Raise My Work Efficiency? (*Wo de gongzuo xiaolü shi zenme tigao de*) [我的工作效率是怎麼提高的]." In Zhongnan People's Press (*Zhongnan renmin chubanshe*) [中南人民出版社], ed. *The Zhang Jiying Typesetting*

Method (Zhang Jiying jianzifa) [張 繼 英 揀 字 法]. Hankou: Zhongnan renmin chuban-she, 1952, 19–22.

Zhang Jiying [張 繼 英]. "I Want to Teach Everyone My Typesetting Method (*Wo yao ba wo de jianzifa jiao gei dajia*) [我 要 把 我 的 揀 字 法 教 給 大 家]." *People's Daily (Renmin ribao)* [人 民 日 报] (June 3, 1952), 2.

Zhang Jiying [張 繼 英]. "Preparing to Issue a Friendly Challenge to Typesetters Across the Country (*Zhunbei xiang quanguo jianzi gongren tichu youyi tiaozhan*) [準 備 向 全 國 揀 字 工 人 提 出 友 誼 挑 戰]." *People's Daily (Renmin ribao)* [人 民 日 报] (June 9, 1952), 2.

Zhang Xiumin [張 秀 民]. *The History of Chinese Printing (Zhongguo yinshuashi)* [中 國 印 刷 史]. Paramas: Homa & Sekey Books, 1989.

Zhang Yuanji [張 元 濟]. *The Complete Works of Zhang Yuanji (Zhang Yuanji quanji)* [張 元 濟 全 集]. Vol. 6. Shanghai: Commercial Press, 2009.

Zhang Yuanji [張 元 濟]. *The Complete Works of Zhang Yuanji (Zhang Yuanji quanji)* [張 元 濟 全 集]. Vol. 7. Shanghai: Commercial Press, 2009.

"Zhang Bangyong: National Phonetic Alphabet Bangyong Shorthand (*Zhang Bangyong xiansheng guoyin bangyong shuji shu*) [張 邦 永 先 生 國 音 邦 永 速 記 術]." *New World (Xin shijie)* [新 世 界] 50 (1934): 57.

Zhongguo dianbao xinbian [中 國 電 報 新 編] 1881. Rigsarkivet [Danish National Archives], Copenhagen, Denmark. 10619 GN Store Nord A/S. 1870–1969 Kode- og telegrafbøger. Kodebøger 1924–1969.

Zhongnan People's Press (*Zhongnan renmin chubanshe*) [中 南 人 民 出 版 社], ed. *The Zhang Jiying Typesetting Method (Zhang Jiying jianzifa)* [張 繼 英 揀 字 法]. Hankou: Zhongnan renmin chubanshe, 1952.

Zhou Daowu [周 道 悟]. "Whatever Work Aims to Complete and Not to Fail the Five Year Plan, All That Work Is Glorious! (*Renhe laodong, dou shi wancheng wunian jihua buke queshaode laodong, dou shi guangrongde laodong!*) [任 何 勞 動,都 是 完 成 五 年 計 畫 不 可 缺 少 的,都 是 光 榮 的 勞 動!]." 1956. PC-1956-013, personal collection, chineseposters.net.

Zhou Houkun [周 厚 坤]. "Diagram Explaining Production of Chinese Typewriter (*Chuangzhi Zhongguo daziji tushuo*) [創 制 中 國 打 字 機 圖 說]." *Eastern Miscellany (Dongfang zazhi)* [東 方 雜 誌] 12, no. 10 (October 1915): 28–31.

Zhou Houkun [周 厚 坤]. "Patent Document for Common Usage Typewriter Tray (*Tongsu dazipan shangqueshu*) [通 俗 打 字 盤 商 榷 書]." *Educational Review (Jiaoyu zazhi)* [字 教 育 雜 誌] 9, no. 3 (March 1917): 12–14.

Zhou Houkun [周 厚 坤] and Chen Tingrui [陳 霆 銳]. "A Newly Invented Typewriter for China (*Xin faming Zhongguo zhi daziji*) [新 發 明 中 國 之 打 字 機]." *Zhonghua xuesheng jie* [中 華 學 生 界] 1, no. 9 (September 25, 1915): 1–11.

Zhou Yukun [周 玉 昆], ed. *Chinese Typing Method (Huawen dazi fa)* [華 文 打 字 法]. Nanjing: Bati yinshuasuo [拔 提 印 刷 所], 1934.

"Zhou and Wang Are Quintessential Scholars (*Zhou Wang liang jun juexue*) [周 王 兩 君 絕 學]." *Shenbao* [申 報] (July 24, 1916), 10.

Zhu Shirong [朱 世 荣], ed. *Manual for Chinese Typists (Zhongwen daziyuan shouce)* [中 文 打 字 员 手 册]. Chongqing: Chongqing chubanshe, 1988. Appendix includes *Chinese Typewriter Tray Bed Comprehensive Character Arrangement Reference Table (Zhongwen daziji zipan zi zonghe pailie cankaobiao)* [中 文 打 字 机 字 盘 字 综 合 排 列 参 考 表].

ENGLISH-LANGUAGE SOURCES

"Accuracy: The First Requirement of a Typewriter." *Dun's Review* 5 (1905): 119.

Adal, Raja. "The Flower of the Office: The Social Life of the Japanese Typewriter in its First Decade." Presentation at the Association for Asian Studies Annual Meeting, March 31–April 3, 2011.

Adas, Michael. *Machines as the Measure of Men: Science, Technology, and Ideologies of Western Dominance*. Ithaca: Cornell University Press, 1989.

"Additional Japanese Typewriters and the Engagement of Typists." Memo from Shanghai Municipal Council Secretary to the Co-ordinating Committee. SMA U1-4-3796 (February 15, 1943), 36.

Adler, Michael H. *The Writing Machine: A History of the Typewriter*. London: Allen and Unwin, 1973.

"An Agreement entered into the 10th day of August, 1887 between the Imperial Chinese Telegraph Company and Great Northern Telegraph Company of Copenhagen and the Eastern Extension, Australasia and China Telegraph Company, Limited." Cable and Wireless Archive DOC/EEACTC/1/304 E.Ex.A&C.T. Co. Ltd Agreements with China and Great Northern Telegraph Co. etc. (August 10, 1887), 185–195.

Ahvenainen, Jorma. *The European Cable Companies in South America before the First World War*. Helsinki: Finnish Academy of Sciences and Letters, 2004.

Allard, J. Frank. "Type-Writing Machine." United States Patent no. 1188875. Filed January 13, 1913; patented June 27, 1916.

Allard, J. Frank. "Typewriting Machine." United States Patent no. 1454613. Filed June 8, 1921; patented May 8, 1923.

Allen, Joseph R. "I Will Speak, Therefore, of a Graph: A Chinese Metalanguage." *Language in Society* 21, no. 2 (June 1992): 189–206.

"An American View of the Chinese Typewriter." *Shanghai Puck* 1, no. 1 (September 1, 1918): 28.

Anderson, Benedict. *Imagined Communities: Reflections on the Origin and Spread of Nationalism*. Rev. ed. New York: Verso, 1991.

Andreas, Joel. *Rise of the Red Engineers: The Cultural Revolution and the Origins of China's New Class*. Stanford: Stanford University Press, 2009.

"Annual Report of the Philadelphia Museums, Commercial Museum." Philadelphia: Commercial Museum, 1923.

Arbisser, Micah Efram. "Lin Yutang and his Chinese Typewriter." Princeton University Senior Thesis no. 13048 (2001).

Arnold, David. *Everyday Technology: Machines and the Making of India's Modernity*. Chicago: University of Chicago, 2013.

"At Last—A Chinese Typewriter—A Remington." *Remington Export Review*, n.d., 7. Hagley Museum and Library. Accession no. 1825. Remington Rand Corporation. Records of the Advertising and Sales Promotion Department. Series I Typewriter Div. Subseries B, Remington Typewriter Company, box 3, vol. 3. [No date appears on the copy housed in the Hagley Museum collection, although the drawing of a Chinese keyboard diagram included within the article is dated February 10, 1921.]

Baark, Erik. *Lightning Wires: The Telegraph and China's Technological Modernization, 1860–1890*. Westport, CT: Greenwood Press, 1997.

Bachrach, Susan. *Dames Employées: The Feminization of Postal Work in Nineteenth-Century France*. London: Routledge, 1984.

Bailey, Paul J. *Reform the People: Changing Attitudes towards Popular Education in Early Twentieth-Century China*. Edinburgh: Edinburgh University Press, 1990.

Barr, John H., and Arthur W. Smith. "Type-Writing Machine." United States Patent no. 1250416. Filed August 4, 1917; patented December 18, 1917.

Bayly, Christopher. *Empire and Information: Intelligence Gathering and Social Communication in India, 1780–1870*. Cambridge: Cambridge University Press, 1999 [1996].

Beeching, Wilfred A. *Century of the Typewriter*. New York: St. Martin's Press, 1974.

Behr, Wolfgang. "Early Medieval Philosophical Crabs." Presentation at the "Literary Forms of Argument in Pre-Modern China" Workshop, Queen's College, University of Oxford, September 16–18, 2009.

Bektas, Yakup. "Displaying the American Genius: The Electromagnetic Telegraph in the Wider World." *British Journal for the History of Science* 34, no. 2 (June 2001): 199–232.

Bektas, Yakup. "The Sultan's Messenger: Cultural Constructions of Ottoman Telegraphy, 1847–1880." *Technology and Culture* 41 (2000): 669–696.

Bellovin, Steve. "Compression, Correction, Confidentiality, and Comprehension: A Modern Look at Commercial Telegraph Codes." Paper presented at the Cryptologic History Symposium (Laurel, MD), 2009.

Benjamin, Walter. *The Work of Art in the Age of Mechanical Reproduction*. In *Illuminations*. Translated by Harry Zohn. New York: Schocken Books, 1968, 217–252.

Bijker, Wiebe E. "Do Not Despair: There Is Life after Constructivism." *Science, Technology and Human Values* 18, no. 1 (Winter 1993): 113–138.

Bijker, Wiebe E. *Of Bicycles, Bakelites, and Bulbs: Toward a Theory of Sociotechnical Change*. Cambridge, MA: MIT Press, 1997 [1995].

Bijker, Wiebe E., and John Law, eds. *Constructing Stable Technologies: Towards a Theory of Sociotechnical Change*. Cambridge, MA: MIT Press, 1992.

Bloom, Alfred H. "The Impact of Chinese Linguistic Structure on Cognitive Style." *Current Anthropology* 20, no. 3 (1979): 585–601.

Bloom, Alfred H. *The Linguistic Shaping of Thought: A Study in the Impact of Language on Thinking in China and the West*. Hillsdale, NJ: L. Erlbaum, 1981.

Bodde, Derk. *Chinese Thought, Society, and Science: The Intellectual and Social Background of Science and Technology in Pre-Modern China*. Honolulu: University of Hawai'i Press, 1991.

Boltz, William G. "Logic, Language, and Grammar in Early China." *Journal of the American Oriental Society* 120, no. 2 (April–June 2000): 218–229.

Bonavia, David. "Coming to Grips with a Chinese Typewriter." *Times* (London) (May 8, 1973), 8.

Borgman, Christine L. *From Gutenberg to the Global Information Infrastructure: Access to Information in the Networked World*. Cambridge, MA: MIT Press, 2003 [2000].

Bourdieu, Pierre. *The Logic of Practice*. Stanford: Stanford University Press, 1992.

Bowker, Geoffrey C. *Memory Practices in the Sciences*. Cambridge, MA: MIT Press, 2005.

Bowker, Geoffrey C., and Susan Leigh Star. *Sorting Things Out: Classification and Its Consequences*. Cambridge, MA: MIT Press, 1999.

Brokaw, Cynthia. "Book History in Premodern China: The State of the Discipline." *Book History* 10 (2007): 253–290.

Brokaw, Cynthia J. "Reading the Best-Sellers of the Nineteenth Century: Commercial Publishers in Sibao." In *Printing and Book Culture in Late Imperial China*, edited by Cynthia Brokaw and Kai-wing Chow. Berkeley: University of California Press, 2005.

Brokaw, Cynthia, and Kai-wing Chow, eds. *Printing and Book Culture in Late Imperial China*. Berkeley: University of California Press, 2005.

Brokaw, Cynthia, and Christopher Reed, eds. *From Woodblocks to the Internet: Chinese Publishing and Print Culture in Transition, Circa 1800 to 2008.* Boston and Leiden: Brill, 2010.

Brook, Timothy. *Collaboration: Japanese Agents and Local Elites in Wartime China.* Cambridge, MA: Harvard University Press, 2007.

Brown, Alexander T. "Type-Writing Machine." United States Patent no. 855832. Filed June 29, 1904/reapplied August 8, 1905; patented June 4, 1907.

Brown, Alexander T. "Type-Writing Machine." United States Patent no. 911198. Filed June 29, 1904; patented February 2, 1909.

Brown, William Norman. "Report on the Chinese Typewriter." May 16, 1948. University of Pennsylvania Archives—W. Norman Brown Papers (UPT 50 B879), box 10, folder 5.

Brumbaugh, Robert S. "Chinese Typewriter." United States Patent no. 2526633. Filed September 25, 1946; patented October 24, 1950.

Bryson, Bill. *Mother Tongue: The English Language.* New York: Penguin, 1999.

Bull, W. "A Short History of the Shanghai Station." Shanghai: n.p., 1893. [Handwritten Manuscript] Cable and Wireless Archive DOC/EEACTC/12/10.

Bunnag, Tej. *The Provincial Administration of Siam, 1892–1915: The Ministry of the Interior under Prince Damrong Rajanubhab.* Kuala Lumpur: Oxford University Press, 1977.

Burgess, Anthony. "Minding the Ps and Qs of our ABCs." *Observer* (April 7, 1991), 63.

Buschmann, Theodor Eugen. "Letter-Width-Spacing Mechanism in Typewriters." United States Patent no. 1472825. Filed March 23, 1921; patented November 6, 1923.

Canales, Jimena. *A Tenth of a Second. A History.* Chicago: University of Chicago Press, 2009.

Carter, John. "The New World Market." *New World Review* 21, no. 9 (October 1953): 38–43.

Carter, Thomas Francis. *The Invention of Printing in China and Its Spread Westward.* New York: Ronald Press Co., 1955.

Cartoon of Chinese Typewriter. *St. Louis Globe-Democrat* (January 11, 1901), 2–3.

Chan, Hok-lam. *Control of Publishing in China, Past and Present.* Canberra: Australian National University Press, 1983.

Chang, C.C. "Heun Chi Invents a Chinese Typewriter." *Chinese Students' Monthly* 10, no. 7 (April 1, 1915): 459.

Chang, Kang-i Sun, Haun Saussy, and Charles Yim-tze Kwong, eds. *Women Writers of Traditional China: An Anthology of Poetry and Criticism*. Stanford: Stanford University Press, 1999.

Characters Formed by the Divisible Type Belonging to the Chinese Mission of the Board of Foreign Missions of the Presbyterian Church in the United States of America. Macao: Presbyterian Press, 1844.

Chartier, Roger. *The Cultural Uses of Print in Early Modern France*. Translated by Lydia G. Cochrane. Princeton: Princeton University Press, 1987.

Chartier, Roger. *Forms and Meanings: Texts, Performances, and Audiences from Codex to Computer*. Philadelphia: University of Philadelphia Press, 1985.

Chartier, Roger. "Gutenberg Revisited from the East." Translated by Jill A. Friedman. *Late Imperial China* 17, no. 1 (1996): 1–9.

Chartier, Roger. "Texts, Printing, Readings." In *The New Cultural History*, edited by Lynn Hunt, 154–175. Berkeley: University of California Press, 1989.

Chen, Jianhua. "Canon Formation and Linguistic Turn: Literary Debates in Republican China, 1919–1949." In *Beyond the May Fourth Paradigm: In Search of Chinese Modernity*, edited by Kai-wing Chow, Tze-ki Hon, Hung-yok Ip, and Don C. Price. Lanham, MD: Lexington Books, 2008, 51–67.

Chen, Li. *Chinese Law in Imperial Eyes: Sovereignty, Justice, and Transcultural Politics*. New York: Columbia University Press, 2015.

Chen Lifu. *Storm Clouds Over China: The Memoirs of Ch'en Li-fu, 1900–1993*. Edited by Sidney Chang and Ramon Myers. Stanford: Hoover Institute Press, 1994.

Cheng, Linsun. *Banking in Modern China: Entrepreneurs, Professional Managers, and the Development of Chinese Banks, 1897–1937*. Cambridge: Cambridge University Press, 2007.

Chi, Heuen [Qi Xuan]. Chinese Exclusion Act File. National Archives and Records Administration, Washington, DC.

Chi, Heuen [Qi Xuan]. "Apparatus for Writing Chinese." United States Patent no. 1260753. Filed April 17, 1915; patented March 26, 1918.

Chia, Lucille. *Printing for Profit. The Commercial Publishers of Jianyang, Fujian*. Cambridge, MA: Harvard University Asia Center, 2003.

Chiang, Yee. *Chinese Calligraphy: An Introduction to Its Aesthetics and Techniques*. Cambridge, MA: Harvard University Press, 1973 [1938].

"Child of the Quarantine: One More Passenger on the Nippon Maru List—Baby Born During Angel Island Stay." *San Francisco Chronicle* (July 11, 1899), 12.

"China." *Atchison Daily Globe* (April 11, 1898), 1.

China as It Really Is. London: Eveleigh Nash, 1912.

"China, Commercial Press Exhibit." City of Philadelphia, Department of Records. Record Group 232 (Sesquicentennial Exhibition Records), 232-4-8.1 "Department of Foreign Participation," box A-1474, folder 8, series folder 29.

"Chinaman Invents Chinese Typewriter Using 4,000 Characters." *New York Times* (July 23, 1916), SM15.

"China Oct 1926." City of Philadelphia, Department of Records. Record Group 232 (Sesquicentennial Exhibition Records), 232-4-8.1 "Department of Foreign Participation," box A-1474, folder 7, series folder 28.

"Chinese Characters Sent by Telegraph Machine." *Los Angeles Times* (November 22, 1936), 5.

"Chinese Divisible Type." *Chinese Repository* 14 (March 1845): 124–129.

"Chinese Language and Dialects." *Missionary Herald* 31 (May 1835): 197–201.

"Chinese Phonetic on a Typewriter." *Popular Science* 97, no. 2 (August 1920): 116.

"Chinese Project: The Lin Yutang Chinese Typewriter." Mergenthaler Linotype Company Records, 1905–1993, Archives Center, National Museum of American History, Smithsonian Institution, box 3628. Multiple Dates in 1950 Listed.

"Chinese Put on Typewriter by Lin Yutang." *Los Angeles Times* (August 22, 1947), 2.

"Chinese Romanized—Keyboard no. 141." Hagley Museum and Library. Accession no. 1825. Remington Rand Corporation. Records of the Advertising and Sales Promotion Department. Series I Typewriter Div. Subseries B, Remington Typewriter Company, box 3, vol. 1.

Chinese Second Historical Archives (*Zhongguo di'er lishi dang'anguan*), ed. *Historical Materials on the Old Chinese Maritime Customs, 1859–1948*. Vol. 112 (1932) (*Zhongguo jiu haiguan shiliao*) [中 國 舊 海 關 史 料]. Beijing: Jinghua Press [京 華 出 版 社], 2001.

Chinese Second Historical Archives (*Zhongguo di'er lishi dang'anguan*), ed. *Historical Materials on the Old Chinese Maritime Customs, 1859–1948*. Vol. 114 (1933) (*Zhongguo jiu haiguan shiliao*) [中 國 舊 海 關 史 料]. Beijing: Jinghua Press [京 華 出 版 社], 2001.

Chinese Second Historical Archives (*Zhongguo di'er lishi dang'anguan*), ed. *Historical Materials on the Old Chinese Maritime Customs, 1859–1948*. Vol. 118 (1935) (*Zhongguo jiu haiguan shiliao*) [中 國 舊 海 關 史 料]. Beijing: Jinghua Press [京 華 出 版 社], 2001.

Chinese Second Historical Archives (*Zhongguo di'er lishi dang'anguan*), ed. *Historical Materials on the Old Chinese Maritime Customs, 1859–1948*. Vol. 122 (1936) (*Zhongguo jiu haiguan shiliao*) [中 國 舊 海 關 史 料]. Beijing: Jinghua Press [京 華 出 版 社], 2001.

Chinese Second Historical Archives (*Zhongguo di'er lishi dang'anguan*), ed. *Historical Materials on the Old Chinese Maritime Customs, 1859–1948*. Vol. 126 (1937) (*Zhongguo jiu haiguan shiliao*) [中 國 舊 海 關 史 料]. Beijing: Jinghua Press [京 華 出 版 社], 2001.

Chinese Second Historical Archives (*Zhongguo di'er lishi dang'anguan*), ed. *Historical Materials on the Old Chinese Maritime Customs, 1859–1948*. Vol. 130 (1938) (*Zhongguo jiu haiguan shiliao*) [中 國 舊 海 關 史 料]. Beijing: Jinghua Press [京 華 出 版 社], 2001.

Chinese Second Historical Archives (*Zhongguo di'er lishi dang'anguan*), ed. *Historical Materials on the Old Chinese Maritime Customs, 1859–1948*. Vol. 134 (1939) (*Zhongguo jiu haiguan shiliao*) [中 國 舊 海 關 史 料]. Beijing: Jinghua Press [京 華 出 版 社], 2001.

Chinese Second Historical Archives (*Zhongguo di'er lishi dang'anguan*), ed. *Historical Materials on the Old Chinese Maritime Customs, 1859–1948*. Vol. 138 (1940) (*Zhongguo jiu haiguan shiliao*) [中 國 舊 海 關 史 料]. Beijing: Jinghua Press [京 華 出 版 社], 2001.

Chinese Second Historical Archives (*Zhongguo di'er lishi dang'anguan*), ed. *Historical Materials on the Old Chinese Maritime Customs, 1859–1948*. Vol. 142 (1941) (*Zhongguo jiu haiguan shiliao*) [中 國 舊 海 關 史 料]. Beijing: Jinghua Press [京 華 出 版 社], 2001.

Chinese Second Historical Archives (*Zhongguo di'er lishi dang'anguan*), ed. *Historical Materials on the Old Chinese Maritime Customs, 1859–1948*. Vol. 144 (1942) (*Zhongguo jiu haiguan shiliao*) [中 國 舊 海 關 史 料]. Beijing: Jinghua Press [京 華 出 版 社], 2001.

"A Chinese Type-Writer." *Chinese Times* (March 1888), 143.

"A Chinese Typewriter." *Peking Gazette* (November 1, 1915), 3.

"A Chinese Typewriter." *San Francisco Examiner* (January 22, 1900).

"A Chinese Type-Writer." *Scientific American* (March 6, 1899), 359.

"A Chinese Typewriter." *Semi-Weekly Tribune* (June 22, 1897), 16.

"A Chinese Typewriter." *Shanghai Times* (November 19, 1915), 1.

The Chinese Typewriter. Written by Stephen J. Cannell. Directed by Lou Antonio. Starring Tom Selleck and James Whitmore, Jr. 78 mins. 1979. Universal City Studios.

"Chinese Typewriter Printing 4,000 Characters." *Chicago Daily Tribune* (June 7, 1899), 6.

"Chinese Typewriters." Memo from "The Secretary's Office, Municipal Council" to "The Director." SMA U1-4-3582 (July 13, 1943): 6–8.

"Chinese Typewriter, Shown to Engineers, Prints 5,400 Characters with Only 36 Keys." *New York Times* (July 1, 1946), 26.

Chow, Kai-wing. *Publishing, Culture, and Power in Early Modern China*. Stanford: Stanford University Press, 2004.

Chow, Rey. "How (the) Inscrutable Chinese Led to Globalized Theory." *PMLA* 116, no. 1 (2001): 69–74.

Chun, Wendy Hui Kyong. "Introduction: Race and/as Technology; or, How to Do Things to Race." *Camera Obscura* 24, no. 170 (2009): 7–35.

"Chu Yin Tzu-mu Keyboard—Keyboard no. 1400." (February 10, 1921.) Hagley Museum and Library. Accession no. 1825. Remington Rand Corporation. Records of the Advertising and Sales Promotion Department. Series I Typewriter Div. Subseries B, Remington Typewriter Company, box 3, vol. 3.

City of Philadelphia, Department of Records. Record Group 232 (Sesquicentennial Exhibition Records), 232-2.6 "Photographs." Photograph 2427.

Clark, Lauren, and Eric Feron. "Development of and Contribution to Aerospace Engineering at MIT." *40th AIAA Aerospace Sciences Meeting and Exhibit* (January 14–17, 2002), 2.

Clarke, Adele E., and Joan Fujimura, eds. *The Right Tools for the Job: At Work in Twentieth-Century Life Sciences.* Princeton: Princeton University Press, 1992.

Clarke, Adele E., and Joan Fujimura. "What Tools? Which Jobs? Why Right?" In *The Right Tools for the Job: At Work in Twentieth-Century Life Sciences,* edited by Adele E. Clarke and Joan Fujimura. Princeton: Princeton University Press, 1992, 3–47.

Clarke, Stephan P. "The Remarkable Sheffield Family of North Gainesville." N.p.: manuscript provided by author.

"Cleaning of Typewriters, Calculators, etc." Memo from Shanghai Municipal Council Secretary to "All Departments and Emergency Offices." SMA U1-4-3586 (April 2, 1943), 35.

Coble, Parks. *Chinese Capitalists in Japan's New Order: The Occupied Lower Yangzi, 1937–1945.* Berkeley: University of California Press, 2003.

Conn, Steven. "An Epistemology for Empire: The Philadelphia Commercial Museum, 1893–1926." *Diplomatic History* 22, no. 4 (1998): 533–563.

Conrad, Frank, and Yasudiro Sakai. "Impedance Device for Use with Current-Rectifiers." United States Patent no. 1075404. Filed January 10, 1912; patented October 14, 1913.

The Cornell University Register 1897–1898. 2nd ed. Ithaca: The University Press of Andrus and Church, 1897–1988, 18.

Cost Estimates for Lin Yutang Typewriter. April 20, 1949. Located within File Marked "Lin Yutang Typewriter." Mergenthaler Linotype Company Records, 1905–1993, Archives Center, National Museum of American History, Smithsonian Institution, box 3628.

"Cost for a Japanese Typewriter." SMA U1-4-3789 (February 25, 1943), 9.

Cousin, A.J.C. "Typewriting Machine." United State Patent no. 1794152. Filed July 13, 1928; patented February 24, 1931.

Cowan, Ruth Schwartz. *A Social History of American Technology.* New York: Oxford University Press, 1997.

Creel, Herrlee Glessner. "On the Nature of Chinese Ideography." *T'oung Pao* 32 (second series), no. 2/3 (1936): 85–161.

Culp, Robert. "Teaching Baihua: Textbook Publishing and the Production of Vernacular Language and a New Literary Canon in Early Twentieth-Century China." *Twentieth-Century China* 34, no. 1 (November 2008): 4–41.

David, Paul A. "Clio and the Economics of QWERTY." *American Economic Review* 75, no. 2 (1985): 332–337.

Davies, E. *Memoir of the Rev. Samuel Dyer; Sixteen Years Missionary to the Chinese.* London: J. Snow, 1846

Davies, Margery W. *A Woman's Place Is at the Typewriter: Office Work and Office Workers 1870–1930.* Philadelphia: Temple University Press, 1982.

Davis, John Francis. *The Chinese: A General Description of the Empire of China and Its Inhabitants.* Vol. 2. London: Charles Knight, 1836.

DeFrancis, John. *Nationalism and Language Reform in China.* New Jersey: Princeton University Press, 1950.

DeFrancis, John. *Visible Speech.* Honolulu: University of Hawai'i Press, 1989.

Derrida, Jacques. *Of Grammatology.* Baltimore: Johns Hopkins University Press, 1976.

"Descriptions of the Commercial Press Exhibit." Shanghai: Commercial Press, n.d. (c. 1926). City of Philadelphia, Department of Records. Record Group 232 (Sesquicentennial Exhibition Records), 232-4-8.1 "Department of Foreign Participation," box A-1474, box folder 8, series folder 29 ("China, Commercial Press Exhibit").

Desnoyers, Charles. *A Journey to the East: Li Gui's A New Account of a Trip Around the Globe.* Ann Arbor: University of Michigan Press, 2004.

"Did NBC Alter the Olympics Opening Ceremony?" *Slashdot* (August 9, 2008) http://news.slashdot.org/story/08/08/09/2231231/did-nbc-alter-the-olympics-opening-ceremony (accessed March 1, 2012).

Dodd, George. *The Curiosities of Industry and the Applied Sciences.* London: George Routledge and Co., 1858.

Dodge, Elbert S. "Typewriting Machine." United States Patent no. 1411238. Filed August 19, 1921; patented March 28, 1922.

"Doings at the Philadelphia Commercial Museum." *Commercial America* 19 (April 1923): 51.

Dolezelová-Velingerová, Milena. "Understanding Chinese Fiction 1900–1949." In *A Selective Guide to Chinese Literature, 1900–1949.* Vol. 1, edited by Milena Dolezelová-Velingerová. Leiden: Brill, 1988.

Dong, Madeleine Yue, and Joshua L. Goldstein, eds. *Everyday Modernity in China.* Seattle: University of Washington Press, 2006.

Douglas, Mary. "Introduction." In *How Classification Works: Nelson Goodman Among the Social Sciences*, edited by Nelson Goodman and Mary Douglas. Edinburgh: Edinburgh University Press, 1992.

Drucker, Johanna. *The Visible Word: Experimental Typography and Modern Art, 1909–1923*. Chicago: University of Chicago Press, 1997.

Du Ponceau, Peter S. "A Dissertation on the Nature and Character of the Chinese System of Writing." *Transactions of the Historical and Literary Committee of the American Philosophical Society* 2 (1838).

"Du Ponceau on the Chinese System of Writing." *North American Review* 48 (January 1839): 271–310.

Duyvendak, J.J.L. "Wong's System for Arranging Chinese Characters. The Revised Four-Corner Numeral System." *T'oung Pao* 28, no. 1/2 (1931): 71–74.

Dyer, Samuel. *A Selection of Three Thousand Characters Being the Most Important in the Chinese Language for the Purpose of Facilitating the Cutting of Punches and Casting Metal Type in Chinese*. Malacca: Anglo-Chinese College, 1834.

Eisenstein, Elizabeth. *The Printing Press as an Agent of Change: Communications and Cultural Transformations in Early-Modern Europe*. Cambridge: Cambridge University Press, 1980.

Engber, Daniel. "What Does a Chinese Keyboard Look Like?" *Slate* (February 21, 2006), http://www.slate.com/articles/news_and_politics/explainer/2006/02/what_does_a_chinese_keyboard_look_like.html

Escayrac de Lauture, Comte d'. *On the Telegraphic Transmission of Chinese Characters*. Paris: E. Brière, 1862.

Esherick, Joseph. *Reform and Revolution in China: The 1911 Revolution in Hunan and Hubei*. Berkeley: University of California Press, 1976.

"Facsimile May Solve Chinese Telegram Problem." *New York Times* (August 8, 1957), 39.

Fan, Fa-ti. "Redrawing the Map: Science in Twentieth-Century China." *Isis* 98 (2007): 524–538.

Fan, Fa-ti. "Science, Earthquake Monitoring, and Everyday Knowledge in Communist China." Paper delivered at Stanford University, History and Philosophy of Science and Technology program, April 22, 2010.

Febvre, Lucien, and Henri-Jean Martin. *The Coming of the Book: The Impact of Printing, 1450–1800*. London: Verso, 2010.

Fenn, Courtenay Hughes. *The Five Thousand Dictionary*. Cambridge, MA: Harvard University Press, 1940.

Ferrier, Claude-Marie, and Sir Hugh Owen. *Exhibition of the Works of Industry of All Nations: 1851 Report of the Juries.* London: William Clowes and Sons, 1852.

Fine, Lisa M. *The Souls of the Skyscraper: Female Clerical Workers in Chicago, 1870–1930.* Philadelphia: Temple University Press, 1990.

Flox, O.D. "That Chinese Type-Writer: An Open Letter to the Hon. Henry C. Newcomb, Agent of the Faroe Islands' Syndicate for the Promotion of Useful Knowledge." *Chinese Times* (March 31, 1888), 199.

Fourteenth Census of the United States, 1920. National Archives and Records Administration, Washington, DC, Records of the Bureau of the Census, Record Group 29, NARA microfilm publication T625.

"4,200 Characters on New Typewriter; Chinese Machine Has Only Three Keys, but There Are 50,000 Combinations. 100 Words in TWO HOURS. Heuen Chi, New York University Student, Patents Device Called the First of Its Kind." *New York Times* (March 23, 1915), 6.

Frazier, Mark W. *The Making of the Chinese Industrial Workplace: State, Revolution, and Labor Management.* Cambridge: Cambridge University Press, 2006.

"Front Views and Profiles: Miss Yin at the Console." *Chicago Daily Tribune* (October 10, 1945), 16.

Fu, Poshek. *Passivity, Resistance, and Collaboration: Intellectual Choices in Occupied Shanghai, 1937–1945.* Stanford: Stanford University Press, 1993.

Fuller, Matthew. *Behind the Blip: Essays on the Culture of Software.* Sagebrush Education Resources, 2003.

Furth, Charlotte. "Culture and Politics in Modern Chinese Conservatism." In *The Limits of Change: Essays on Conservative Alternatives in Republican China*, 22–56. Cambridge, MA: Harvard University Press, 1976.

Galison, Peter. *Einstein's Clocks, Poincaré's Maps: Empires of Time.* New York: W.W. Norton and Co., 2003.

Gamble, William. *List of Chinese Characters in the New Testament and Other Books.* N.p., 1861. Library of Congress. G/C175.1/G15.

Gamble, William. *1878 Chinese Characters in William Gamble's List which can be Formed by Divisible Type (Liangbian pin xiaozi)* [兩 邊 拼 小 字]. Manuscript. N.p., 1863. Library of Congress. G/C175.1/G18.

Gamble, William. *Two Lists of Selected Characters Containing All in the Bible.* Shanghai: n.p., 1861.

Gellner, Ernest. *Nations and Nationalism.* Ithaca: Cornell University Press, 1983.

Gerth, Karl. *China Made: Consumer Culture and the Creation of the Nation.* Cambridge, MA: Harvard Asia Center, 2003.

Gilbert, Paul T. "Putting Ideographs on Typewriter." *Nation's Business* 17, no. 2 (February 1929): 156.

Gitelman, Lisa. *Scripts, Grooves, and Writing Machines: Representing Technology in the Edison Era.* Stanford: Stanford University Press, 2000.

Goldman, Merle, and Leo Ou-fan Lee, eds. *An Intellectual History of Modern China.* Cambridge: Cambridge University Press, 2001.

Goodman, Nelson. *Ways of Worldmaking.* Indianapolis: Hackett Publishing, 1978.

Goodman, Nelson, Mary Douglas, and David L. Hull, eds. *How Classification Works: Nelson Goodman Among the Social Sciences.* Edinburgh: Edinburgh University Press, 1992.

Goody, Jack. *The Interface between the Written and the Oral.* Cambridge: Cambridge University Press, 1987.

Goody, Jack. "Technologies of the Intellect: Writing and the Written Word." In *The Power of the Written Tradition.* Washington: Smithsonian Institution Press, 2000: 133–138.

Gottlieb, Nanette. "The Rōmaji Movement in Japan." *Journal of the Royal Asiatic Society* 20, no. 1 (2010): 75–88.

Grant, John Cameron, and Lucien Alphonse Legros. "A Method and Means for Adapting Certain Chinese Characters, Syllabaries or Alphabets for use in Type-casting or Composing Machines, Typewriters and the Like." Great Britain Patent Application no. 2483. Filed January 30, 1913; patented October 30, 1913.

Greene, Stephen Lyon Wakeman. *Absolute Dreams. Thai Government Under Rama VI, 1910–1925.* Bangkok: White Lotus, 1999.

Grose, Thomas K. "London Admits It Can't Top Lavish Beijing Olympics When It Hosts 2012 Games." *U.S. News* (August 22, 2008), http://www.usnews.com/news/world/articles/2008/08/22/london-admits-it-cant-top-lavish-beijing-olympics-when-it-hosts-2012-games

Gunn, Edward. *Rewriting Chinese: Style and Innovation in Twentieth-Century Chinese Prose.* Stanford: Stanford University Press, 1991.

Guy, R. Kent. *The Emperor's Four Treasuries: Scholars and the State in the Late Ch'ien-lung Era.* Cambridge, MA: Council on East Asian Studies, Harvard University, 1987.

Haddad, Selim S. "Types for Type-Writers or Printing-Presses." United States Patent no. 637109. Filed October 13, 1899; patented November 14, 1899.

"The Hall Typewriter." *Scientific American* (July 10, 1886), 24.

Hannas, William C. *Asia's Orthographic Dilemma.* Honolulu: University of Hawai'i Press, 1996.

Hannas, William C. *The Writing on the Wall: How Asian Orthography Curbs Creativity.* Philadelphia: University of Pennsylvania Press, 2003.

Hansen, Harry. "How Can Lin Yutang Make His New Typewriter Sing?" *Chicago Daily Tribune* (August 24, 1947), C4.

Harrison, Samuel A. "Oriental Type-Writer." United States Patent no. 977448. Filed December 15, 1909; patented December 6, 1910.

Harrist, Robert E., and Wen Fong. *The Embodied Image: Chinese Calligraphy from the John B. Elliott Collection.* Princeton: Art Museum, Princeton University in association with Harry N. Abrams, 1999.

Havelock, Eric A. *The Literate Revolution in Greece and Its Cultural Consequences.* Princeton, NJ: Princeton University Press, 1981.

Havelock, Eric A. *The Muse Learns to Write: Reflections on Orality and Literacy from Antiquity to the Present.* New Haven: Yale University Press, 1986.

Havelock, Eric A. *Origins of Western Literacy.* Toronto: Ontario Institute for Studies in Education, 1976.

Hay, John. "The Human Body as a Microcosmic source of Macrocosmic Values in Calligraphy." In *Self as Body in Asian Theory and Practice,* edited by Thomas Kasulis, Roger Ames, and Wimal Dissanayake. Albany: State University of New York Press, 1993, 179–212.

Hayford, Charles W. *To the People: James Yen and Village China.* New York: Columbia University Press, 1990.

Hayles, N. Katherine. *How We Think: Digital Media and Contemporary Technogenesis.* Chicago: University of Chicago Press, 2012.

Headrick, Daniel. *The Invisible Weapon: Telecommunications and International Politics, 1851–1945.* Oxford: Oxford University Press, 1991.

Headrick, Daniel. *The Tentacles of Progress: Technology Transfer in the Age of Imperialism, 1850–1940.* Oxford: Oxford University Press, 1988.

Headrick, Daniel. *The Tools of Empire: Technology and European Imperialism in the Nineteenth Century.* Oxford: Oxford University Press, 1981.

Headrick, Daniel, and Pascal Griset. "Submarine Telegraph Cables: Business and Politics, 1838–1939." *Business History Review* 75, no. 3 (2001): 543–578.

Hearn, Maxwell, and Judith Smith. eds. *Arts of the Sung and Yuan.* New York: Metropolitan Museum of Art, 1996.

Hedtke, Charles H. "The Sichuanese Railway Protection Movement: Themes of Change and Conflict." *Bulletin of the Institute of Modern History, Academia Sinica (Zhongyang yanjiuyuan jindaishi yanjiusuo jikan)* [中央研究院近代史研究所集刊] 6 (1977): 353–407.

Hegel, Georg Wilhelm Friedrich. *The Philosophy of History*. Translated by John Sibree. New York: Wiley Book Co., 1900.

Heijdra, Martin J. "The Development of Modern Typography in East Asia, 1850–2000." *East Asia Library Journal* 11, no. 2 (Autumn 2004): 100–168.

Hernisz, Stanislas. *A Guide to Conversation in the English and Chinese Languages for the Use of Americans and Chinese in California and Elsewhere*. Boston: John P. Jewett and Co., 1854.

"Highlights of Syracuse Decade by Decade." *Syracuse Journal* (March 20, 1939), E2.

Hill, Michael G. "National Classicism: Lin Shu as Textbook Writer and Anthologist, 1908–1924." *Twentieth-Century China* 33, no. 1 (November 2007): 27–52.

"Hiring Telegraphers for China." *New York Times* (September 30, 1887), 1.

Hirth, Friedrich. "Western Appliances in the Chinese Printing Industry." *Journal of the China Branch of the Royal Asiatic Society* (Shanghai) 20 (1885): 163–177.

"The History of the Typewriter Recited by Michael Winslow," http://www.filmjunk. com/2010/06/20/the-history-of-the-typewriter-recited-by-michael-winslow/ (accessed September 5, 2010).

Hoare, R. "Keyboard Diagram for Chinese Phonetic." Mergenthaler Linotype Collection. Museum of Printing, North Andover, Massachusetts, February 4, 1921.

Hoare, R. "Keyboard Diagram for Chinese Phonetic Amended." Mergenthaler Linotype Collection. Museum of Printing, North Andover, Massachusetts, March 3, 1921.

H.O. Fuchs Engineering Case Program: Case Files. Department of Special Collections, Stanford University Archives, SC 269, box 1—J.E. Arnold, A.T. Ling, MIT, "Chinese Typewriter."

Holcombe, Charles. *In the Shadow of the Han: Literati Thought and Society at the Beginning of the Southern Dynasties*. Honolulu: University of Hawai'i Press, 1995.

Hon, Tze-ki, and Robert Culp, eds. *The Politics of Historical Production in Late Qing and Republican China*. Leiden: Brill, 2007.

Honolulu, Hawaii Passenger and Crew Lists, 1900–1959. National Archives and Records Administration, Washington, DC, Series A3422, Roll 49.

"How Can the Chinese Use Computers Since Their Language Contains So Many Characters?" *Straight Dope* (December 8, 1995), http://www.straightdope.com/ columns/read/1138/how-can-the-chinese-use-computers-since-their-language -contains-so-many-characters (accessed January 7, 2010).

Howland, Douglas. *Borders of Chinese Civilization: Geography and History at Empire's End*. Durham: Duke University Press, 1996.

H.R.H. The Crown Prince of Siam. *The War of the Polish Succession.* Oxford: Black-well, 1901.

Hughes, Thomas P. "The Evolution of Large Technical Systems." In *The Social Construction of Technical Systems: New Directions in the Sociology and History of Technology,* edited by Wiebe E. Bijker, Thomas P. Hughes, and Trevor Pinch, 51–82. Cambridge, MA: MIT Press, 1989.

Hull, Matthew. *Government of Paper: The Materiality of Bureaucracy in Urban Pakistan.* Berkeley: University of California, 2012.

Humphrey, Henry Noel. *The Origin and Progress of the Art of Writing: A Connected Narrative of the Development of the Art, its Primeval Phases in Egypt, China, Mexico, etc.* London: Ingram, Cooke, and Co., 1853.

Hunter, Edward. "Increasing Program of China Foundation." *China Weekly Review* (August 8, 1931), 379.

Hunter, Janet. "Technology Transfer and the Gendering of Communications Work: Meiji Japan in Comparative Historical Perspective." *Social Science Japan Journal* 14, no. 1 (Winter 2011): 1–20.

Huters, Theodore. *Bringing the World Home: Appropriating the West in Late Qing and Early Republican China.* Honolulu: University of Hawai'i Press, 2005.

Innis, Harold. *Empire and Communications.* Toronto: University of Toronto Press, 1972.

Inoue, Miyako. "Stenography and Ventriloquism in Late Nineteenth-Century Japan." *Language and Communication* 31 (2011): 181–190.

International Telegraph Convention of Saint-Petersburg and Service Regulations Annexed (1925). London: His Majesty's Stationery Office, 1926.

International Telegraph Convention with Berlin Revision of Service Regulations and Tariffs (1885). London: Blackfriars Printing and Publishing Co., 1885.

"Invents Typewriter for Chinese Language; First Machine of the Kind Ever Built Is Announced by the Underwood Company." *New York Times* (May 16, 1926), 5.

Ip, Manying. *Life and Times of Zhang Yuanji, 1867–1959.* Beijing: Commercial Press, 1985.

Ishii, Kae. "The Gendering of Workplace Culture: An Example from Japanese Telegraph Operators." *Bulletin of the Health Science University* (*Kenkō kagaku daigaku kiyō*) [健康科学大学紀要] 2 (2005): 37–48.

Ismail, Ibrahim bin. "Samuel Dyer and His Contributions to Chinese Typography." *Library Quarterly* 54, no. 2 (April 1984): 157–169.

"It Takes Four Thousand Characters to Typewrite in Chinese." *Popular Science Monthly* 90, no. 4 (April 1917): 599.

Jacobowitz, Seth. *Writing Technology in Meiji Japan: A Media History of Modern Japanese Literature and Visual Culture*. Cambridge, MA: Harvard Asia Center, 2015.

Jacobsen, Kurt. "A Danish Watchmaker Created the Chinese Morse System." *NIAS-nytt (Nordic Institute of Asian Studies) Nordic Newsletter* 2 (July 2001): 17–21.

"Japanese Typewriters Cleaning and Repair Service." Memo from Sanwa Shoji Company. SMA U1-4-3789 (February 12, 1943), 6.

Jin Jian. *A Chinese Printing Manual*. Translated by Richard C. Rudolph. Los Angeles: Ward Ritchie Press, 1954 [1776].

Johns, Adrian. *The Nature of the Book: Print and Knowledge in the Making*. Chicago: University of Chicago Press, 1998.

Jones, Robert McKean. "Typewriting Machine." United States Patent no. 1687939. Filed May 19, 1927; patented October 16, 1928.

Jones, Robert McKean. "Urdu—Keyboard no. 1130—No. 4 Monarch." (March 13, 1918) Hagley Museum and Library. Accession no. 1825.

Jones, Stacy V. "Telegraph Printer in Japanese with 2,300 Symbols Patented." *New York Times* (December 31, 1955), 19.

Judge, Joan. *Print and Politics: Shibao and the Culture of Reform in Late Qing China*. Stanford: Stanford University Press, 1996.

"Judging Eastern Things from Western Point of View." *Chinese Students' Monthly* 8, no. 3 (1913): 154.

"Just How Smart Are We?" *Daily News New York* (September 2, 1947). Clipping included in Archives of John Day Co. Princeton University. Box 236, folder 14, call no. CO123.

Kadry, Vassaf. "Type Writing Machine." United States Patent no. 1212880. Filed January 15, 1914; patented January 30, 1917.

"Kamani Eng. Corporation Ltd. Agent of the Olivetti in India." *Rivista Olivetti* 6 (December 1951): 12–13.

Kara, György. "Aramaic Scripts for Altaic Languages." In *The World's Writing Systems*, edited by Peter T. Daniels and William Bright. New York: Oxford University Press, 1994, 536–558.

Kaske, Elisabeth. *The Politics of Language in Chinese Education, 1895–1919*. Leiden: Brill, 2008.

Keenan, Barry C. *Imperial China's Last Classical Academies: Social Change in the Lower Yangzi, 1864–1911*. Berkeley: China Research Monographs, University of California, 1994.

Keep, Christopher. "The Cultural Work of the Type-Writer Girl." *Victorian Studies* 40, no. 3 (Spring 1997): 401–426.

Kennedy, George A., ed. *Minimum Vocabularies of Written Chinese*. New Haven: Far Eastern Publications, 1966.

Khalil, Seyed. "Typewriting Machine." United States Patent no. 1403329. Filed April 14, 1917; patented January 10, 1922.

Kittler, Friedrich A. *Gramophone, Film, Typewriter*. Translated by Geoffrey Winthrop-Young and Michael Wautz. Stanford: Stanford University Press, 1999.

Kline, Ronald. *Consumers in the Country: Technology and Social Change in Rural America*. Baltimore: Johns Hopkins University Press, 2000.

Kline, Ronald, and Trevor Pinch. "Users as Agents of Technological Change: The Social Construction of the Automobile in the Rural United States." *Technology and Culture* 37 (1996): 763–795.

Knorr-Cetina, Karin. *Epistemic Cultures: How the Sciences Make Knowledge*. Cambridge, MA: Harvard University Press, 1999.

Ko, Dorothy. *Teachers of the Inner Chambers: Women and Culture in Seventeenth-Century China*. Stanford: Stanford University Press, 2004.

Kraus, Richard Kurt. *Brushes with Power: Modern Politics and the Chinese Art of Calligraphy*. Berkeley: University of California Press, 1991.

"Kurita, Shunjiro." *Who's Who in Japan* 13–14 (1930): 8.

Labor and Cost Estimates Associated with Lin Yutang Typewriter, n.d. (circa late 1948/early 1949). Located within File Marked "Lin Yutang Typewriter." Mergenthaler Linotype Company Records, 1905–1993, Archives Center, National Museum of American History, Smithsonian Institution, box 3628.

Latour, Bruno. *Reassembling the Social: An Introduction to Actor-Network-Theory*. Oxford: Oxford University Press, 2007 [2005].

Latour, Bruno. *Science in Action: How to Follow Scientists and Engineers through Society*. Cambridge, MA: Harvard University Press, 1987.

Latour, Bruno, and Steve Woolgar. *Laboratory Life: The Construction of Scientific Facts*. Princeton: Princeton University Press, 1986 [1979].

Law, John. "Technology and Heterogeneous Engineering: The Case of Portuguese Expansion." In *The Social Construction of Technological Systems: New Directions in the Sociology and History of Technology*, edited by Wiebe E. Bijker, Thomas P. Hughes, and Trevor Pinch. Cambridge, MA: MIT Press, 1989 [1987]: 111–134.

Lee, En-han. "China's Response to Foreign Investment in Her Mining Industry." *Journal of Asian Studies* 28, no. 1 (November 1968): 55–76.

Lee, Leo Ou-fan. *Shanghai Modern: The Flowering of a New Urban Culture in China 1930–1945*. Cambridge, MA: Harvard University Press, 1999.

Lee-Wong, Song Mei. "Coherence, Focus and Structure: The Role of Discourse Particle *ne*." *Pragmatics* 11, no. 2 (2001): 139–153.

Lent, John A., and Ying Xu. "Chinese Animation Film: From Experimentation to Digitalization." In *Art, Politics, and Commerce in Chinese Cinema*, edited by Ying Zhu and Stanley Rosen. Hong Kong: Hong Kong University Press, 2010.

Leslie, Stuart. "Exporting MIT." *Osiris* 21 (2006): 110–130.

Letter from Lin Yutang to M.M. Reed, circa February 23, 1949. Mergenthaler Linotype Company Records, 1905–1993, Archives Center, National Museum of American History, Smithsonian Institution, box 3628.

Letter from Lin Yutang to M.M. Reed, n.d. (Precedes/Prompts March 10, 1949 Response—Likely Date of February 23, 1949). Found within file marked "Lin Yutang Typewriter." Mergenthaler Linotype Company Records, 1905–1993, Archives Center, National Museum of American History, Smithsonian Institution, box 3628.

Letter from Lin Yutang to Richard Walsh and Pearl S. Buck, December 16, 1937. Archives of John Day Co. Princeton University. Box 144, folder 6, call no. C0123.

Letter from Lin Yutang to Richard Walsh and Pearl S. Buck, December 13, 1938, sent from Paris. Archives of John Day Co. Princeton University. Box 144, folder 6, call no. C0123.

Letter from Mirovitch to Chung-yuan Chang. August 12, 1949. Mergenthaler Linotype Company Records, 1905–1993, Archives Center, National Museum of American History, Smithsonian Institution, box 3628.

Letter from Mirovitch to Chung-yuan Chang. October 1, 1949. Mergenthaler Linotype Company Records, 1905–1993, Archives Center, National Museum of American History, Smithsonian Institution, box 3628.

Letter from Mirovitch to G.B. Welch. March 8, 1949. Mergenthaler Linotype Company Records, 1905–1993, Archives Center, National Museum of American History, Smithsonian Institution, box 3628.

Letter from M.M. Reed to Lin Yu-tang [Yutang]. March 10, 1949. Found within File Marked "Lin Yutang Typewriter." Mergenthaler Linotype Company Records, 1905–1993, Archives Center, National Museum of American History, Smithsonian Institution, box 3628.

Letter from Pearl S. Buck to Lin Yutang. May 4, 1947. Pearl S. Buck International Archive, record group 6, box 3, folder 29, item 10.

Letter from Tao Minzhi to author, February 11, 2010.

Letter from Yu's Chinese Typewriter Mfg. Co. to the Shanghai Municipal Police Administration (*Shanghai shi jingchaju*) [上海市警察局]. SMA Q131-7-1368 (December 13, 1945), 4.

Levering, Gilbert. "Chinese Language Typewriter." *Life* 2311 (February 17, 1927): 4.

Li Yu. "Character Recognition: A New Method of Learning to Read in Late Imperial China." *Late Imperial China* 33, no. 2 (December 2012): 1–39.

Li Yu. "Learning to Read in Late Imperial China." *Studies on Asia: Series II* 1, no. 1 (2004): 7–29.

Lichtentag, Alexander. *Lichtentag Paragon Shorthand. A Vast Improvement in the Art of Shorthand.* New York: Paragon Institute Home Study Department, 1918.

Lin Yutang. "At Last: A Chinese Typewriter." Reprint from *New York Post,* n.d. ("May 24, 1946 Received" stamped on top). Mergenthaler Linotype Company Records, 1905–1993, Archives Center, National Museum of American History, Smithsonian Institution, box 3628.

Lin Yutang. "Chinese Typewriter." United States Patent no. 2613795. Filed April 17, 1946; patented October 14, 1952.

Lin Yutang. "Features of the Invention." Archives of John Day Co. Princeton University. Box/folder 14416, call no. CO123 (circa October 14, 1931).

Lin Yutang. *My Country and My People.* New York: Reynal and Hitchcock, 1935.

Lin Yutang. "Newly Invented Chinese Typewriter Has Sixty-Four Keys." *Washington Post* (December 5, 1945), C1.

"A Line O'Type or Two." *Chicago Daily Tribune* (August 31, 1949), 16.

"The Lin Yutang Chinese Typewriter." N.d. Mergenthaler Linotype Company Records, 1905–1993, Archives Center, National Museum of American History, Smithsonian Institution, box 3628.

"Lin Yutang Invents Chinese Typewriter: Will Do in an Hour What Now Takes a Day." *New York Herald Tribune* (August 22, 1947), 13.

"Lin Yutang Solves an Oriental Puzzle: Newly Invented Chinese Typewriter Has Sixty-Four Keys." *Washington Post* (December 5, 1945), C1.

"Lin Yutang Typewriter." Mergenthaler Linotype Company Records, 1905–1993, Archives Center, National Museum of American History, Smithsonian Institution, box 3628. Dates include January 14, 1949; January 19, 1949.

"List of Awards-General, n.d." City of Philadelphia, Department of Records. Record Group 232 (Sesquicentennial Exhibition Records), 232-4-6.4 (Jury of Awards-Files), box a-1472, folder 17, series folder 1.

"List of Chinese Characters Formed by the Combination of the Divisible Type of the Berlin Font Used at the Shanghai Mission Press of the Board of Foreign Missions of the Presbyterian Church in the United States." Shanghai: n.p., 1862.

"List of Vendors Who Submitted Quotations." February 2, 1949. Found within File Marked "Lin Yutang Typewriter." Mergenthaler Linotype Company Records, 1905–

1993, Archives Center, National Museum of American History, Smithsonian Institution, box 3628.

Littell, John Stockton. *Some Great Christian Jews*. 2nd ed. N.p., 1913.

Liu, Alan P.L. *Communications and National Integration in Communist China*. Berkeley: University of California Press, 1971.

Liu, James T.C. "The Classical Chinese Primer: Its Three-Character Style and Authorship." *Journal of the American Oriental Society* 105, no. 2 (April–June 1985): 191–196.

Liu, Lydia. *The Clash of Empires: The Invention of China in Modern World Making*. Cambridge, MA: Harvard University Press, 2004.

Liu, Lydia. *Translingual Practice: Literature, National Culture, and Translated Modernity—China, 1900–1937*. Stanford: Stanford University Press, 1995.

Logan, Robert K. *The Alphabet Effect: The Impact of the Phonetic Alphabet on the Development of Western Civilization*. New York: William Morrow & Co., Inc., 1986.

Lovett, R. *History of the London Missionary Society, 1795–1895*. 2 vols. London: Henry Frowde, 1899.

Ma, Sheng-mei. *Immigrant Subjectivities in Asian American and Asian Diaspora Literatures*. Albany: State University of New York Press, 1998.

Maclachlan, Patricia L. *The People's Post Office: The History and Politics of the Japanese Postal System, 1871–2010*. Cambridge, MA: Harvard Asia Center, 2012.

Maddox, Brenda. "Women and the Switchboard." In *The Social History of the Telephone*, edited by Ithiel de Sola Pool. Cambridge, MA: MIT Press, 1977, 262–280.

Maher, John Peter. "More on the History of the Comparative Methods: The Tradition of Darwinism in August Schleicher's Work." *Anthropological Linguistics* 8 (1966): 1–12.

Markwyn, Abigail. "Economic Partner and Exotic Other: China and Japan at San Francisco's Panama-Pacific International Exposition." *Western Historical Quarterly* 39, no. 4 (2008): 439–465.

Marshall, John. Email communication, October 21, 2011.

Marshman, Joshua. *Elements of Chinese Grammar, with a Preliminary Dissertation on the Characters and the Colloquial Medium of the Chinese*. Serampore: Mission Press, 1814.

Martin, W.A. *The History of the Art of Writing*. New York: Macmillan, 1920.

Marvin, Carolyn. *When Old Technologies Were New: Thinking about Electric Communication in the Late Nineteenth Century*. Oxford: Oxford University Press, 1998.

Massachusetts Institute of Technology Alumni Association, ed. *Technology Review* 18, nos. 7–12 (1916). Cambridge: Association of Alumni and Alumnae of the Massachusetts Institute of Technology, 1916.

Mathews, Jay. "The Chinese Language: Sounds and Fury." *Washington Post* (December 28, 1980), C1.

Mathias, Jim, and Thomas L. Kennedy, eds. *Computers, Language Reform, and Lexicography in China. A Report by the CETA Delegation.* Pullman, WA: Washington State University Press, 1980.

Matsusaka, Y. Tak. "Managing Occupied Manchuria." In *Japan's Wartime Empire*, edited by Peter Duus, Ramon H. Myers, and Mark R. Peattie. Princeton: Princeton University Press, 1996, 112–120.

McDermott, Joseph P. *A Social History of the Chinese Book: Books and Literati Culture in Late Imperial China.* Hong Kong: Hong Kong University Press, 2006.

McFarland, George B. *Reminiscences of Twelve Decades of Service to Siam, 1860–1936.* Bancroft Library. BANCMSS 2007/104, box 4, folder 14, George Bradley McFarland 1866–1942.

McNair, Amy. "Engraved Calligraphy in China: Recension and Reception." *Art Bulletin* 77, no. 1 (March 1995): 106–114.

Medhurst, Walter Henry. *China: Its State and Prospects, with Especial Reference to the Spread of the Gospel.* London: John Snow, 1842.

Memo from N. Inagaki, Commissioner, Commodity Control Department to "The Secretary, Shanghai Municipal Council." SMA U1-4-3796 (February 4, 1943), 43.

Meng, Liansu. "The Inferno Tango: Gender Politics and Modern Chinese Poetry, 1917–1980." PhD diss., University of Michigan, 2010.

Mergenthaler Linotype Company. *China's Phonetic Script and the Linotype.* Brooklyn: Mergenthaler Linotype Co., April 1922. Smithsonian National Museum of American History Archives Center. Collection no. 666, box LIZ0589 ("History—Non-Roman Faces"), folder "Chinese," subfolder "Chinese Typewriter."

Meyer-Fong, Tobie. "The Printed World: Books, Publishing Culture, and Society in Late Imperial China." *Journal of Asian Studies* 66, no. 3 (August 2007): 787–817.

Milne, W. *A Retrospect of the First Ten Years of the Protestant Mission to China.* Malacca: Anglo-Chinese Press, 1820.

"Missionaries of the American Board." *Congregationalist* (September 26, 1872), 3.

Mittler, Barbara. *A Newspaper for China? Power, Identity, and Change in Shanghai's Mass Media, 1872–1912.* Cambridge, MA: Harvard University Asia Center, 2004.

Mizuno, Hiromi. *Science for the Empire: Scientific Nationalism in Modern Japan.* Stanford: Stanford University Press, 2009.

"Monarch Arabic Keyboard—Haddad System—Keyboard no. 724." (October 16, 1913.) Hagley Museum and Library. Accession no. 1825. Remington Rand Corporation. Records of the Advertising and Sales Promotion Department. Series I Typewriter Div. Subseries B, Remington Typewriter Company, box 3, vol. 2.

Morrison, E. *Memoirs of the Life and Labours of Robert Morrison, D.C.* 2 vols. London: Orme, Brown, Green and Longmans, 1839.

Morrison, Robert, comp. *A Dictionary of the Chinese Language, in Three Parts.* 6 vols. Macao: East India Co., 1815–1823.

Mullaney, Thomas S. *Coming to Terms with the Nation: Ethnic Classification in Modern China.* Berkeley: University of California Press, 2010.

Mullaney, Thomas S. "Controlling the Kanjisphere: The Rise of the Sino-Japanese Typewriter and the Birth of CJK." *Journal of Asian Studies* 75, no. 3 (August 2016): 725–753.

Mullaney, Thomas S. "How to Spy on 600 Million People: The Hidden Vulnerabilities in Chinese Information Technology." *Foreign Affairs* (June 5, 2016), https://www.foreignaffairs.com/articles/china/2016-06-05/how-spy-600-million-people

Mullaney, Thomas S. "The Movable Typewriter: How Chinese Typists Developed Predictive Text during the Height of Maoism." *Technology and Culture* 53, no. 4 (October 2012): 777–814.

Mullaney, Thomas S. "Semiotic Sovereignty: The 1871 Chinese Telegraph Code in Historical Perspective." In *Science and Technology in Modern China, 1880s–1940s,* edited by Jing Tsu and Benjamin Elman. Leiden: Brill, 2014, 153–184.

Mullaney, Thomas S. "'Ten Characters per Minute': The Discourse of the Chinese Typewriter and the Persistence of Orientalist Thought." Association for Asian Studies Annual Meeting 2010.

Müller, Friedrich Wilhelm. "Typewriter." United States Patent no. 1686627. Filed January 22, 1925; patented October 9, 1928.

Needham, Joseph. "Poverties and Triumphs of the Chinese Scientific Tradition." In *Scientific Change (Report of History of Science Symposium, Oxford, 1961),* edited by A.C. Crombie. London: Heinemann, 1963.

Needham, Joseph. *Science and Civilisation in China.* Vol. 2. Cambridge: Cambridge University Press, 1956.

Needham, Joseph. *Science and Civilisation in China.* Vol. 5, part 1. Cambridge: Cambridge University Press, 1985.

"New Chinese Typewriter." *China Trade News* (July 1946), 5. Mergenthaler Linotype Company Records, 1905–1993, Archives Center, National Museum of American History, Smithsonian Institution, box 3628.

"New Chinese Typewriter Triumphs over Language of 43,000 Symbols." *New York Times* (October 18, 1952), 26, 30.

Newcomb, Henry C. "Letter to the Editor: That Chinese Type-writer." *Chinese Times* [Tianjin] (March 17, 1888), 171–172

"The Newest Inventions." *Washington Post* (March 21, 1917), 6.

"New Typewriter Will Aid Chinese. Invention of Dr. Lin Yutang Can Do a Secretary's Day's Work in an Hour." *New York Times* (August 22, 1947), 17.

Nineteenth Annual Report of the American Tract Society. Boston: Perkins and Marvin, May 29, 1833.

"No Chinese Typewriters." *Gregg Writer* 15 (1912): 382.

"Nothing Serious." *Utica Observer* (April 10, 1900), 1.

"Obituary: Robert McKean Jones. Inventor of Chinese Typewriter Was Able Linguist." *New York Times* (June 21, 1933), 18.

Official Congressional Directory. Washington, DC: United States Congress, 1916 (December).

Ogasawara, Yuko. *Office Ladies and Salaried Men: Power, Gender, and Work in Japanese Companies.* Berkeley: University of California Press, 1998.

Olwell, Victoria. "The Body Types: Corporeal Documents and Body Politics Circa 1900." In *Literary Secretaries/Secretarial Culture,* edited by Leah Price and Pamela Thurschwell. Aldershot: Ashgate, 2005.

Ong, Walter J. *Orality and Literacy.* New York: Routledge, 2013 [1982].

Oudshoorn, Nelly, and Trevor Pinch, eds. *How Users Matter: The Co-Construction of Users and Technology.* Cambridge, MA: MIT Press, 2005 [2003].

"Our Benevolent Causes." *Southwestern Christian Advocate* (July 8, 1897), 6.

Passenger and Crew Lists of Vessels Arriving at Seattle, Washington, 1890–1957. National Archives and Records Administration, Washington, DC, Record Group 85, NARA microfilm publication M1383_109.

"Passenger Lists of Vessels Arriving or Departing at Honolulu, Hawaii, 1900–1954." National Archives and Records Administration, Washington, DC, Records of the Immigration and Naturalization Service, Record Group 85. Series/roll no. m1412:6.

Passport Applications January 2, 1906–March 31, 1925. National Archives and Records Administration, Washington, DC, ARC Identifier 583830, MLR, Number A1534, NARA Series M1490, Roll 109.

Paterno, Roberto. "Devello Z. Sheffield and the Founding of the North China College." In *American Missionaries in China,* edited by Kwang-ching Liu. Cambridge, MA: Harvard East Asian Monographs, 1966, 42–92.

Payment Slip for Chung-yuan Chang. Mergenthaler Linotype Company Records, 1905–1993, Archives Center, National Museum of American History, Smithsonian Institution, box 3628, January 17, 1950.

Peeters, Henry. "Typewriter." United States Patent no. 1528846. Filed February 20, 1924; patented March 20, 1925.

Peeters, Henry. "Typewriter." United States Patent no. 1634042. Filed February 20, 1924; patented June 28, 1927.

"Phonetic Chinese." Letter from R. Hoare (Foreign Department) to Chauncey Griffith (January 7, 1921). Mergenthaler Linotype Collection. Museum of Printing, North Andover, Massachusetts.

Photograph of Fong Sec. *Asia: Journal of the American Asiatic Association* 19, no. 11 (November 1919): front matter.

Photograph of Lin Yutang and Lin Taiyi. From "Inventor Shows His Chinese Typewriter." Acme News Pictures—New York Bureau (August 21, 1947).

Photograph of Woman Using Chinese Typewriter at Trade Fair in Munich. November 25, 1953. Author's personal collection.

Photographs. George Bradley McFarland Papers, box 3, folder 15, October 23, 1938. Bancroft Library. University of California, Berkeley.

Poletti, Pietro. *A Chinese and English Dictionary Arranged According to Radicals and Sub-Radicals*. Shanghai: American Presbyterian Mission Press, 1896.

Price Quotes from Typewriter Companies to the General Office, First District of the Shanghai Government. SMA R22-2-776 (circa December 21, 1943), 1–28.

"Psychological Warfare, EUSAK Compound, Seoul, Korea (1952)." National Archives and Records Administration, Washington, DC, ARC Identifier 25967, Local Identifier 111-LC-31798.

"Public Works Department—Chinese Typewriters." Memo from Treasurer to Secretary General. SMA U1-4-3582 (August 12, 1943), 9.

Qi Xuan [Heuen Chi]. "The Principle of My Chinese Typewriter." *Chinese Students' Monthly* 10, no. 8 (May 1, 1915): 513–514.

Rankin, Mary Backus. "Nationalistic Contestation and Mobilization Politics: Practice and Rhetoric of Railway-Rights Recovery at the End of the Qing." *Modern China* 28, no. 3 (July 2002): 315–361.

Rawski, Evelyn. *Education and Popular Literacy in Ch'ing China*. Ann Arbor: University of Michigan Press, 1979.

"A Reason Why the Chinese Business Man May Soon Be Tired." *Life* 68 (1916): 272.

Receipt from C.Y. Chao for Cleaning Services Sent to Secretariat Office. SMA U1-4-3582 (October 12, 1943), 5.

"Reducing Chinese Letters from 40,000 Symbols to 40: New Typesetting Machine Expected Greatly to Facilitate Elimination of Illiteracy in China—American Invention Will Take Place of Twelve Men." *New York Times* (March 2, 1923), X3.

Reed, Christopher A. *Gutenberg in Shanghai: Chinese Print Capitalism, 1876–1937*. Honolulu: University of Hawai'i Press, 2004.

Reed, Martin W. "Lin Yutang Typewriter." Mergenthaler Linotype Company Records, 1905–1993, Archives Center, National Museum of American History, Smithsonian Institution.

Reich, Donald. "Freezing Assets of Nationalistic China." June 3, 1949. Mergenthaler Linotype Company Records, 1905–1993, Archives Center, National Museum of American History, Smithsonian Institution, box 3628.

Reid, Robert A. *The Panama-Pacific International Exposition*. San Francisco: Panama-Pacific International Exposition Co., 1915.

Reply to Tore Hellstrom. January 20, 1944. Mergenthaler Linotype Company Records, 1905–1993, Archives Center, National Museum of American History, Smithsonian Institution, box 3628.

Review of *Dianbao xinshu*. *Chinese Recorder and Missionary Journal* 5 (February 1874): 53–55.

Richards, G. Tilghman. *The History and Development of Typewriters: Handbook of the Collection Illustrating Typewriters*. London: His Majesty's Stationery Office, 1938.

Richardson, Ingrid. "Faces, Interfaces, Screens: Relational Ontologies of Framing, Attention and Distraction." *Transformations* 18 (2010).

Richardson, Ingrid. "Mobile Technosoma: Some Phenomenological Reflections on Itinerant Media Devices." *fibreculture* 6 (December 10, 2005).

Robbins, Bruce. "Commodity Histories." *PMLA* 120, no. 2 (2005): 454–463.

Rogers, Everett M. *Diffusion of Innovations*. New York: Free Press, 2003 [1962].

Rotman, Brian. *Becoming Beside Ourselves: The Alphabet, Ghosts, and Distributed Human Being*. Durham: Duke University Press, 2008.

Said, Edward W. *Orientalism*. New York: Vintage Books, 1979.

Sakai, Yasudiro. "Armature Winding." United States Patent no. 1156711. Filed February 3, 1910; patented October 12, 1915.

Sakai, Yasudiro. "Electrical Terminal." United States Patent no. 1049404. Filed January 7, 1911; patented January 7, 1913.

Sakai, Yasudiro. "Vapor Electric Apparatus." United States Patent no. 1148628. Filed June 14, 1912; patented August 3, 1915.

Sakai, Yasudiro. "Vapor Electric Device." United States Patent no. 1101665. Filed December 30, 1910; patented June 30, 1914.

"Salmis Journalier." *Milwaukee Journal* (May 3, 1897), 4.

Sammons, Thomas. "Chinese Typewriter of Unique Design." Department of Commerce Bureau of Foreign and Domestic Commerce. *Commerce Reports* 3, nos. 154–230 (May 24, 1916): 20.

Sampson, Geoffrey. *Writing Systems*. Stanford: Stanford University Press, 1985.

Schleicher, August. "Darwinism Tested by the Science of Language." Translated by Max Müller. *Nature* 1, no. 10 (1870): 256–259.

Schmalzer, Sigrid. *The People's Peking Man: Popular Science and Human Identity in Twentieth-Century China*. Chicago: University of Chicago Press, 2008.

Schmid, Andre. *Korea between Empires, 1895–1919*. New York: Columbia University Press, 2002.

Schoenhals, Michael. *Doing Things with Words in Chinese Politics*. Berkeley: Institute of East Asian Studies, University of California, Berkeley, 1992.

Schurmann, Franz. *Ideology and Organization in Communist China*. Berkeley: University of California Press, 1966.

Schwarcz, Vera. "A Curse on the Great Wall: The Problem of Enlightenment in Modern China." *Theory and Society* 13, no. 3 (May 1984): 455–470.

"Science and Industry." *Arkansas Democrat* (October 10, 1898), 7.

Scott, Edward Benjamin. *Sixpenny Telegrams. Scott's Concise Commercial Code of General Business Phrases*. London: n.p., 1885.

"Secretariat Purchase of Japanese Typewriter." Memo from "S. Ozawa Secretary" to "The Treasurer." SMA U1-4-3582 (February 15, 1943), 3.

Seeley, Christopher. *A History of Writing in Japan*. Leiden: Brill, 1991.

Seybolt, Peter J., and Gregory Kuei-ke Chiang. *Language Reform in China: Documents and Commentary*. White Plains, NY: M.E. Sharpe, 1979.

Shah, Pan Francis. "Type-Writing Machine." United States Patent no. 1247585. Filed October 20, 1916; patented November 20, 1917.

Shay, Frank. *Cipher Book for the Use of Merchants, Stock Operators, Stock Brokers, Miners, Mining Men, Railroad Men, Real Estate Dealers, and Business Men Generally*. Chicago: Rand McNally and Co., 1922.

Sheehan, Brett. *Trust in Troubled Times: Money, Banks, and State-Society Relations in Republican Tianjin*. Cambridge, MA: Harvard University Press, 2003.

Sheffield, Devello Z. "The Chinese Type-writer, Its Practicability and Value." In *Actes du onzième Congrès International des Orientalistes*, vol. 2. Paris: Imprimerie Nationale, 1898.

Sheffield, Devello Z. *Selected Lists of Chinese Characters, Arranged According of Frequency of their Recurrence*. Shanghai: American Presbyterian Mission Press, 1903.

Sheridan, Dave. Memo to Sales Staff regarding Remington Japanese Typewriter. Hagley Museum and Library. Accession no. 1825. Remington Rand Corporation. Records of the Advertising and Sales Promotion Department. Series I Typewriter Div. Subseries B, Remington Typewriter Company, box 3, folder 6 "Keyboards and Typestyles–Correspondence, 1906."

"The Shrewd Buyer Investigates." *New Metropolitan* 21, no. 5 (1905): 662.

"A Siamese Typewriter." *School Journal* (July 3, 1897), 12.

"Siam's Future King Guest in Syracuse." *Syracuse Post-Standard* (November 4, 1902), 5.

Siegert, Bernard. *Cultural Techniques: Grids, Filters, Doors, and Other Articulations of the Real*. Translated by Geoffrey Winthrop-Young. New York: Fordham University Press, 2015.

"Simplified Chinese." *The Far Eastern Republic* 1, no. 6 (March 1920): 47.

The Simpsons. Season 13, episode 1304. "A Hunka Hunka Burns in Love." December 2, 2001.

Sinensis, Typographus. "Initial Notes on Estimate of Proportionate Expense of Xylography, Lithography, and Typography." *Chinese Repository* 3 (May 1834–April 1835).

Slater, Robert. *Telegraphic Code to Ensure Secresy [sic] in the Transmission of Telegrams*. London: W.R. Gray, 1870.

Smith, A.H. "In Memoriam. Dr. Devello Z. Sheffield." *Chinese Recorder* (September 1913), 564–568.

So, Richard Jean. "Chinese Exclusion Fiction and Global Histories of Race: H.T. Tsiang and Theodore Dreiser, 1930." *Genre* 34 (2006): 1–21.

So, Richard Jean. "Collaboration and Translation: Lin Yutang and the Archive of Asian American Literature." *Modern Fiction Studies* 56, no. 1 (2010): 40–62.

Soothill, William Edward. *Student's Four Thousand 字 and General Pocket Dictionary*. Shanghai: American Presbyterian Mission Press, 1908.

Specimen of Cuts and Types in the Printing Office of the Shanghai Mission of the Board of Foreign Missions of the Presbyterian Church in the United States. Shanghai: n.p., 1865. Library of Congress. G/C175.6/P92S2.

Spurgin, Richard A. "Type Writer." United States Patent no. 1055679. Filed August 11, 1911; patented March 11, 1913.

Standage, Tom. *The Victorian Internet: The Remarkable Story of the Telegraph and the Nineteenth Century's On-line Pioneers.* New York: Berkeley Books, 1999.

Star, Susan Leigh. "Introduction: The Sociology of Science and Technology." *Social Problems* 35, no. 3 (June 1988): 197–205.

Staunton, George Thomas. *Ta Tsing Leu Lee: Being the Fundamental Laws, and a Selection from the Supplementary Statutes, of the Penal Code of China.* London: Printed for T. Cadell and W. Davies, in the Strand, 1810.

Steele, H.H. "Arabic Typewriter." United States Patent no. 1044285. Filed October 24, 1910; patented November 12, 1912.

Stellman, Louis John. *Said the Observer.* San Francisco: The Whitaker & Ray Co., 1903.

"Stenographer Has a Tough Job." *Ludington Daily News* (April 8, 1937), 5.

Steward, J. *The Stranger's Guide to Paris.* Paris: Baudry's European Library, 1837.

Stewart, Neil. "China at the Leipzig Fair." *Eastern World* 7, no. 10 (October 1953): 42–44.

Stickney, Burnham. "Typewriting Machine." United States Patent no. 1549622. Filed February 9, 1923; patented August 11, 1925.

Strom, Sharon Hartman. *Beyond the Typewriter: Gender, Class, and the Origins of Modern American Office Work, 1900–1930.* Chicago: University of Illinois Press, 1992.

Su, Ching. "The Printing Presses of the London Missionary Society among the Chinese." PhD diss., University College London, 1996.

Suchman, Lucy. *Human-Machine Reconfigurations: Plans and Situated Actions.* Cambridge: Cambridge University Press, 2006.

Su Tow, Julius. *The Real Chinese in America.* New York: Academy Press, 1923.

Tai, Evelyn. Interview. July 11, 2010. London, United Kingdom.

Tao, Wen Tsing. "Mr. H. Chi's New Contribution." *Chinese Students' Monthly* 12 (1916): 101–105.

Tcherkassov, Baron Paul, and Robert Erwin Hill. "Type for Type-Writing or Printing." United States Patent no. 714621. Filed November 21, 1900; patented November 25, 1902.

"Telegraphy of the Chinese." *San Francisco Chronicle* (July 5, 1896), 14.

Temporary Catalogue of the Department of Fine Arts Panama-Pacific International Exposition: Official Catalogue of Exhibitors. Rev. ed. San Francisco: The Wahlgreen Co., 1915.

Tsu, Jing. *Sound and Script in Chinese Diaspora.* Cambridge, MA: Harvard University Press, 2011.

Turkle, Sherry, ed. *Evocative Objects: Things We Think With.* Cambridge, MA: MIT Press, 2007.

Turkle, Sherry. "Inner History." In *The Inner History of Devices,* edited by Sherry Turkle, pp. 2–31. Cambridge, MA: MIT Press, 2008.

Twelfth Annual Report of the American Tract Society. Boston: Perkins and Marvin, 1837.

"Typewriter in Chinese." *Denver Evening Post* (May 29, 1897), 1.

"Typewriter Made for Chinese After 20 Years of Toil." *Washington Post* (April 18, 1937), F2.

"Typewriter Notes." *Phonographic Magazine and National Shorthand Reporter* 18 (1904): 322.

"Typewriters Built to Correspond with Merchants of China, Servia [*sic*], Armenia, Russia and Other Countries." *Washington Post* (November 17, 1912), M2.

"Typewriters to Orient: Remington Rand Sends Consignment of 500 in the Mongolian Language." *Wall Street Journal* (April 26, 1930), 3.

"Typewrites in Chinese; Oriental Student at New York Invents Machine." *Washington Post* (March 28, 1915), B2.

"Typewriting in Chinese. Machine Developed upon Which Forty Words a Minute May Be Written by an Expert Operator." *Washington Post* (May 23, 1915), III, 17.

Unger, J. Marshall. "The Very Idea: The Notion of Ideogram in China and Japan." *Monumenta Nipponica* 45, no 4 (Winter 1990): 391–411.

University of Illinois Urbana-Champaign, ed. *University of Illinois Directory: Listing the 35,000 Persons Who Have Ever Been Connected with the Urbana-Champaign Departments, Including Officers of Instruction and Administration and 1397 Deceased.* Urbana-Champaign, 1916.

Vella, Walter Francis. *Chaiyo! King Vajiravudh and the Development of Thai Nationalism.* Honolulu: University Press of Hawai'i, 1978.

Wager, Franz X. "Type-Writing Machine." United States Patent no. 829494. Filed November 9, 1905; patented August 28, 1906.

Wagner, Rudolph G. "The Early Chinese Newspapers and the Chinese Public Sphere." *European Journal of East Asian Studies* 1, no. 1 (2001): 1–33.

Wang, Chih-ming. "Writing Across the Pacific: Chinese Student Writing, Reflexive Poetics, and Transpacific Modernity." *Amerasia Journal* 38, no. 2 (2012): 136–154.

Wang, Chin-chun [王 景 春]. "The New Phonetic System of Writing Chinese Characters." *Chinese Social and Political Science Review* 13 (1929): 144–160.

Wang Hui. "Discursive Community and the Genealogy of Scientific Categories." In *Everyday Modernity in China*, edited by Madeleine Yue Dong and Joshua L. Goldstein. Seattle: University of Washington Press, 2006, 80–120.

Wang, John [H. L. Huang]. "Technical Education in China." *Chinese Students' Monthly* 11, no. 3 (January 1, 1916): 209–214.

Wang Kuoyee. "Chinese Typewriter." United States Patent no. 2534330. Filed March 26, 1948; patented December 19, 1950.

Wang Zheng. *Women in the Chinese Enlightenment: Oral and Textual Histories*. Berkeley: University of California Press, 1999.

Wang Zuoyue. "Saving China through Science: The Science Society of China, Scientific Nationalism, and Civil Society in Republican China." *Osiris* 17 (2002): 291–322.

Wasserstrom, Jeffrey N. *Student Protests in Twentieth-Century China: The View from Shanghai*. Stanford: Stanford University Press, 1997.

Waterman, T.T., and W.H. Mitchell, Jr. "An Alphabet for China." *Mid-Pacific Magazine* 43, no. 4 (April 1932): 343–352.

Watt, Lori. *When Empire Comes Home: Repatriates and Reintegration in Postwar Japan*. Cambridge, MA: Harvard University Asia Center, 2009.

Wershler-Henry, Darren. *The Iron Whim: A Fragmented History of Typewriting*. Ithaca: Cornell University Press, 2007.

Weston, Timothy B. "Minding the Newspaper Business: The Theory and Practice of Journalism in 1920s China." *Twentieth-Century China* 31, no. 2 (April 2006): 4–31.

"William P. Fenn, 90, Protestant Missionary." *New York Times* (April 25, 1993), A52.

Williams, R. John. "The Technê-Whim: Lin Yutang and the Invention of the Chinese Typewriter." *American Literature* 82, no. 2 (2010): 389–419.

Williams, Samuel Wells. "Draft of General Article on the Chinese Language." Samuel Wells Williams Family Papers, box 13, folder 38. Yale University Library, n.d.

Williams, Samuel Wells. *The Middle Kingdom: A Survey of the Chinese Empire and Its Inhabitants*. New York: Wiley & Putnam, 1848.

Williams, Samuel Wells. "Movable Types for Printing Chinese." *Chinese Recorder and Missionary Journal* 6 (1875): 22–30.

Williams, Samuel Wells, family papers. Yale University Library Manuscripts and Archives. MS 547 Location LSF, series II, box 13.

"Will Typewrite Chinese." *Atchison Daily Globe* (June 1, 1897), 3.

Wilson, Mary Badger. "Fleet-Fingered Typist." *New York Times* (December 2, 1923), SM2.

World War I Draft Registration Card. United States, Selective Service System. World War I Selective Service System Draft Registration Cards, 1917–1918. National Archives and Records Administration, Washington, DC, M1509.

World War II Draft Registration Card. United States, Selective Service System. Selective Service Registration Cards, World War II: Fourth Registration. National Archives and Records Administration Branch locations: National Archives and Records Administration Region Branches.

Wu, K.T. "The Development of Typography in China During the Nineteenth Century." *Library Quarterly* 22, no. 3 (July 1952): 288–301.

Wyckoff, Seamans & Benedict. *The Remington Standard Typewriter*. Boston: Wyckoff, Seamans & Benedict (Remington Typewriter Co.), 1897.

Yamada, Hisao. "A Historical Study of Typewriters and Typing Methods; from the Position of Planning Japanese Parallels." *Journal of Information Processing* 2, no. 4 (February 1980): 175–202.

Yamada, Hisao, and Jiro Tanaka. "A Human Factors Study of Input Keyboard for Japanese Text." *Proceedings of the International Computer Symposium* [Taipei] (1977), 47–64.

Yanagiwara, Sukeshige. "Type-writing Machine." United States Patent no. 1206072. Filed February 1, 1915; patented November 28, 1916.

Yang, Daqing. *Technology of Empire: Telecommunications and Japanese Expansion in Asia, 1883–1945*. Cambridge, MA: Harvard University Asia Center, 2011.

Yang, Daqing. "Telecommunication and the Japanese Empire: A Preliminary Analysis of Telegraphic Traffic." *Historical Social Research* 35, no. 1 (2010): 66–89.

Ye, Weili. *Seeking Modernity in China's Name: Chinese Students in the United States, 1900–1927*. Stanford: Stanford University Press, 2001.

Yee, James. Email communication, July 6, 2009.

Yen, Tisheng. "Typewriter for Writing the Chinese Language." United States Patent no. 2471807. Filed August 2, 1945; patented May 31, 1949.

Yu, Pauline, Peter Bol, Stephen Owen, and Willard Peterson, eds. *Ways with Words: Writing about Reading Texts from Early China*. Berkeley: University of California Press, 2000.

Zacharias, Yvonne. "Longest Olympic Torch Relay Ends in Vancouver." *Vancouver Sun* (February 12, 2010).

Zhang Longxi. *The Dao and the Logos*. Durham: Duke University Press, 1992.

Zheng, Xiaowei. "The Making of Modern Chinese Politics: Political Culture, Protest Repertoires, and Nationalism in the Sichuan Railway Protection Movement in China." PhD diss. University of California, San Diego, 2009.

Zhou Houkun [Chow, Houkun]. "The Problem of a Typewriter for the Chinese Language." *Chinese Students' Monthly* (April 1, 1915), 435–443.

Zhu Jiahua [Chu Chia-hua]. *China's Postal and Other Communications Services.* Shanghai: China United Press, 1937.

JAPANESE-LANGUAGE SOURCES

Advertisement for *The Army Typist (Jūgun taipisuto)* [從軍タイピスト], by Sakurada Tsunehisa [櫻田常久]. *Taipisuto* [タイピスト] 18, no. 3 (March 1943) 35.

"Headed Valiantly to the Continent as Typists *(Taipisuto to shite ooshiku tairiku e)* [タイピストとして雄雄しく大陸へ]." *Taipisuto* [タイピスト] 17, no. 5 (May 1942): 36.

"Heading to the South Seas as a Typist *(Taipisuto toshite nanyō e)* [タイピストとして南洋へ]." *Asahi Shinbun* [朝日新聞] (August 24, 1939), E6.

Investigation of Women's Occupations in Tokyo and Osaka (Tōkyō Ōsaka ryōshi ni okeru shokugyō fujo chōsa) [東京大阪両市に於ける職業婦女調査]. Tokyo: n.p., 1927.

Japanese Typewriter Company Divisions. *Taipisuto* [タイピスト] 17, no. 10 (October 1942—Showa 16).

Kurosawa Sumiko [黒澤澄子]. "A Snowy Night in Nanjing *(Nankin no yuki no yoru)* [南京の雪の夜]." *Taipisuto* [タイピスト] 16, no. 4 (April 1940): 20–21.

Makimasa [牧正]. "Diary of a Garrison in South China *(Nanshi chūgun ki)* [南支駐軍記]." *Taipisuto* [タイピスト] 17, no. 1 (February 1942): 16–25.

Manchukuo Capacity Research Committee *(Manshū nōryoku kenkyūkai)* [滿洲能力研究會]. *Efficiency of the Japanese Typewriter (Hōbun taipuraitā no nōritsu)* [邦文タイプライターの能率]. N.p., 1936.

"Mantetsu Type Competition Results *(Zen Mantetsu jousho kyōgi taikai no seiseki)* [全滿鉄淨書競技大會の成績]." *Taipisuto* [タイピスト] 17, no. 10 (October 1942—Showa 16): 6–11.

Morita Torao [森田虎雄]. *Japanese Typewriter Textbook (Hōbun taipuraitā kyokasho)* [邦文タイプライター教科書]. Tokyo: Tokyo Women's Foreign Language School [東京女子外國語學校], 1934.

"Newly Invented Chinese Character Typewriter *(Saishin hatsumei kanbun taipuraitā)* [最新発明漢文タイプライター]." *New Youth (Shin Seinen)* [新青年] (June 1927—Showa 2). Reprint of Gilbert Levering. "Chinese Language Typewriter." *Life* 2311 (February 17, 1927): 4.

Nippon Typewriter Company [日本イプライター株式會社], ed. *Character Index for Japanese Typewriter* [邦文タイプライター用文字の索引]. Tokyo: n.p., 1917.

Nippon Typewriter Company Divisions. *Taipisuto* [タ イ ピ ス ト] 17, no. 10 (October 1942—Showa 16): 54.

Nishida Masaaki [西 田 正 秋]. "Japanese-Style Female Beauty of Today (*Kyō no nipponteki joseibi*) [今 日 の 日 本 的 女 性 美]." *Taipisuto* [タ イ ピ ス ト] 17, no. 7 (July 1942): 2–5.

Omi Hironobu [小 見 博 信]. "Japanese Culture and the Mission of the Japanese Typewriter (*Nipponbunka to Hōbun taipuraitā no shimei*) [日 本 文 化 と 邦 文 タ イ プ ラ イ タ ー の 使 命]." *Taipisuto* [タ イ ピ ス ト] 17, no. 11 (November 1942): 12–13.

Oriental Typewriter Character Handbook (Tōyō taipuraitā moji binran) [東 洋 タ イ プ ラ イ タ ー 文 字 便 覧: 弐 号 機 用]. Tokyo: Oriental Typewriter Co. [東 洋 タ イ プ ラ イ タ ー], 1923.

Photographs of Japanese Typing School Graduates. *Taipisuto* [タ イ ピ ス ト] 17, no. 1 (February 1942—Showa 16): 27.

Photographs of Japanese Typing School Graduates. *Taipisuto* [タ イ ピ ス ト] 17, no. 3 (March 1942—Showa 16): 31.

Photographs of Japanese Typing School Graduates. *Taipisuto* [タ イ ピ ス ト] 17, no. 10 (October 1942—Showa 16): 27.

"Recommended Words for Common Usage Kanji (*Jōyō Kanji amane mi suisen no kotoba*) [常 用 漢 字 遍 覽 推 薦 の 言 葉]." *Taipisuto* [タ イ ピ ス ト] 16, no. 5 (May 1940): 4.

"A Sailor's Inspection of the Typewriter (*Suihei-san no taipuraitā kengaku*) [水 兵 さ ん の タ イ プ ラ イ タ ー 見 学]." *Taipisuto* [タ イ ピ ス ト] 16, no. 7 (July 1940): 16.

"The Six Patriotic Women Arrive in Tianjin (*Tenshin e tsuita aikoku roku josei*) [天 津 へ 着 い た 愛 国 六 女 性]." *Asahi Shinbun* [朝 日 新 聞] (January 4, 1938), 10.

"Spring Meeting of the Shanghai Branch (*Shanhai shibu shunki taikai*) [上 海 支 部 春 季 大 會]." *Taipisuto* [タ イ ピ ス ト] 16, no. 7 (July 1940): 21.

"Telegraph Code (*Denshin jigō*) [電 信 字 号]." "Extension Selskabet—Japansk Telegrafnøgle." 1871. Arkiv nr. 10.619. In "Love og vedtægter med anordninger." GN Store Nord A/S SN China and Japan Extension Telegraf. Rigsarkivet [Danish National Archives]. Copenhagen, Denmark.

Watabe Hisako [渡 部 久 子]. *Japanese Typewriter Textbook (Hōbun taipuraitā tokuhon)* [邦 文 タ イ プ ラ イ タ ー 讀 本]. Tokyo: Sūbundō [崇 文 堂], 1929.

Yuji Riichi [湯 地 利 市]. "On the Mantetsu Type Competition (*Mantetsu no taipu kyōgi ni tsuite*) [満 鉄 の タ イ プ 競 技 に 就 て]." *Taipisuto* [タ イ ピ ス ト] 18, no. 10 (October 1943—Showa 17): 2–3.

FRENCH-LANGUAGE SOURCES

Bembanaste, V. "Turquie d'hier ..." *Rivista Olivetti* 2 (July 1948): 56–58.

"A Beyrouth la Olivetti." *Rivista Olivetti* 5 (November 1950): 54.

"Cérémonie d'ouverture des jeux olympiques" in "Règlements et Protocole de la Célébration des Olympiades Modernes et des Jeux Olympiques Quadriennaux" (1921), 10.

Châh Nameh, sous le titre: Le livre des rois par Aboul'kasim Firdousi, publié, traduit et commenté par M. Jules Mohl. 7 vols. Paris: Jean Maisonneuve, 1838–1878; "Le Clavier Arabe." *Rivista Olivetti* 2 (July 1948): 26–28.

Christian, Arthur. *Débuts de l'Imprimerie en France.* Paris: G. Roustan and H. Champion, 1905.

Convention télégraphique internationale de Berlin (1885): Publiée par le Bureau International des Administrations Télégraphiques. Bern: Imprimerie Rieder & Simmen, 1886.

Convention télégraphique internationale de Paris, révisée à Vienne (1868) et Règlement de service international (1868)—Extraits de la publication: Documents de la conférence télégraphique internationale de Vienne. Vienna: Imprimerie Impériale et Royale de la Cour et de l'Etat, 1868.

Convention télégraphique internationale de Saint-Pétersbourg et Règlement et tarifs y annexés (1875). Extraits de la publication—Documents de la Conférence télégraphique internationale de St-Pétersbourg: Publiés par le Bureau International des Administrations Télégraphiques. Bern: Imprimerie Rieder & Simmen, 1876.

Convention télégraphique internationale et règlement et tarifs y annexés révision de Londres (1903). London: The Electrician Printing and Publishing Co., 1903.

Convention télégraphique internationale et règlement y annexé—Révision de Paris (1925). Bern: Bureau International de l'Union Télégraphique, 1926.

Dictionnaire télégraphique officiel chinois en français (Fawen yi Huayu dianma zihui) [法文譯華語電碼字彙]. Shanghai: Dianhouzhai [點后齋], n.d. Rigsarkivet [Danish National Archives]. Copenhagen, Denmark.10619 GN Store Nord A/S. 1870–1969 Kode- og telegrafbøger. Kodebøger 1924–1969.

Documents de la conférence télégraphique internationale de Berlin: Bureau International des Administrations Télégraphiques. Bern: Imprimerie Rieder & Simmen, 1886.

Documents de la conférence télégraphique internationale de Lisbonne. Bern: Bureau International de l'Union Télégraphique, 1909.

Documents de la conférence télégraphique internationale de Madrid (1932). Vol. 1. Bern: Bureau International de l'Union Télégraphique, 1933.

Documents de la conférence télégraphique internationale de Paris. Bern: Imprimerie Rieder & Simmen, 1891.

Documents de la conférence télégraphique internationale de Paris (1925). Vol. 1. Bern: Bureau International de l'Union Télégraphique, 1925.

Documents de la conférence télégraphique internationale de Paris (1925). Vol. 2. Bern: Bureau International de l'Union Télégraphique, 1925.

Drège, Jean-Pierre. *La Commercial Press de Shanghai, 1897–1949*. Paris: Publications Orientalistes de France, 1979.

Escayrac de Lauture, Comte d'. *Grammaire du télégraphe: Histoire et lois du langage, hypothèse d'une langue analytique et méthodique, grammaire analytique universelle des signaux*. Paris: J. Best, 1862.

Imprimerie Nationale. *Catalogue des caractères chinois de l'Imprimerie Nationale, fondus sur le corps de 24 points*. Paris: Imprimerie Nationale, 1851.

Legrand, Marcellin. *Spécimen de caractères chinois gravés sur acier et fondus en types mobiles par Marcellin Legrand*. Paris: n.p., 1859.

Legrand, Marcellin. *Tableau des 214 clefs et de leurs variantes*. Paris: Plon frères, 1845.

"La Olivetti au Viet-Nam, au Cambodge et au Laos." *Rivista Olivetti* 5 (November 1950): 70–72.

Pauthier, Jean-Pierre Guillaume. *Chine ou Description historique, géographique et litté-raire de ce vaste empire, d'après des documents chinois. Première partie*. Paris: Firmin Didot Frères, Fils, et Cie., 1838.

Pauthier, Jean-Pierre Guillaume. *Foe Koue Ki ou Relation des royaumes bouddhiques*. Paris: Imprimerie Royale, 1836.

Pauthier, Jean-Pierre Guillaume. *Sinico-Aegyptiaca. Essai sur l'origine et la formation des écritures chinoise et égyptienne*. Paris: F. Didot Frères, 1842.

Pauthier, Jean-Pierre Guillaume. *Le Ta-Hio ou la Grande Étude, ouvrage de Confucius et de ses disciples, en chinois, en latin et en français, avec le commentaire de Tchou-hi*. Paris: n.p., 1837.

Pauthier, Jean-Pierre Guillaume. *Le Tào-te-Kîng, ou le Livre de la Raison Suprême et de la Vertu, par Lao-Tseu, en chinois, en latin et en français, avec le commentaire de Sie-Hoèi, etc.* Paris: F. Didot Frères, Libraires, 1838.

Pauthier, Jean-Pierre Guillaume. *Ta thsîn Kîng-Kiao; l'Inscription Syro-chinoise de Singan-fou, monument nestorien élevé en Chine l'an 781 de notre ère et découvert en 1625. En chinois, en latin et en français, avec la prononciation figurée, etc.* Paris: Librairie de Firmin Didot Frères, Fils, et Cie, 1858.

Pelliot, Paul. *Les débuts de l'imprimerie en Chine*. Paris: Librairie d'Amérique et d'Orient Adrien-Maisonneuve, 1953.

"Règle 69, 'Cérémonies d'ouverture et de clôture,'" *Charte Olympique* (1991), n.p.

Rosny, L. Léon de. *Table des principales phonétiques chinoises*. Paris: Maisonneuve et Cie, 1857.

Viguier, Septime Auguste. *Memoir on the Establishment of Telegraph Lines in China* (*Mémoire sur l'établissement de lignes télégraphiques en Chine*). Shanghai: Imprimerie Carvalho & Cie., 1875.

ITALIAN-LANGUAGE SOURCES

"Diagramma per Tastiera M. 80: Hindi (Diagram for Keyboard M. 80: Hindi)." August 19, 1954. Simbolo 205-B (46 Tasti). Fase 220. Olivetti Historical Archives.

"Diagramma per Tastiera M. 80: Inglese per Shanghai (Diagram for Keyboard M. 80: English for Shanghai)." July 12, 1950. Simbolo -B. DCUS. Fase 220. Olivetti Historical Archives.

"Diagramma per Tastiera M. 80: Italia (Diagram for Keyboard M. 80: Italy)." December 17, 1953. Simbolo 1-B. DCUS. Fase 220. Olivetti Historical Archives.

"Diagramma per Tastiera M. 80: Londra (Diagram for Keyboard M. 80: London)." October 13, 1948. Simbolo 118-B. DCUS. Fase 220. Olivetti Historical Archives.

"In India con l'Olivetti (In India with Olivetti)." *Giornale di fabbrica* 4–5 (August–September 1949): 8.

"La lexicon oltre il Circolo Polare." *Rivista Olivetti* 5 (November 1950): 16–17.

"Le macchine arabe scrivono a ritroso: A Beirut l'Olivetti vince in arabo e in francese." *Rivista Olivetti* 5 (November 1950): 52–53.

"Le macchine Olivetti scrivono in tutte le lingue." *Notizie Olivetti* 55 (March 1958): 1–4.

"Notizie dall'estero (News from abroad)." *Notizie Olivetti* 36 (April 1956): 13–15.

"Notizie dall'estero (News from abroad)." *Notizie Olivetti* 38 (June 1956): 14–16.

"La Olivetti nei mercati del Medio Oriente: Incontro con gli Arabi." *Notizie Olivetti* 11 (November 1953): 8–9.

"La Olivetti nel mondo." *Notizie Olivetti* 21 (November 1954): 3.

"Radio Olivetti." *Rivista Olivetti* 4 (April 1950): 78–101.

"Un po' d'Europa nel cuore dell'Africa: il Congo (A Little Bit of Europe in the Heart of Africa: the Congo)." *Notizie Olivetti* 32 (December 1955): 6–8.

OTHER LANGUAGES

Marakueff, Aleksandr Vladimirovitch. *Chinese Typewriter* (*Kitaisckaya pishutcaya mashina*). *Memoirs of the Far Eastern State University* (Vladivostok) 1 (1932).

INDEX

Page numbers followed by the letter "f" refer to figures illustrating the text. Page numbers followed by the letter "t" refer to tables. Page numbers containing the letter "n" refer to endnotes, with the number of the note following.

"abject form," 65
abolition of characters, in Korea and
  Japan, 201–202
abolition of Chinese characters, 15, 22,
  44
  as "easy iconoclasm," 14, 26
  lack of typewriter as argument for, 124
  proponents of, 13–14, 190
  rejection of, 125, 138, 156, 171–172
advertising, Chinese, 153–154
A.E.G. Company, 343n37
aesthetics
  of divisible type, 96–103, 154
  machinic, 151, 153
  of Oriental "essences" in Occident, 92
  politics of, 156
  traditional Chinese orthographic, 93–
    95, 96, 130, 132, 151, 154
  twentieth-century transformation of,
    153
agonistic methodology, 30, 31–32

Ai Weiwei, 3
aikoku taipisuto (patriotic typists), 222
algorithmic reading, 76, 84, 85, 88, 91,
  347n3
  by Chinese researchers, 141, 361n63
Allard, J. Frank, 185
Allen, Joseph, 261
alphabet, Cyrillic, 60
alphabet, fetishization of, 67–68
alphabet, Greek, 2, 186
alphabet, Latin
  on stenotype machine, 317–318
  used as paratextual technology in
    Chinese, 8–9, 255–256
  used in Chinese telegraph code, 117,
    118f, 120
alphabet, Phoenician, 2, 186
alphabetic order
  as analogy for character organization
    methods, 249
  Chinese telegraph code and, 120

alphabetic order (cont.)
  lack of in Chinese script, 3–9, 250
  in Olympic Parade of Nations, 2, 4,
    338n9
  as stable, 250, 252, 253
alphacentric technologies, 9–10, 25, 46,
  60, 352n48
alphasyllabaries (abugidas), 11
Analytic Universal Telegraphy (Escayrac
  de Lauture), 106, 108–109
Anderson, Leroy, 27–28, 29, 31
Anglo-Chinese College, 102, 147
Anglo-French expedition to China
  (1859), 106
animation, Chinese, 170
anti-reading. *See* algorithmic reading
Apple (computer company), 241
Arabic numerals
  adoption of, in Chinese, 140
  on Shu-style typewriter, 366n19
Arabic script, and typewriter, 62–63,
  344n52
Army Typist, The (Sakurada Tsunehisa),
  222
Asahi shinbun (periodical), 222, 375n7
ASIA (magazine), 188, 278
"auto-telecommunication," 318

Baidu, 320
baihua. See vernacularization of
  Chinese
Bangkok, 50
Baoshan Chinese Typewriting
  Supplementary School, 228–229
Barr, John H., 345n54
Beijing, 227, 228, 278, 284. *See also*
  Olympic Games
beiyongzi. *See* infrequently used
  characters
Bell, Alexander Graham, 55
Berry, Mary Elizabeth, 248
Beyerhaus, Auguste, 102, 149, 265, 325,
  351n38, 351n39

Bi Sheng, 81, 169, 325
Bible, 87, 89, 135, 349n19, 351n38,
  358n20
Blair, Ann, 247–248
blind typing, 179
Bloom, Alfred, 70, 72
Boas, Franz, 68
Bodde, Derk, 70, 72
bodies, human
  attuned to Communist rhetoric, 292,
    304
  Chinese typewriting and, 134, 172,
    182, 371n55
  in Chinese versus alphabetic typing,
    178–180
  natural-language experimentation
    and, 290, 300, 304
  standardization and, 303
  technolinguistic history and, 178,
    290
Bollywood, 29
Boltz, William, 70
Bombay Talkie (film), 28–29, 30
Book from the Sky (Xu Bing), 16–18, 22
Boulder Dam, as analogy for Chinese
  typewriter, 244, 245
Boxer Indemnity Scholarship Program,
  137, 138
boycott of Japanese goods in China,
  218, 219
Boyfriend (magazine), 215
Breteuil, 355n74
Brook, Timothy, 225
Brown, Alexander T., 47
Brumbaugh, Robert, 151
brushtalk, 209–210, 211
Bryans, William Remington, 148,
  363n78
Bryson, Bill, 35, 40, 41, 43
Buck, Pearl S., 272, 278
Burgess, Anthony, 35
bushou. See radicals
Byrne, David, 319

Cable and Wireless, 201, 315
Cadmus, 186–187, 279
Cai Yuanpei, 185, 249
Caligraph (typewriter model), 50
calligraphy, Chinese, 16, 96, 281,
    399n66
  aesthetics of, 94–95, 99
  orthographic principles of, 5, 94–95,
    256
candidacy (in input), 240, 281
Cang Jie, 170
carbon-paper copying, 170, 180, 220
carriage advance mechanism, 59, 60, 63
  reversed in Arabic and Hebrew
    typewriters, 60–61, 343n45, 344n52
Carter, Harry, 17, 18, 19
cartoons, mocking Chinese typewriters,
    35–38, 42, 192–193
cenemic script, 11–12, 68
Chang, C.C., 157, 363n78
Chang, Chung-yuan, 279
Changsha, 142
character cylinder (typewriter part),
    166, 167, 265, 363n79
  in Japanese typewriter, 205
"character retrieval problem." *See under*
  retrieval of Chinese characters
characters, Chinese
  abundance of, 77, 88, 264–265
  in computing, 239–243
  described verbally by components,
    261
  as elemental unit of script, 79–80
  encounter with Western information
    technologies, 9–10, 44
  fake, 16–18
  "fundamental method" of organizing,
    250–253, 256–258
  as hindrance to technology, 184, 316
  as "ideographic," 66, 320
  invention of, 170
  as irresolvable individuals, 131, 132
  "natural laws" of, 99

not made to fit the typewriter, 156
"puzzling" of, 75–121, 211
as semi-phoneticized "half-breed," 67
simplification of, 15, 19, 20, 301
spatial archetypes of, 262
"spelled" from components, 80, 90,
    94, 102, 104, 131, 149
structural balance of, 75
as ultimate Other of world scripts,
    11–12, 64–65, 317
used for transnational
    communication, 209–210
Western denigration of, 38, 43–44,
    65–74, 81
See also abolition of Chinese
    characters; radicals
character selector (typewriter part), 143,
    197, 199
character slugs (typewriter part), 30,
    169, 226
  lacking for infrequent characters, 284
  not fixed, 163, 295
  on Shu-style typewriter, 163–164, 167,
    364n1
  Yu Binqi and, 219
  See also infrequently used characters
Chartier, Roger, 178
Chen Changgeng, 230, 231, 233
Chen Duxiu, 12, 15, 252
Chen Guangyao, 296, 298
Chen Heqin, 141, 142, 361n58
Chen Lifu, 253–258, 325, 390n23,
    390n25
  invention of Five-Stroke system, 254
  as "righting the wrongs of the
    ancients," 256–258
Chen Pengnian, 77
Chen Songling, 175
Chiang Kai-shek, 254, 278, 371n55
Chicago Daily Tribune, 244, 245
China Standard Typewriter
    Manufacturing Company, 226
Chinatown, Los Angeles, 161

Chinatown, San Francisco, 35
China Typewriter Company, 233
"Chinese alphabet" (Western
    misconception), 183–187
Chinese-American Typing Institute, 174
Chinese Civil War, 278
Chinese Exclusion Act, 362n76
Chinese Inventors Association, 220
Chinese National Physical Education
    Federation, 215
Chinese Phonetic Alphabet. *See zhuyin
    fuhao*
Chinese Railway Institute, 165
Chinese Students' Monthly, 125
Chinese Telegraph Administration,
    114
Chinese Times, 126
"Chinese Typewriter" (dance), 40
Chinese Typewriter, The (film), 38–39
chinmun. See "truth script"
Chu Yin Tzu-mu Keyboard. *See*
    typewriter models, Chinese:
    Remington Chinese Phonetic
    Typewriter
Civil Service Examination, Chinese, 21
Civil War, U.S., 127
CJK (Chinese-Japanese-Korean
    information technology), 200, 210
    in typewriter market, 209
Clarke, Adele, 307
cleaning services for typewriters, 225
cliché, 291, 295
cloud, the, 320
Coble, Parks, 225
"code consciousness," 240
codex, 178
Cold War, 196
colonialism, Western, 9, 12, 107
combinatorialism, 79, 121, 159, 316
    in Chinese-language printing, 89–103,
    106
    criticism of in typewriting, 157–158,
    264

definition of, 80
    in MingKwai, 264–265, 266f
    in Qi Xuan typewriter, 149–154, 150f
    in telegraphy, 103–104
    See also divisible type
Commercial Press, 142, 250, 252, 274,
    296, 364n104
    competition to, 212, 217, 223, 224,
    229
    and formation of typewriter industry,
    170–172, 175, 182
    implements *lianchuanzi*, 292
    interest in Zhou Houkun typewriter,
    157, 159, 165
    Japanese bombing of, 212–213
    mass-produces first Chinese
    typewriter, 161, 164, 166–169, 168f,
    169f
    presents Chinese typewriter to West,
    188–190
*Commercial Press Characters in the
    Chinese Spoken Language* (Chen
    Heqin), 141
common usage, 17–18, 121, 316
    approach to typewriting, 155–156
    in Chinese-language printing, 81–88,
    89, 106, 128
    in Chinese telegraph code, 110
    criticism of in typewriting, 158, 264,
    274
    definition of, 79
    "discovery" of by Westerners, 85
    Japanese kanji and, 204–205
    in MingKwai, 264, 267, 268
    and number of characters used in
    texts, 84–86
    politics of exclusion in, 87, 88, 100,
    103, 198
    in Qi Xuan typewriter, 148–149
    in Sheffield typewriter, 132–134
    in Shu-style typewriter, 163
    vibrancy in early twentieth-century
    China, 140–142

in Zhou Houkun typewriter, 137, 140, 143, 159, 166

See also descriptive imperative; exclusion of characters; prescriptive imperative

Communist Party, Chinese (CCP)
alliance with Nationalists, 254
attitude toward typing experiments, 300–301
in Chinese Civil War, 278–279
language reform and, 12, 142, 301
Zhang Jiying and, 292

Communist revolution (1949), 8, 232, 278–279
key political changes caused by, 299–300

competitions, typing
Manchuria-Wide Typing Competition, 224
North China Military Region Typing Competition, 298

"complete keyboard" typewriter. *See* double-keyboard typewriter

Computer History Museum, 237

computing
Chinese, 9, 14, 19, 24, 238–243, 288, 315–321
history of Western, 237–238
Western misconceptions of Chinese, 321

conceptual algorithms, 42–44

Conference on Unification of Pronunciation, 185

Confucianism, 12, 15, 202

Confucius, 86

Connected Language Tray Bed. *See* lianchuanzi

continuity, 18, 21–23

correspondence (of keystrokes to inscription), 239, 245, 317

Costas, Bob, 3–4, 6, 9

Cousin, Abel Joseph Constant, 53–54

Creel, Herrlee Glessner, 68–69, 70

criteria (in computing), 239, 240, 281

C.Y. Chao Typewriting Maintenance Department, 225

Daily Hiragana News, 202

Daily Picayune-New Orleans, 136

Damrong Rajanubhab, Prince, 45

dao (of organizing Chinese characters), 252, 253, 258

Daodejing, 89, 92

Daoism, 15, 89, 202

dayinben. See type-and-mimeograph editions

daziji. See typewriter, Chinese

daziyuan. See typists, Chinese

decentralization of tray bed changes, 288, 290, 300, 304, 308, 311, 397n55

deconstruction, 31–32

De la transmission télégraphique des caractères chinois (Escayrac de Lauture), 103–104

Denshin jigō. See kana: in Japanese telegraph code

Densmore Typewriter Company, 46, 50

descriptive imperative, 87, 134

Detroit, 137

"dialects" of Chinese, 13, 21, 40, 187, 243, 356n75
and "national language," 19
in popular culture, 39

dictionaries, Chinese, 15, 77, 82–83, 95, 270
foundational character dictionaries, 142, 204
See also Kangxi Dictionary

Ding, Cook, 99

directionality of script, in Hebrew and Arabic, 60

discipline, physical and mental, 178

divisible type, 80, 90–103, 106, 131, 351n37, 351n38
aesthetic politics of, 96–103, 154, 198

divisible type (cont.)
  compared to full-body type, 97–100,
    98f, 101f
  number of radicals needed in, 158
  Qi Xuan's typewriter and, 147, 149,
    151
  telegraphy and, 103
  widespread adoption of, 102
Dodge, Elbert S., 343n45
Dong Jing'an, 143
double-keyboard typewriter, 47–49, 48f,
    342n27
  disappearance of, 51, 56
Du Dingyou, 249, 250, 252, 260–263,
    270, 325, 390n29
Du Ponceau, Peter S., 66–67
Dyer, Samuel, 102, 325, 349n20

East Asia Japanese-Chinese Typewriting
    Professional Supplementary School,
    228, 229, 386n109
Eastern Extension A&C Telegraph
    Company, 110
East India Company, 84
edicts, imperial (*zhaoshu*), 213
Egyptian script, 66
eight fundamental strokes, 5, 5f, 94–95,
    256, 257f
electronic music, 318–319
embedded messages, 217
embodied practice. *See* bodies, human
"enciphered" transmissions
    (telegraphy), 112–113, 115
  Chinese considered to be, 114, 355n74
  during Nationalist military campaign,
    254
engineers
  duty of according to Zhou Houkun,
    138–139
  role in Chinese technolinguistic
    modernity, 10, 14
Escayrac de Lauture, Pierre Henri
    Stanislas d', 104–105, 110, 114, 325

Chinese telegraph code proposal,
    103–104, 105f, 109
  and universal telegraphic language,
    106, 108–109, 110, 112, 353n59
Ethnic Classification project (*minzu
    shibie*), 283
"evocative objects," 198
evolutionism
  in character retrieval debate, 252
  questioning of, 68–70
  in Western critiques of Chinese,
    43–44, 65–67, 71, 72
"examination Chinese," 202
exclusion of characters
  in Sheffield's typewriter, 132, 133,
    134–135
  in typewriters in general, 155–156
  in Zhou Houkun's typewriter, 166
"exotic type," 91

factories, typewriter, 59, 61, 63,
    167–169, 168f, 169f
false universalisms, 9, 10–12, 64, 317
  in Olympic Games, 4
Fan Jiling, 232, 233
Far East Trade Monthly, 224
feminization of clerical workforce,
    24
  in Japan, 205
  worldwide, 172–173, 175, 177, 274,
    392n47
Fengtian Japanese-Chinese Typing
    Institute, 229
Fengtian Typing Professional School,
    213
finding rod. *See* character selector
Five-Stroke (Wubi) retrieval system,
    254–258, 259f, 390n25. *See also*
    Chen Lifu
Flox, O.D., 126–128
Foguoji (A Record of Buddhist Kingdoms),
    89, 102, 351n39
Fong Sec, 188

fonts, Chinese, 89, 90, 92–93, 96, 102,
  348n16, 351n39
  radical new fonts, 153–154
  Song-style font (*songti*), 18
  See also divisible type; movable type
Forbidden City, 36, 147
foundational character sets, 141–142,
  155, 166, 361n63
Four Treasuries, The (Siku quanshu), 82
France, rationality of, 90, 93
frequency analysis of alphabetic scripts,
  60
frequency analysis of Chinese characters
  by Chinese researchers, 140–142,
    361n58
  in computer input, 19, 240
  in corpus of Chinese texts, 84, 89
  by Gamble, 75–76
  in natural-language arrangement, 305
  by Sheffield, 132
  in Shu-style typewriter tray bed, 163
  for typesetting, 82
Fujimura, Joan, 307
Fukuzawa Yukichi, 204
Fuller, Matthew, 239

gada gada gada (sound of Chinese
  typewriter), 30, 31
Gamble, William, 75–76, 134, 141, 325,
  349n19
  analysis of Chinese common usage,
    75–76, 79, 86, 87, 89, 127, 361n58
  criticism of Chinese movable type, 81,
    83, 347n4
Gao Mengdan, 249, 250
Gattopardo, Il (Lampedusa), 22
Gellner, Ernest, 202
Gestalt pattern-finding, 262, 271
Gilbert, Paul T., 183
Ginsberg, Allen, 71
Girouard, Gerry, 319
Glidden, Carlos, 46
Goody, Jack, 69

Google, 241, 281, 319
Grant, John Cameron, 184–185
"graphology" ("metalanguage"), in
  Chinese, 261
Great Asia Pictorial (periodical), 176
Greater East Asia Co-Prosperity Sphere,
  224
Great Leap Forward, 79, 300
Great Learning, The (Daxue), 89
Great Northern Telegraph Company,
  109, 201
Great Qing Legal Code, 84–85, 89
Greece, Western reverence for, 2
"Greek Miracle," 2, 69, 186
Guangde Chinese Typewriting
  Supplementary School, 227, 228
Guomindang, 195, 253
  Central Executive Committee, 253,
    255
  in Chinese Civil War, 278
  military campaign in warlord period,
    254
  retreat from Japanese forces, 220, 225
  urban development campaign, 255
Gutenberg, Johannes, 81

Haddad, Salim, 62
Hall, Thomas, 46, 48, 325
  attempt to develop Chinese
    typewriter, 131–132
Hammond Typewriter Company, 61,
  185, 186
Han Zonghai, 233
Han dynasty, 94
hangul (Korean phonetic script), 3
hanja (Korean characters), 201–202, 210
Hannas, William, 70
Hanyang factory, 167
Hanyeping Steel Company, 167
Hanyu pinyin. See pinyin
hanzi. See characters, Chinese
Harbin Municipal Bureau of Public
  Security, 284

Harrison, Samuel A., 60–61
Harvard University, 244, 273
Havelock, Eric, 2, 69, 72, 186
Hayford, Charles, 142
heat maps, 308, 309f, 310f, 398n63
Hebei, 284
Hegel, Georg Wilhelm Friedrich, 65–66, 72
hegemony, alphabetic, 9–12, 23, 26
Henan Provincial Government, 175
Hepburn, James Curtis, 376n28
Hernisz, Stanislas, 351n41
Herodotus, 186
heteropraxy and political orthodoxy, 292, 295
high modernism, 153
Hill, Robert Erwin, 62
Hillier, Walter, 372n65
History of the Art of Writing, The (Mason), 67
Holly, Carlos, 325
Hong, Y.C. (You Chung), 161, 164
Hongye Company, 219
Howl (Ginsberg), 71
Howland, Douglas, 209
Hu Shi, 138, 156–157, 326
Hu Zhixiang, 233
Huadong Machinery Factory, 171
Huang Xisheng, 250
Huanqiu Chinese Typewriter Manufacturing Company, 225–226
Huanqiu Typing Academy, 174–175, 368n34
Hull, Matthew, 339n23
human-machine interactionism (HCI), 260
Humphrey, Henry Noel, 67
Hung Shen, 142
Huntington Library, 161, 163
Hwangsong sinmun (periodical), 375n14
Hwa-yin Type-writing School, 176
hypermediation. *See* mediation

IBM (International Business Machines Corporation), 272, 315, 391n38
"ideographic" scripts, 66
  Chinese rejected as, 67
Ilion, New York, 46
imitation (*fangzhi*), of Japanese typewriters by Chinese companies, 230–232, 233
immediacy of script, myth of, 316–317
imperialism, Japanese, 200, 201, 223, 229
imperialism, Western, 10, 11, 12
Imperial Printing Office (*Wuying dian*), 82, 348n10
Importance of Living, The (Lin Yutang), 244
imports of typewriters to China, 220–221, 221f, 232
Imprimerie Royale, 91, 348n16
Indo-European language family, 66, 95
inflectional (agglutinative) languages, 96
"information crisis" in China, 10, 248–249, 250
information environment, early modern, 248
information infrastructure, Chinese, 19–21
"information overload," 247–248
information technology, Chinese
  current-day robustness of, 14, 26
  lack of archives on, 32
  "technology transfer" and, 25
infrequently used characters
  in movable type, 82, 359n36
  in "secondary usage character box," 140, 197, 371n55
  in telegraphy, 110
  in typewriting, 191, 284
input, 237–243, 286, 318
  in Chinese computing, 238–243, 315–320

cloud input (*yun shuru*), 319–320, 399n3
MingKwai and birth of, 243, 245–247, 280–281
input method editors (IMEs), 239–243, 318, 319, 388n2
Cangjie input, 243, 281
Sougou pinyin input, 240–241, 242f, 281
structure-based, 243
Interim Committee on the National Language (Japan), 204
International Olympic Committee, 2, 4, 8, 9, 11
International Telegraphic Union, 107, 114, 355n74
Iroha organizational system, 201, 205, 212
"Italian engineer" of MingKwai, 276, 392–393n54

Japan
invasion of China (1937), 220, 228
invasion of northeast China (1931), 213, 218, 263
military occupation of Southeast Asia, 221
neologisms from, 86
See also imperialism, Japanese; typewriter, Japanese
Japanese Business Machines, Ltd., 198, 199
Japanese Imperial Army, 230
Japanese Typewriter Company, 232
Japanese Typist Association, 207
Jesus
placement on Sheffield's typewriter, 135–136
in typewriter art, 203
Jiang Yiqian, 250–252
Jiangsu Province Education Committee Summer Supplementary School, 165

Jiaotong Engineering University, 137, 360n49
Jin Jian, 82–84, 134, 326, 348n10, 348n14, 348n16
Jin Shuqing, 216, 380n59
Jinggangshan, 294
Jinggangshan Newspaper Printing House, 292–294
Jingyi Typewriter Company, 233
Jiyang Chinese Typewriting Supplementary School, 227–228
Jones, Robert McKean, 265, 326
creation of Chinese Phonetic Typewriter, 182–183, 185–186, 187
and kana typewriter, 376n23
Lin Yutang and, 391n37

Kadry, Vassaf, 344n50
Kafka, Ben, 339n23
Kaifeng, 278, 291, 294
kana (Japanese phonetic writing), 3, 202, 203
hiragana, 3, 200, 202
in Japanese telegraph code, 201
katakana, 3, 200, 201, 203, 212
in typewriting, 59, 200, 202–204, 212
See also typewriter, Japanese: kana-based
Kangxi Dictionary, 18, 77, 82–83, 91, 149, 256, 270, 348n11
criticisms of, 190, 249, 299
number of characters in, 85
Kangxi emperor, 82
kanji (Japanese characters), 3, 202, 210
common usage and, 204
in typewriting, 199, 200, 205
See also typewriter, Japanese: kanji-based
kanjisphere, 210, 211, 212, 213
Kanto earthquake, 204
Karlgren, Bernhard, 67

Kataoka Kotarō, 205, 326
Kennedy, George, 141, 142, 143
keys, computer, 239
keys, typewriter
  analogous to piano keys, 179
  "dead keys," 48, 59, 63
  definitive of typewriter's essence,
    41–45
  on double-keyboard machine, 47
  on imaginary Chinese typewriter,
    35–37, 40–44, 157
  lack of on Zhou Houkun machine,
    145
  as metaphor for life, 28
  on MingKwai, 243, 245, 246f, 265,
    271–272
  Western typewriters without, 46
Khalil, Seyed, 344n52
Kittler, Friedrich, 28
Klaproth, Julius Heinrich, 102
Korean War, 279–280, 292, 297
Kurosawa Teijirō, 202, 375n17

Lampedusa, Giuseppe Tomasi di, 22
language reform in China, 12–16,
  19–23, 138, 186
  ethnographic notions in, 253, 262,
    263
  and reaction to typewriters, 156
  See also abolition of Chinese
    characters; characters, Chinese:
    simplification of; retrieval of
    Chinese characters; Romanization
    of Chinese; vernacularization of
    Chinese
language reform in Japan and Korea,
  201–202, 204
laser printer, 285
Lauer, Matt, 3, 9
L.C. Smith & Brothers Typewriters Inc.,
  51
leapfrogging, technological, 24, 375n7
Lee, Tuh-Yueh, 273

Legrand, Marcellin, 75, 91–103, 151,
  154, 326, 351n38, 351n39, 351n41.
  See also divisible type
Legros, Lucien Alphonse, 184–185
Levering, Gilbert, 193
Lewis, Jerry, 28
Lewis, Samuel, 46
lexicon, Chinese, size and expansion of,
  77, 78f, 85, 86
Li Chih, 142–143
Li Fuguang, 94
Li Gui, 24–25
Li Xianyan, 213–214, 229
Li Youtang, 228
Li Zuhui, 174
lianchuanzi organization, 291, 292–294,
  295, 298
Liangyou (periodical), 176
Liaoning Chinese Typing Institute, 176
Life (periodical), 193
Lin Gensheng, 305–306, 398n58
Lin Taiyi, 237, 272, 277f, 326,
  392–393n54
  demonstrates MingKwai, 273–276
Lin Yutang, 235, 237, 244–247, 277f,
  326, 391n32, 392–393n54
  early development of typewriter,
    263–265, 391n36, 391–392n38
  failure to mass-produce MingKwai,
    278–281
  Jones and, 391n37
  legacy in transforming technology,
    280–281, 316
  and mechanical design of MingKwai,
    246, 267
  and taxonomic system for MingKwai,
    267, 268–272
  unveils and promotes MingKwai
    prototype, 272–276
  work with character retrieval methods,
    250, 252
  See also typewriter models, Chinese:
    MingKwai

linguistic fitness, of Chinese, 65–72
Linotype, 360n54, 392n47
literacy, mass, 15, 19, 166
   campaigns for, 260
   in current-day China, 14
   information technology and, 21
   supposed lack of in China, 69
Literacy in Traditional Societies (Goody), 69
literary (Classical) Chinese, 20, 140, 211, 347n4
Logan, Robert, 69
London, 196, 198
Los Angeles Times, 276
Lu Xun, 13, 15, 29
Lu-Ho Rural Service Bureau, 143
Luoyang, 288

Macartney, George, 84
machine gun, compared to typewriter, 28, 29
machine translation, 106, 315
Madagascar, 104
Maejima Hisoka, 202
Magic Eye viewfinder (in MingKwai), 245, 246, 270
Maha Vajiravudh. *See* Rama VI, King
Maha Vajirunhis, Crown Prince, 49
Malaysia, 196, 197
Malling-Hansen, Rasmus, 42
Manchukuo, 195, 213–214, 222, 223, 224, 230
Manchu script, 224
Mao Zedong, 279, 298
   and language reform, 142, 279
   name on typewriter tray beds, 294, 295, 302–303, 307–308, 307f
   poetry of, 285
   speeches of, 285
Marshall, John, 391n38
Marshman, Joshua, 86
Maschinenfabrik Augsburg-Nürnberg (MAN), 167

Mason, W.A., 67
Massachusetts Institute of Technology (MIT), 138, 139, 145, 315, 391n38
"masses," Chinese, 260–261, 292, 311
   Chinese common usage and, 141, 142, 143
   how characters are found by, 253, 260–261, 270
   MingKwai advertised as usable by, 274
   as projections of imagination, 262–263
   and proletarian knowledge, 299
mass production
   of Chinese typewriters, 161, 164, 167–170
   failure of Chinese phonetic typewriters to achieve, 187
   failure of MingKwai to achieve, 277
   failure of Sheffield typewriter to achieve, 137
   of images, 37
McFarland, Edwin Hunter, 45–50, 56, 59, 326, 341n11, 342n27
McFarland, George, 49–50, 51, 54, 56, 326, 341n11, 341n13, 342n22, 342n27, 342n31
McFarland, Jane Hays, 45
McFarland, Samuel Gamble, 45, 341n10
McFarland Siamese Typewriter. *See* typewriter, Siamese
MC Hammer, 40, 43
mechanization, industrial, 177
mediation
   in Chinese computing, 239, 240
   of Chinese in telegraph code, 112, 116–121, 118f, 119f, 356n76
   and hypermediation in Chinese telegraph code, 117–121, 356n75
Meeting for the Improvement of the Typewriter Character Chart, 301–302, 303
memory practices
   in Chinese typing, 178, 180

memory practices (cont.)
  and Japanese tray beds, 211
  and natural-language tray beds, 311
  and telegraph code books, 120
Mergenthaler Linotype, 10, 199, 272,
    273, 277, 278–279, 392n47
metadata, 104
"Mexican" script, 66
Microsoft Word, 318, 320, 399n3
MIDI (Musical Instrument Digital
    Interface), 317, 318–319
Ming dynasty, 12, 78, 82
Ming Kee Typewriter Company, 226
MingKwai Chinese Typewriter
  mechanism of, 245, 265–268
  retrieval system used in, 267, 270,
    279, 280
  See also Lin Yutang; typewriter models,
    Chinese: MingKwai
minimal modification, 63, 123, 345n54
Ministry of Communications (China),
    114
Ministry of Education (Chinese
    Republic), 142, 185
Ministry of Foreign Affairs (France), 104
Minsheng Chinese Typewriter
    Company, 233
missionaries in China, 45, 64, 86, 102,
    127, 341n10, 349n20
  and hope of introducing new
    vocabulary, 87, 134–135, 349n19
  See also Gamble, William; Sheffield,
    Devello Z.
Missipi (Edna Eagle Feather), 343n36
model workers, 288, 292, 298, 300, 304
modernity
  Chinese "incompatibility" with,
    65–72, 172, 248
  of Chinese women, 176
  compatibility of languages with, 10,
    11, 60, 66, 70
  how to achieve in China, 13, 15, 20
  information technology as test of, 16

Japanese, premised on cutting ties
    with China, 203
  Japanese and Korean exclusion from,
    210
modernity, Chinese technolinguistic
  between mimicry and independence,
    165
  character organization systems hinder,
    249
  intensity of engagement in, 26
  methodological issues, 29–32
modernization initiatives in Siam, 45,
    48
Monarch Typewriter Company, 344n52
Mongolia, 56
Mongolian script, 224
monoculture, technolinguistic, 44, 56,
    58, 140
Monotype, 139, 360n54
Morse, Samuel, 106, 326
Morse code, 25, 113f, 117
  authorized symbols in, 107
  Chinese and, 108, 110, 112, 115–116
  Continental, 107, 354n66
  as inherently English-centered,
    106–107, 352n48
  See also telegraph code of 1871,
    Chinese
movable type, 8, 18, 74, 96, 169, 281,
    349n20, 359n36
  and common usage in Chinese
    printing, 84–88
  compared to hot metal printing,
    360n54
  invention of in China, 81
  and "Oriental" type design by
    Westerners, 90–94
  resolving "incompatibility" of Chinese
    with, 75, 81–84, 92, 106, 347n4
  See also divisible type
Mow, Pang-Tsu, 272
Mr. Hui's Chinese-English Typing
    Institute, 174

"Mr. Tsiang" and "Mr. Cü" (assistants to William Gamble), 76
multiple-character sequences, 241, 290, 291, 294, 295–296, 311
Musée de la Machine à Écrire, 59
Museo della Macchina da Scrivere, 59
Museum of the History of Science (Oxford), 320
My Country and My People (Lin Yutang), 244

Nagasaki, 110, 230
Nanbu Yoshikazu, 376n28
Nanjing, 159, 220, 254–255
Nanjing, Treaty of, 127, 357n14
Nanjing Normal College, 159
Nanyang College, 137, 360n49
National Conference of Youth Activists in Social Construction, 298
Nationalist Party. *See* Guomindang
National Language Monthly, 186
National Ping-Pong Association (China), 215
National University of Commerce, 215
Nation's Business, 183
natural-language arrangement, 286–311, 289f, 396n35, 398n62
  attempts to centralize, 301–303, 311
  in Chinese typewriters, 286–288, 295–311
  failure to adopt in Republican era, 296–297, 298
  individual styles of, 306, 313f, 398n58
  limited uses of, 295, 297, 297f
  spread of in Communist era, 297–300, 308–311, 399n66
  strategies for setting up, 305–306, 311, 312f
  in typesetting, 291–295
NBC, 6
Needham, Joseph, 346n77
New China Press, 291, 295
Newcomb, Henry C., 126

New Culture Movement, 156
newspaper industry, Japanese, 201, 204, 375n7
New Typing Method, 298
New York Times, 143, 154, 210, 276
New York University (NYU), 147, 148, 154, 157
New York World Tribune, 276
Nietzsche, Friedrich, 42
Ningbo, 74, 75
Nipon o daï itsi ran (Annales des empereurs du Japon), 102, 351n39
Nippon Typewriter Company, 207, 223, 226, 228
  introduction of Wanneng model, 223–224
North China Language School, 142
North China Military Region Headquarters, 297
Notizie Olivetti, 59
"numbered languages" (*langues chiffrées*). *See* "enciphered" transmissions

Oba Sachiko, 222
occupation of China, Japanese, 221–223, 235
  collaboration with, 214, 225, 228
  resistance to, 225
  offset printing, 279
Olivetti, 9–10, 27, 42, 71, 199
  globalization of, 56–58
Olympic Games (Beijing, 2008), 1–9, 10, 11, 94. *See also* Parade of Nations, Olympic
Ong, Walter, 2, 40, 186
"On the Nature of Chinese Ideography" (Creel), 68–69
"On the Psychology of How the Masses Search for Characters" (Du Dingyou), 260–261
Opium Wars, 105, 127
oracle bones, 71

Orality and Literacy (Ong), 40
organizational methods for characters.
   See retrieval of Chinese characters
Orientalism, 44, 91–92
Oriental Library (Shanghai), 260
Origin and Progress of the Art of Writing,
   The (Humphrey), 67
Otani Typewriter Company, 205

Palace of Fine Arts, San Francisco, 147
Panama-Pacific International Exposition
   (1915), 56, 147, 204
Parade of Nations, Olympic, 2–9, 11,
   337n6, 338n9
   in Beijing Olympics (2008), 2–9, 7t, 94
   regulations of, 2
   in Tokyo and Seoul Olympics, 3
particles (in Chinese language), 136,
   140, 296, 398n62
patent rights, in Communist China,
   279
Pauthier, Jean-Pierre Guillaume, 89–103,
   151, 154, 326, 351n38, 351n41. *See*
   also divisible type
Peking Syllabary, 128
People's Daily, 290, 297
People's Liberation Army, 297
People's Republic of China, 79, 196,
   279
   campaigns in, 284, 300
   "golden age" of typewriting in,
   284–286
   in Korean War, 279–280
   model workers in, 288
   natural-language tray bed experiments
   and, 286–288
   Reform era, 79, 286
   seizure of Japanese interests, 232
   typewriter models in, 195
People's University, 294
Peter Mitterhofer
   Schreibmaschinenmuseum, 58–59,
   341n8

Philadelphia Centennial Exposition
   (1876), 24, 55
Philadelphia Commercial Museum, 188,
   189
Philadelphia Sesquicentennial
   International Exposition (1926), 161,
   188–190, 192
Philology and Ancient China (Karlgren),
   67
Philosophy of History, The (Hegel), 65
phoneticization of Chinese, 8, 13,
   264–265
   Western dreams of, 183–187
   See also zhuyin fuhao
phoneticization of Japanese, 203
phonetics (structural components of
   Chinese script), 265
piano playing, compared to typewriting,
   179
pictography, 68, 320
pinyin, 8, 281, 338n12
   in Chinese computer input, 240–241
place names, on typewriter tray bed,
   135, 295–296, 306, 396n35
plaintext, 113–116, 318, 355n74
Playing the Building (Byrne), 319
pleremic script, 11–12
point (*dian*), 5
   as fundamental element of characters,
   258, 259f
politics of objects, 196, 198–199
   kana typewriter and, 203
Popular Science Monthly, 145
predictive text, 319
   analog roots of in typewriting,
   286–288, 308
   "predictive turn," 308, 309f
Presbyterian Mission Press, 75
prescriptive imperative, 87
   in Sheffield's typewriter, 135–136
primary and secondary transcripts,
   318
primers, character, 88, 142

printing industry, Japanese, 201
printing press, 170, 171. *See also*
Linotype; Monotype; movable type
pronunciation of Chinese, 9, 243
propaganda, Chinese Communist, 284,
292
Provençal troubadour poetry, 396n31
"pseudo-radicals" (in MingKwai),
271
Psychological Warfare Unit, Eighth
United States Army in Korea, 280
publishers, Chinese, 154
punctuation, Western, 19
on typewriter tray beds, 366n19,
398n62
Puyi (Kangde emperor), 213, 214

Qi Xuan, 147–159, 148f, 204, 265, 267,
326, 362n76
debates Zhou Houkun, 157–158
design of combinatorial typewriter,
149
reception of typewriter, 154–159
transnational ambitions of, 379n39
See also typewriter models, Chinese:
Qi Xuan typewriter
Qian Xuantong, 13, 186, 264, 326
criticism of Chinese typewriter,
190–192
Qianlong emperor, 82, 84
Qing dynasty, 127, 147, 244
printing during, 82
restored in Manchukuo, 213
telegraphy and, 108, 115
QQ, 241
QWERTY keyboard, 238
training regimen on, 179
used in Chinese computing, 9,
238–243, 315, 316, 319

race science, refutation of, 68, 70
racism, 40
"radiating compounds," 298, 301–302

radicals (structural components of
Chinese script), 83, 197, 256, 262,
348n12
in divisible type, 90–100, 99
likened to prefixes, 265
in MingKwai, 267, 271
new character retrieval methods and,
250
reimagined as equivalent of letters, 80,
94–96, 149, 156
with shared orthographic features,
271
size and placement, 91, 92, 157–158
as taxonomic entity, 95–96, 348n11
in telegraphy, 104
used in IMEs, 243
radical-stroke organization system,
82–83, 256, 348n13, 349n19
in Chinese telegraph code, 110
criticisms of, 190, 249, 296, 299
departure from in Mao era, 286, 288,
290, 291, 298, 303, 304
exceptions to in tray beds, 295–296
in movable type, 348n14
reaffirmed by typing reform
committee, 302, 303
relaxed in tray beds, 302–303, 397n50
typewriter tray beds using, 143, 212,
307–308
Rama V, King, 45, 49, 55, 341n13,
342n22
Rama VI, King, 49, 341n13, 342n21,
342n22, 342n23
Red Star Typewriter Company, 232
Reed, Martin, 279
reference materials
anxiety about in China, 248–249
Chinese, 15, 255, 258, 389n9
made possible by codex form, 178
refugee crisis in northeast China, 219
Remington, Eliphalet, 46
Remington Export Review, 182
Remington News, 59

Remington Rand. *See* Remington
  Typewriter Company
Remington Typewriter Company, 28,
  36, 42, 44–45, 48, 50, 343n33,
  343n36, 343n45
  Arabic typewriter and, 345n54
  argues for Japanese language reform,
    203
  arrival in Siam, 51–55, 52f
  Chinese Phonetic Typewriter and,
    182–183, 185–187, 265
  claims of universality by, 9–10
  as constitutive of modern
    technolinguistic consciousness,
    31–32, 74
  failure to encompass Chinese, 10, 64,
    71, 159, 199, 200
  feminization of typing and, 177
  founding of, 46
  globalization of, 55–57, 57f, 60, 123,
    182
  Lin Yutang and, 272–275, 391n37
  purchases Smith Premier company, 51
  sale of Japanese kana typewriter,
    202–203, 376n23
Rémusat, Abel, 351n39
repatriation of Japanese after Second
  World War, 230
Republican era
  character retrieval and reference
    materials in, 247, 253–256
  iconoclasm in, 153, 299
  telegraphy in, 110, 115
Resist Japan Association, 218
retrieval of Chinese characters, 15
  and "character retrieval problem,"
    247–249, 250, 252–253, 255, 258,
    260–261
  Communist push for proletarian
    taxonomy in, 299
  in computing, 316
  ethnographic viewpoint on, 253,
    260–263, 270

input and, 280
MingKwai and, 245–246, 247
new experimental methods of, 248,
  249, 250–263, 251f, 270, 286,
  299
predictive text and, 286
slowness of, 248–249
surrogacy and, 81
in telegraphy, 104
transposed to inscription, 247, 267
in typing school curricula, 179–180
See also Five-Stroke retrieval system;
  Lin Yutang; Shape-Position retrieval
  system
revolution of 1911, 79, 147, 185, 213,
  244, 371n55
rhetoric, Communist, 286, 291, 292,
  299–300
  ingested into typist's body, 304
  metacognitive distance from, 294–295
Richardson, Ingrid, 290
Rivista Olivetti, 56
Robbins, Bruce, 23
Rockefeller Foundation, 391n38
Romance of the Three Kingdoms, 86
Romanization of Chinese, 14, 19, 40,
  128
  as bound to fail, 184–185
  call for in Mao era, 279
Romanization of Japanese, 376–377n28
Ryōshin Minami, 201

Sakai Yasujiro, 378n38
Sakurada Tsunehisa, 222
"same script, same race," 224
Sammons, Thomas, 145
Sampson, Geoffrey, 69
San Francisco Chronicle, 137, 276
San Francisco Examiner, 35
Schjellerup, H.C.F.C., 110
Schurmann, Franz, 299–300
Science and Civilization (Needham),
  346n77

scroll (textual form), 178
searching. *See* retrieval of Chinese
    characters
Second World War, 200, 221, 229–230
Selleck, Tom, 38, 40, 43
semiotics, 317
semiotic sovereignty, 115, 121, 355n74
Semi-Weekly Tribute, 136
Shang dynasty, 71
Shanghai, 109, 110, 145, 157, 216, 226,
    244, 260
    fall to Japanese forces, 220
    Japanese bombing of, 212, 218, 260
    typewriter manufacturing in, 161, 230
    typing schools in, 174–175
    Zhou Houkun in, 159, 165
Shanghai Baptist College and Seminary,
    143
Shanghai Calculator and Typewriter
    Factory, 195, 233
Shanghai Central Stadium, 215
Shanghai Chamber of Commerce, 189,
    218
Shanghai Chinese Typewriter
    Manufacturers Association, 233,
    235
Shanghai Machinery Import-Export
    Company, 233
Shanghai Press, 188
Shape-Position retrieval system,
    260–263, 262f. *See also* Du Dingyou
Sheffield, Devello Z., 123, 126–137, 139,
    140, 143, 151, 326, 357n14, 358n20,
    359n38, 362n68
    Chinese clerks and, 129–130
    early version of invention, 127–128
    fate of typewriter, 136–137
    invention of Chinese typewriter
        prototype, 132–133
    missionary work, 129, 134–136
    See also typewriter models, Chinese:
        Sheffield typewriter
Sheffield, Eleanor, 127

Shen Yunfen, 297–298, 300
Shenbao (periodical), 190, 214, 218, 219,
    220
Sheng Yaozhang, 228, 229
Shibao (periodical), 176
shift key, 46–47, 48, 129
shift-keyboard typewriter, 46, 47, 48
    dominance of, 44–45, 51, 56–59, 63
    encounter with foreign scripts, 59–63,
        182
    inadequacy of for Chinese, 124
    Japanese typewriters and, 202
    MingKwai's similarity to, 245
    as typewriter par excellence, 42
    See also typewriter, Western; typewriter
        form, universal
Shimada Minokichi, 205, 326
Shimizu Usaburō, 202
Shinozawa Yūsaku, 209
Sholes, Christopher Latham, 46, 123
Shu Changyu. *See* Shu Zhendong
Shu Zhendong, 167–169, 171, 229,
    247, 326, 365n12, 366n14. *See also*
    typewriter models, Chinese:
    Shu-style typewriter
shuru. See input
"Shu Zhendong Chinese Typewriter,
    The" (animated film), 170, 171
Siam, 45, 48–55, 342n23
Siamese script, 45, 46, 47, 53–54, 64
    changed to fit typewriter, 49
Silicon Valley, 237, 238, 315
Simpsons, The, 321
Singapore, 196, 197, 199
single-shift typewriter. *See* shift-
    keyboard typewriter
Sino-Japanese War, First, 244
Sino-Japanese War, Second. *See* War of
    Resistance Against Japan
Smith, Arthur W., 345n54
Smith, Lyman C., 47
Smith Premier Typewriter Company,
    47–48, 49, 342n23

Smith Premier Typewriter Company
(cont.)
demise of, 50–51
presence in Bangkok, 50
Snowden, Edward, 399n3
social Darwinism. *See* evolutionism
Société Asiatique de Paris, 89
Song Mingde, 170–171
Sonic Banana project, 319
Southeast University of Commerce, 215
Soviet Union, 230
space, and spatial recognition of
Chinese characters, 262
within divisible type, 99–100, 103,
149
speed of typesetting, using natural-
language arrangement, 290–291,
292, 294, 395n26
speed of typing, 27
average, in Mao era, 288
on index typewriter, 343n37
on Qi Xuan typewriter, 154
on Shu-style typewriter, 170, 190
using natural-language tray bed, 286,
288, 298, 311
Western notions of, on Chinese
typewriter, 320
on Zhou Houkun typewriter, 165
Spurgin, Richard A., 61
State Department, U.S., 279
Staunton, George, 84–85, 88, 89, 100,
132, 327
Steele, Herbert H., 344n52
Stellman, Louis John, 36, 41, 43
stenography, 317–318
Stickney, Burnham, 204
St. John's University, 244
St. Louis Globe-Democrat, 36
Strauss Festival Orchestra Vienna, 27
stroke count, materiality of in character
slugs, 180
stroke-count organization, 4, 8, 83, 302
mocked by Olympic commentators, 3

stroke ranking, 6
strokes (*bihua*), 131–132, 243, 256,
262
as elemental unit of script, 94–95
See also eight fundamental strokes
"study sessions," 284, 285
Suchman, Lucy, 307
Sugimoto Kyōta, 204, 209, 211, 327
Suiyuan province, 220
Sun Yat-sen, 230, 371n55
surrogacy, 79, 121, 316
in Chinese telegraphy, 103–121
definition of, 80–81
MingKwai and, 265, 270
surveillance, 256, 284, 399n3
Sushan, 215
synaesthesia, 261
Syracuse, New York, 47, 48
relationship with Bangkok, 50,
342n23

"Table of 214 Keys and Their Variants"
(Tableau des 214 clefs et leurs
variants) (Legrand), 92, 93f
Tai, Evelyn (pseudonym), 196–199, 200,
233, 374n2
Tai, Maria (pseudonym), 197–199,
374n2
Taipisuto (magazine), 205, 207, 208f,
222–223
taipuraitā. See typewriter, Japanese
Taiwan, 195–196, 223, 244
Tanakadate Aikitsu, 377n28
Tang Chongli, 171
Tao Minzhi, 233, 384n93
Tao Xingshi, 142
Tap-Key (imaginary inventor), 36–37,
40, 41, 42, 45, 65, 74, 193, 316, 320,
327
taxonomic evenness, 270
Tcherkassov, Baron Paul, 62
Teaching of Words, The (Fukuzawa),
204

technolinguistic imagination, collapse of, 42–45, 54, 316
technolinguistic realm, 18–23, 72, 339n23
definition of, 17
"technological abyss," 24, 26, 243
technological fitness, denigration of Chinese according to, 44, 71–72, 346n77
technosomatic complex. *See* bodies, human
Tekniska Museet, 320
telecommunication, 317, 318
telegraph code books, 81, 104, 113, 355n74
changes to, 117, 120
for Chinese telegraphy, 110, 121
during Nationalist military campaign, 254
telegraph code of 1871, Chinese, 110–121, 111f, 201, 254, 265, 354n66
adjustments to, 116, 354n70
and Chinese as inherently secret language, 114
disadvantages of, 112, 114–115, 355n74
exempted as plaintext, 115
taking "symbolic possession" of, 121
Telegraph-Herald, 185
"telegraph sovereignty," 115
telegraphy, 9
causes sense of crisis in China, 248
Chinese, 24, 25, 74, 80, 103–121, 131, 178, 240, 265
global expansion of, 106, 107–108, 109–110, 114, 354n73
international regulations, 107, 113, 355n74, 356n76
Japanese, 201
Japanese domination of in China, 221–222

MingKwai system and, 279
and non-Western scripts, 108
structural inequality of China within, 114, 116, 121, 355–356n74
telephone, 55
terrain, Chinese language portrayed as, 83, 85, 191–192
Tewksbury, Reverend E.G., 185
textbooks, typing, 180, 301, 303
thimble linking (*dingzhen xuxian*), 294, 395–396n31
Thirteen Classics, 86
"Three People's Principles" (Sun Yat-sen), 230
Tianjin, 128, 133, 222, 232, 278
typing reform committee in, 301–302
Tianjin, Treaty of, 127
Times (London), 320
tinkering, 307
Titsingh, Isaac, 351n39
Today! (journal), 286
Tokugawa period, Japan, 248
Tokyo, firebombing of, 229
tones (in Chinese), 185
Tong Lisheng, 233
Tongji (periodical), 171
Tongzhou, 126, 127, 132
Too Much to Know (Blair), 247–248
Toshiba, 205
Touch Method, 51
training drills. *See under* tray bed
"transnational culturalism" in East Asia, 209–210, 211
tray bed (typewriter part), 30, 169, 199, 226, 295, 298
absence of in MingKwai, 265, 267
adjacency of characters on, 288–290, 296
Communist terms on, 294–295
democratization of, 300, 304, 308, 398n62
how to rearrange, 305–306

tray bed (typewriter part) (cont.)
on Japanese-built Chinese typewriters, 212
on Japanese typewriter, 205
memorization of characters on, 178, 197, 211, 305–306, 311
move to standardize, 301–303
with "out-of-the-box" natural-language arrangement, 303
with relaxed radical-stroke system, 302–303, 304
reorganized by Mao-era typists, 286, 288–290, 300, 307–308
on Shu-style typewriter, 163–164, 364n1, 366n19
sold blank by manufacturers, 311
special usage region of, 163, 295, 297, 298, 303
as territorial expanse, 191–192, 320
training drills on, 179–180, 181f, 370n51, 371n55
visibility of characters on, 270
on Yu Binqi typewriter, 217
Zhou Houkun's "popular," 143
treaty ports in China, 64, 127
trigraph coded transmission (telegraphy), 117, 120, 356n76
"truth script," 201–202
Tsinghua University, 244
Tsugi Kitahara, 204
Turkle, Sherry, 198
Twain, Mark, 177
two-character compounds (ci), 180, 241, 286, 289, 291, 295–296, 306, 397n50
Two Lists of Selected Characters Containing All in the Bible and Twenty Seven Other Books (Gamble), 81, 87
type, materiality of, 17
type-and-copy shops (dazi tengxieshe), 284
type-and-mimeograph editions (dayinben), 285–286, 287f

type design. See under movable type
type lever (typewriter part), 30, 148, 180
"Types de Charles X" (Legrand), 91
typesetters, Chinese, 82–84, 132, 139
natural-language experimentation and, 290–295
as "nomadic," 83–84, 85
typesetters, Western, 81–88, 102, 348n16
typewriter, Arabic, 60, 61–63, 64, 344n52, 345n54
Jones keyboard for, 182, 187
Olivetti model, 56, 58f
typewriter, Chinese
aurality of, 27, 29–30, 198
challenge of creating, 25, 138–140
functional mechanics of, 30, 199
history of textual reproduction and, 170–171
how to identify documents typed on, 283–284
impact of, 24
linguistic ambidexterity of, 226, 379n39
measured against Western typewriter, 23–24, 27–32, 125, 164–165, 188, 190
merits of, 170–171
as monstrous imagined object, 35–44, 36f, 37f, 38f, 45, 65, 71–72, 192f, 192–193, 203, 316, 320–321
national identity and, 229
and personal relationship with user, 163–164, 198
in popular culture, 35–40, 321
as symbol of modernity, 124–125, 229
transnational nature of, 33
used for state documents, 283–284
used in Korean war, 280, 280f
used to reproduce books, 285–286

Western criticism of, 188
See also typewriter models, Chinese
typewriter, Hebrew, 60–61, 64,
    343n45
typewriter, Japanese, 200, 206f
adaptability to Chinese, 209, 211
dual modes of, 200, 209
H-Style, 218
Japanese Smith Typewriter (*sumisu
    taipuraitā*), 202
kana-based, 202–204
kanji-based, 204–205, 226, 378n38
market for in occupied China, 222
Remington and Underwood models,
    202–204
retrofitted to handle Chinese, 226,
    385n100
sound of, 223
Sugimoto kanji typewriter, 204–205
training regimens on, 203
used in occupied China, 213
Western denigration of kanji-based,
    210–211
typewriter, Romance language, 60
typewriter, Russian, 60, 64
typewriter, Siamese, 45–55, 49f, 53f
change to shift-keyboard design,
    51–53
invention and design of, 45–50
"Typewriter, The" (musical piece,
    Anderson), 27
typewriter, Turkish, 182
typewriter, Urdu, 182, 187
typewriter, Western
aurality of, 27–30, 58, 340n32
Chinese response to, 24–25
compared to machine gun, 28
early diversity of, 42, 44, 45, 56
encounter with Chinese script, 44,
    63–65
essence of, 41
expansion to foreign scripts, 44, 46,
    53–54, 63, 123

in film, 28
functional mechanics of, 41, 59
globalization of, 55–64, 124
iconic nature of, 24, 28, 71, 74,
    203
inability to encompass Chinese,
    123–124, 129, 139, 200
revolutionary nature of, 23
as symbol of modernity, 124
as "true" typewriter, 124–125,
    164–165, 188
universality claimed for, 64, 71, 73f
See also shift-keyboard typewriter;
    typewriter form, universal;
    typewriter models, Western
typewriter art, 203
"typewriter boy," nonexistent, 177
typewriter form, universal, 59
confronted with Chinese, 64–65,
    124
as constitutive of technolinguistic
    consciousness, 74
stretched by foreign scripts, 60, 61
See also shift-keyboard typewriter
"typewriter girl," 173, 176–177
typewriter industry, Chinese
formation of, 170–172
Japanese domination of, 200, 212,
    214, 217, 220–221, 223–225, 232,
    235
in postwar period, 230–233, 387n115
thriving in 1940s, 229
typewriter models, Chinese
Brumbaugh patent, 151, 152f
Double Pigeon, 195, 199, 233–235,
    234f
Grant and Legros Chinese phonetic
    typewriter, 183–184
Hammond Multiplex, 185
Horizontal-style machine, 228
improved Shu-style machine, 224
Japanese-built Chinese machines, 212,
    213, 228, 274, 378n38

typewriter models, Chinese (cont.)
  MingKwai, 237, 243, 244–247,
    263–281, 268–269f, 273f, 275f, 316,
    392–393n54 (see also Lin Yutang)
  "Mr. Fan Wanneng Chinese
    Typewriter," 232
  "People's Welfare Typewriter," 230,
    231f
  pre-MingKwai prototype, 391–392n38
  Qi Xuan typewriter, 147–159, 148f,
    204, 265
  "Reformed" typewriter of 1956, 303
  Remington Chinese Phonetic
    Typewriter, 182–183, 184f, 187, 265
  Sheffield typewriter, 126–137, 133f,
    140, 166, 167, 359n36, 359n38
  Shu-style typewriter, 161–164, 162f,
    167–169, 170–172, 175, 182,
    188–191, 189f, 191f, 197, 212, 224,
    265, 274, 308, 364n1, 366n14,
    366n19
  Standard Horizontal-Vertical-style
    Chinese-Japanese Typewriter, 226
  Standard-style machine, 228
  Suganuma-style machine, 228
  Superwriter, 196–199, 200, 233–235,
    374n2
  Underwood Chinese National
    Phonetic Typewriter, 185, 186–187
  Wang Kuoyee patent, 151, 153f
  Wanneng ("All-Purpose") typewriter,
    223–224, 224f, 228, 230–232, 233,
    235, 302–303
  Yu Binqi typewriter, 216–220, 224,
    380n61, 381n71
  Zhou Houkun typewriter, 138–145,
    144f, 146f, 148, 149, 156–159,
    165–167
typewriter models, multi-language
  "Oriental Type-Writer," 60–61
  "Universal Eastern alphabet
    typewriter" (Tcherkassov and Hill),
    62–63
typewriter models, Western
  American Visible Typewriter, 42
  Circular Index, 46
  Hall Typewriter, 46
  Hammond models, 391n36
  Hughes Typewriter for the Blind, 46
  index typewriters, 46, 56, 131, 343n37
  Lambert typewriter, 42
  L.C. Smith & Brothers "Standard," 51
  Malling-Hansen Writing Ball, 42
  Mignon, 343n37
  Olivetti Lettera 22, 27, 71, 72f
  Remington Number 7, 55
  Remington Number 10, 55
  Sholes and Glidden Type-Writer, 46
  Smith-Corona Elliott model, 202
  Smith Premier Number 4, 47, 51
  Underwood Number 1, 342n30
  See also double-keyboard typewriter;
    shift-keyboard typewriter
"Typewriter Tip Tip Tip" (song), 29
typing classes, given by Commercial
  Press, 167. See also typing schools,
  Chinese
typing schools, Chinese, 164, 182, 301,
  303, 380n61
  curricula in, 178–180, 197, 213
  early entrepreneurial era of, 174–175
  enrollment data from, 174, 227–228
  establishment of first, 172
  during Japanese occupation, 226–229
  Latin-alphabet typing in, 179
  in Manchukuo, 213
  See also names of individual schools
typing schools, Japanese, 205
  in occupied China, 213, 223
typing schools, Siamese, 51
"typist fever," 223
typists, Chinese, 164, 172
  and allegiance in Manchukuo,
    213–214
  demands for "production" by, 299,
    300

employment of in Communist
   government, 300
gender of, 172–174, 175–177, 227, 274
media representations of, 172–177,
   173f
in occupied China, 229
patriotic images of, 284, 285f
and sensitivity to machine's
   materiality, 180
Sheffield as first, 134
spread of in government and business,
   175
tray bed reorganization experiments
   by, 235, 286–290, 304–311, 398n62
typists, Japanese, 207, 207f
in occupied China, 213, 222–223

"U Can't Touch This" (MC Hammer), 40
Underwood, John T., 342n30
Underwood Typewriter Company, 42
   Arabic typewriter and, 344n50,
      344n52
   Chinese phonetic typewriter and, 185,
      186
   failure to encompass Chinese, 71, 159,
      199
   at Panama-Pacific exposition, 147
   sale of Japanese kana typewriter,
      203–204
   Siamese typewriter and, 53, 54
   visible typewriting and, 50, 342n30
UNESCO, Chinese typewriter used in,
   306–308, 398n62
Union Typewriter Company, 50
United Front alliance, 254
United Nations, 279
   Chinese typewriter used in, 306–308,
      398n62
Universal History (Sheffield), 129
universal telegraphic language, 106,
   108–109, 110, 353n59
University of Illinois at Urbana-
   Champaign, 138

University of the Philippines, 260
upper-case letters, frequency of in
   English, 47
user-driven change, 304, 311
user experience, 260, 270
users of information technology, 21
   and hypothetical "average Chinese
      user," 253, 262–263, 303

vernacularization of Chinese, 15, 19,
   20–21
   state-sponsored movement, 140, 142,
      303
vernacular taxonomy. See natural-
   language arrangement
vertigo zoom, 389n10
Victory and Success Typing Academy,
   174
Viguier, Septime Auguste, 110, 327,
   354n70
visible typing, 50–51, 55, 342n30

Wade, Sir Thomas Francis, 128
Wagner, Franz X., 342n30
Wall Street Journal, 56
Walsh, Richard, 272, 278
Wan Guchan, 170
Wan Laiming, 170
Wang Chao, 185
Wang Guihua, 305–306, 398n58
Wang, H.L., 389n9
Wang Jialong, 298
Wang Jingchun, 327
Wang Kuoyee, 151
Wang Xinshun, 294
Wang Xizhi, 5, 94, 183, 256–258
Wang Yi, 216
Wang Yunwu, 250, 252, 296, 327
Wang Zhen, 82
Warlord period, 254, 371n55
War of Resistance Against Japan
   (1937–1945), 200, 220, 229, 230
Waseda University, 215

Washington Post, 155
Watt, Lori, 230
wax duplicating paper, 219
Wei Geng, 227
Wei Shuo (Lady Wei), 94–95
weights and measures, 140
Wen, King, 99
Wenhua Chinese Typewriter Company, 233
Western Civilization Union, 126, 128
White Terror, 254
Who's Minding the Store? (film), 28
Williams, Samuel Wells, 66
Wilson, Mary Badger, 210–211
women, professional, 173, 176, 205, 366n29
women's magazines, 141, 207
woodblock printing, 82, 96
word processing, 315
Writing Systems (Sampson), 69
Wu Zhihui, 185
Wubi. See Five-Stroke retrieval system
Wuxi, 137
Wyckoff, Seamans, and Benedict, 55, 343n33

Xia Liang, 175
Xu Bing, 16–18, 22
Xu Shen, 77

Yamagata Girls' Professional School, 222
Yana Fumio, 204
Yanagiwara Sukeshige, 203, 204
Yang, Daqing, 221
Yang Yuying, 148, 154
Yasujiro Sakai, 327
Ye Shuyi, 176
Yen, James, 142, 327
Yesu. See Jesus
yin-yi-xing triad, 16–20, 22, 23
Yost (typewriter company), 46, 50, 60–61, 342n30

Yu Binqi, 215–220, 215f, 230, 232, 233, 327, 380n59, 384n93
and anti-Japanese politics, 218–220
decline of business, 223, 224–225
modifies Japanese typewriter, 217–218
and typewriting enterprise, 216–217, 380n60
See also typewriter models, Chinese: Yu Binqi typewriter
Yu Shuolin, 216, 384n93
Yuan Shikai, 371n55
Yu Binqi Chinese Typing Professional School, 216–217, 232, 380n60, 380n61
Yucai Chinese Typing School, 228
Yunnan University Mao Zedong-ism Artillery Regiment Foreign Language Division Propaganda Group, 285
Yu-Style Chinese Typewriter Company, 233

Zhang Jiying, 283, 290–293, 293f, 294, 295, 297, 298, 300, 302, 327, 395n26
Zhang Xiangling, 188–189, 190, 373n74, 373n77
Zhang Yimou, 1
Zhang Yuanji, 157, 159, 166–167, 171, 250, 365n12
Zhanwang (periodical), 176
Zhao Yuanren (Yuen Ren Chao), 138, 186, 237, 273, 327
Zhongshan University, 260
Zhou Houkun, 123, 137–146, 146f, 148, 265, 267, 327, 359n49, 362n68, 365n5
attempt to mass-produce typewriter, 145
Commercial Press and, 159, 165–167, 364n104
debates Qi Xuan, 157–158
on duty of engineers, 138–139

mechanics of typewriter built by, 143
reception of typewriter built by,
    143–145, 156–159
See also typewriter models, Chinese:
    Zhou Houkun typewriter
Zhou Yaru, 228
Zhu Yunming, 95
Zhuangzi, 76, 99
zhuyin fuhao (Chinese Phonetic
    Alphabet), 183–187, 279, 366n19.
    See also typewriter models, Chinese:
    Remington Chinese Phonetic
    Typewriter

STUDIES OF THE WEATHERHEAD EAST ASIAN INSTITUTE
COLUMBIA UNIVERSITY

SELECTED TITLES

(Complete list at: http://weai.columbia.edu/publications/studies-weai/)

The Chinese Typewriter: A History, by Thomas S. Mullaney. MIT Press, 2017.

Darwin, Dharma, and the Divine: Evolutionary Theory and Religion in Modern Japan, by G. Clinton Godart. University of Hawai'i Press, 2017.

The Cultural Revolution on Trial: Mao and the Gang of Four, by Alexander C. Cook. Cambridge University Press, 2016.

Dictators and Their Secret Police: Coercive Institutions and State Violence, by Sheena Chestnut Greitens. Cambridge University Press, 2016.

Inheritance of Loss: China, Japan, and the Political Economy of Redemption after Empire, by Yukiko Koga. University of Chicago Press, 2016.

Homecomings: The Belated Return of Japan's Lost Soldiers, by Yoshikuni Igarashi. Columbia University Press, 2016.

The Social Life of Inkstones: Artisans and Scholars in Early Qing China, by Dorothy Ko. University of Washington Press, 2016.

Samurai to Soldier: Remaking Military Service in Nineteenth-Century Japan, by D. Colin Jaundrill. Cornell University Press, 2016.

The Red Guard Generation and Political Activism in China, by Guobin Yang. Columbia University Press, 2016.

Accidental Activists: Victim Movements and Government Accountability in Japan and South Korea, by Celeste L. Arrington. Cornell University Press, 2016.

Negotiating Rural Land Ownership in Southwest China: State, Village, Family, by Yi Wu. University of Hawai'i Press, 2016.

Ming China and Vietnam: Negotiating Borders in Early Modern Asia, by Kathlene Baldanza. Cambridge University Press, 2016.

Ethnic Conflict and Protest in Tibet and Xinjiang: Unrest in China's West, coedited by Ben Hillman and Gray Tuttle. Columbia University Press, 2016.

One Hundred Million Philosophers: Science of Thought and the Culture of Democracy in Postwar Japan, by Adam Bronson. University of Hawai'i Press, 2016.

Conflict and Commerce in Maritime East Asia: The Zheng Family and the Shaping of the Modern World, c. 1620–1720, by Xing Hang. Cambridge University Press, 2016.

Chinese Law in Imperial Eyes: Sovereignty, Justice, and Transcultural Politics, by Li Chen. Columbia University Press, 2015.

Imperial Genus: The Formation and Limits of the Human in Modern Korea and Japan, by Travis Workman. University of California Press, 2015.

Yasukuni Shrine: History, Memory, and Japan's Unending Postwar, by Akiko Takenaka. University of Hawai'i Press, 2015.

The Age of Irreverence: A New History of Laughter in China, by Christopher Rea. University of California Press, 2015.

The Nature of Knowledge and the Knowledge of Nature in Early Modern Japan, by Federico Marcon. University of Chicago Press, 2015.

The Fascist Effect: Japan and Italy, 1915–1952, by Reto Hofmann. Cornell University Press, 2015.

The International Minimum: Creativity and Contradiction in Japan's Global Engagement, 1933–1964, by Jessamyn R. Abel. University of Hawai'i Press, 2015.

Empires of Coal: Fueling China's Entry into the Modern World Order, 1860–1920, by Shellen Xiao Wu. Stanford University Press, 2015.

Casualties of History: Wounded Japanese Servicemen and the Second World War, by Lee K. Pennington. Cornell University Press, 2015.

City of Virtues: Nanjing in an Age of Utopian Visions, by Chuck Wooldridge. University of Washington Press, 2015.

The Proletarian Wave: Literature and Leftist Culture in Colonial Korea, 1910–1945, by Sunyoung Park. Harvard University Asia Center, 2015.

Neither Donkey nor Horse: Medicine in the Struggle over China's Modernity, by Sean Hsiang-lin Lei. University of Chicago Press, 2014.

When the Future Disappears: The Modernist Imagination in Late Colonial Korea, by Janet Poole. Columbia University Press, 2014.

Bad Water: Nature, Pollution, & Politics in Japan, 1870–1950, by Robert Stolz. Duke University Press, 2014.

Rise of a Japanese Chinatown: Yokohama, 1894–1972, by Eric C. Han. Harvard University Asia Center, 2014.

Beyond the Metropolis: Second Cities and Modern Life in Interwar Japan, by Louise Young. University of California Press, 2013.

From Cultures of War to Cultures of Peace: War and Peace Museums in Japan, China, and South Korea, by Takashi Yoshida. MerwinAsia, 2013.

Imperial Eclipse: Japan's Strategic Thinking about Continental Asia before August 1945, by Yukiko Koshiro. Cornell University Press, 2013.

The Nature of the Beasts: Empire and Exhibition at the Tokyo Imperial Zoo, by Ian J. Miller. University of California Press, 2013.

Public Properties: Museums in Imperial Japan, by Noriko Aso. Duke University Press, 2013.

Reconstructing Bodies: Biomedicine, Health, and Nation-Building in South Korea since 1945, by John P. DiMoia. Stanford University Press, 2013.

Taming Tibet: Landscape Transformation and the Gift of Chinese Development, by Emily T. Yeh. Cornell University Press, 2013.

Tyranny of the Weak: North Korea and the World, 1950–1992, by Charles K. Armstrong. Cornell University Press, 2013.

The Art of Censorship in Postwar Japan, by Kirsten Cather. University of Hawai'i Press, 2012.

Asia for the Asians: China in the Lives of Five Meiji Japanese, by Paula Harrell. MerwinAsia, 2012.

Lin Shu, Inc.: Translation and the Making of Modern Chinese Culture, by Michael Gibbs Hill. Oxford University Press, 2012.

Redacted: The Archives of Censorship in Postwar Japan, by Jonathan E. Abel. University of California Press, 2012.